AN INTR
TO IDENTIFICATION

J. P. NORTON

Integrated Catchment Assessment & Management Centre
Fenner School of Environment & Society
The Australian National University, Canberra
and
School of Electronic, Electrical & Computer Engineering
The University of Birmingham, Birmingham, UK

DOVER PUBLICATIONS, INC.
Mineola, New York

Bibliographical Note

This Dover edition, first published in 2009, is an unabridged republication of the second printing of the work (1988) originally published by the Academic Press, London and New York, in 1986.

Library of Congress Cataloging-in-Publication Data

Norton, J. P.
An introduction to identification / John Norton — Dover ed.
p. cm.
Originally published: London ; New York : Academic Press, 1986.
Includes index.
ISBN-13: 978-0-486-46935-5 (pbk.)
ISBN-10: 0-486-46935-2 (pbk.)
1. System identification. I. Title.

QA402.N667 2009
003'.1—dc22

2009002969

Manufactured in the United States of America
Dover Publications, Inc., 31 East 2nd Street, Mineola, N.Y. 11501

Preface

Mathematical models of dynamical systems (systems with links between past history and present behaviour) are required in engineering, physics, medicine, economics, ecology and in most areas of scientific enquiry. In control engineering, model-building from measurements on a dynamical system is known as identification, and has enjoyed a sustained boom as a research topic for a decade and a half. In that time technique and theory have developed at such a pace that, although there are a number of good advanced or specialised textbooks on identification, a gap has opened in the coverage at the undergraduate and introductory graduate level. This book is aimed at that gap. As the book gives a broad view at a fairly modest mathematical level, it should also suit the reader who, with a particular modelling problem in mind, needs a quick appraisal of established methods and their limitations.

A serious attempt has been made to recognise that identification, like any engineering activity, is an art backed up by some science, and not a branch of applied mathematics. The presentation is therefore informal and more concerned with usefulness than with elegance. It is also highly selective, inevitably so in such a diverse and eclectic field.

The mathematical requisites increase gradually from Chapter 1 to Chapter 7, but never go far beyond what appears in most first-degree courses in electrical or control engineering. All necessary topics are well covered in many textbooks, so brief reviews and references in the text are given rather than additional mathematical appendices. With few exceptions results are derived in full, but questions of rigour are mentioned only when absolutely necessary. Chapters 2 and 3 use classical single-input-single-output linear-system methods embracing superposition, Fourier and Laplace transforms, z transforms, transfer functions, correlation functions, power spectra and a minute amount of probability theory. Matrix algebra first appears in Chapter 4, including inner products, quadratic forms, gradient vectors, Jacobian and Hessian matrices, inverses and singularity, positive definiteness, rank and linear dependence, Euclidean norm, orthogonality of vectors, orthogonal matrices, eigenvalues and eigenvectors. Probability and statistics are required from Chapter 5 on but they are introduced in an elementary way, that is, intuitively rather than axiomatically. Convergence of random sequences is discussed in connection with asymptotic properties of estimators, but no background in analysis is needed. Some acquaintance with state-variable methods will help in Sections 7.3, 7.5, 8.2, 8.6 and 8.7.

Selections for special purposes could be made as follows: basic topics for an undergraduate course, introducing classical methods with modest mathematical requirements: Chapters 1 to 4 as far as Section 4.2.1; estimation theory, suitable for state estimation as well as identification: Chapters 4, 5, 6 and 7 up to Section 7.3; computational methods for parameter estimation: Chapters 3, 4, 7, 8 and 10 (presupposing some background in probability); review of some recent and current areas of activity: Section 7.5, Chapters 8 and 9; practice of identification for linear, lumped systems: Chapters 2, 3, 4, 7, 9 and 10. The most specialised sections are starred.

My debt to a handful of prominent workers in identification is obvious from the text and is shared by everyone in the field. Less obvious but no less appreciated are debts to friends and collaborators over the years, which range from odds and ends of technique to shaping of an attitude to the subject. Rather than attempting a list, I shall just mention two people whose influence at crucial times has been especially valued: Percy Hammond, in whose group at the NPL I encountered identification as an inexhaustible source of interest, and Keith Godfrey, whose stimulus is responsible for this book being written. Thanks are also due to Graham Goodwin for facilities and encouragement during a sabbatical at the University of Newcastle, New South Wales, in which early sections were written, and to my long-suffering family.

Provision of records by the Hydro-Electric Commission of Tasmania, the Institute of Hydrology and Dr. Alan Knell of Warwick Hospital is gratefully acknowledged, as is the provision of Exercise 9.2.1 by Dr. Alan Robins of British Aerospace, Dynamics Group (Bristol).

Finally, I must thank Mandy Dunn for her cheerful good nature and efficiency throughout the preparation of the typescript, a far longer job than either of us expected.

Birmingham
November 1985

J. P. NORTON

Contents

List of Abbreviations

a.c.f.	Autocorrelation function
a.m.l.	Approximate maximum likelihood (algorithm)
a.r.	Autoregression, autoregressive
a.r.m.a.	Autoregressive moving-average
a.r.m.a.x.	Autoregressive moving-average exogenous
a.s.	Almost surely
c.c.f.	Cross-correlation function
d.c.	Direct current
e.l.s.	Extended least-squares (algorithm)
e.m.m.	Extended matrix method
g.l.s.	Generalised least-squares
i.c.	Initial condition
i.r.w.	Integrated random walk
m.a.	Moving average
m.f.d.	Matrix-fraction description
m.i.m.o.	Multi-input–multi-output
m.l.	Maximum-likelihood
m.m.s.e.	Minimum-mean-square-error
m.s.	Mean-square
m.s.e.	Mean-square error
o.l.s.	Ordinary least-squares
p.d.f.	Probability density function
p.e.	Persistently exciting
p.r.b.s.	Pseudo-random binary sequence
p.s.d.	Power spectral density
q.r.	Quadratic residue
r.i.v.	Recursive instrumental-variable (algorithm)
r.m.l. 1	Recursive maximum-likelihood 1 (algorithm)
r.m.l. 2	Recursive maximum-likelihood 2 (algorithm)
r.m.s.	Root-mean-square
s.a.	Stochastic approximation
s.i.s.o.	Single-input–single-output
s.n.r.	Signal-to-noise ratio
s.r.w.	Simple random walk
s.s.i.	Strongly system-identifiable
u.p.r.	Unit-impulse response
w.l.s.	Weighted least-squares
w.p. 1	With probability 1

Chapter 1

Introduction

1.1 WHAT IS IDENTIFICATION?

Identification is the process of constructing a mathematical model of a dynamical system from observations and prior knowledge. This definition raises quite a few questions. Why should we want a model? What exactly does "dynamical" signify? What sort of mathematical model? What sort of prior knowledge and observations? How do you construct such a model? How do you decide whether it is any good? The first two questions can be answered fairly quickly, as we shall soon see, but the others will take the rest of the book to answer even partially.

1.2 WHY SHOULD WE WANT A MODEL?

To make any sense, identification must have some definite purpose, although it is sometimes not clearly stated. Dynamical models find application in areas as diverse as engineering and the hard sciences, economics, medicine and the life sciences, ecology and agriculture; the references at the start of Chapter 7 give some idea of the range. The same few basic purposes underlie identification in all these fields.

1.2.1 Models to Satisfy Scientific Curiosity

A characteristic of science is its use of mathematical models to extract the essentials from complicated evidence and to quantify the implications. Identification has a long history in science.

Example 1.2.1 Halley conducted an identification exercise in 1704 when, realising that reports of comets in 1531, 1607 and 1682 related to a single object, he calculated the parameters of its orbit from the limited observations, and hence predicted its return in 1758 (Asimov, 1983; Murdin and Allen, 1979). The orbit determination relied on prior knowledge, from Newtonian dynamics and gravitational theory, that the orbit would be an ellipse.

We shall return to this example more than once, appropriately enough. △

The aim in scientific modelling is to increase understanding of some mechanism by finding the connections between observations relating to it. Any predictive ability of the model is an incidental benefit, valuable as a means of testing the model.

Example 1.2.2 Halley would have been pleased, no doubt, to know that the comet did return in 1758, sixteen years after he died, and pleased that his model was the reason for people reaching for their binoculars in 1985/6, but his immediate satisfaction came from understanding better how the comet behaved. △

1.2.2 Models for Prediction and Control

A wish to predict is a common and powerful motive for dynamical modelling. On a utilitarian view, a prediction model should be judged solely on the accuracy of its predictions. The plausibility and simplicity of the prediction model and its power to give insight are all incidental, although they help in model construction and acceptance. The narrowness of prediction as an aim paradoxically makes the choice of model wider.

Example 1.2.3 Hydrologists have a range of techniques for predicting river flow from measurements of rainfall and flow (Kitanidis and Bras, 1980; Kashyap and Rao, 1976). At one extreme a runoff-routing model represents in detail the physical features of areas within the catchment, and traces the passage of water through all the areas. At the other extreme a black-box model is estimated from measurements of flow at one place and rainfall at one place or averaged over the catchment. It makes no attempt to depict the internal workings of the catchment, but aggregates the catchment dynamics as they affect that particular flow. Runoff-routing models require much more field measurement to construct. They force a detailed examination of catchment peculiarities and, perhaps for that reason, inspire confidence in spite of the difficulty of testing such a large model. Black-box models have the advantages of a simple and standard form, e.g. linear difference equations, and fairly standard estimation techniques. They are simple enough to be updated continually according to recent prediction performance, though they may not be flexible enough, even with updating, to match the complicated non-linear dynamics of the catchment. △

Prediction by a dynamical model is important in control-system design. Design of any control scheme more ambitious than the traditional trial-and-error-tuned two-term controller requires a model (D'Azzo and Houpis, 1981; Richards, 1979). To keep the design procedure tractable, the model must be simple, even if that means rough. An accurate model may not be a realistic aim in any case, because of variability in the system to be controlled. Ideally the model would indicate the extent of the variability, so that the controller could be designed to have acceptable worst-case performance. This is less than straightforward, as disturbances, measurement inaccuracy and the limitations of the model structure also contribute uncertainty to the model (Ashworth, 1982).

Most control-design models employ a model for prediction only in the sense of saying how the system will respond to a standard input such as a step in the desired output value, or a specified disturbance. Prediction is more directly involved in two control techniques, feed-forward and self-tuning control, which have been underexploited through lack of reliable models. In feed-forward control, a disturbance is detected early in its progress through the system and fed forward, suitably shaped and inverted, to cancel its own effects further on. Self-tuning control recalculates the controlling input to the system periodically by reference to a periodically updated prediction model of the effect of that input. The identification aspects of self-tuning control are discussed briefly in Section 8.3.3.

1.2.3 Models for State Estimation

The object of state estimation is to track variables which characterise some dynamical behaviour, by processing observations afflicted by errors, wholly or partly random. The 1960's and early 1970's saw strikingly successful examples of state estimation in space navigation, including the Apollo moon landings, Mariner Mars orbiter and flybys of Venus and Mercury. The state estimated then was the position and velocity of a space vehicle, or equivalently its orbital parameters. The range of applications of state estimation has expanded rapidly, now embracing radar tracking, terrestrial navigation, stabilisation of ship motion, remote sensing, geophysical exploration, regulation of various industrial processes, monitoring of power systems and applications in demography (*IEEE Transactions on Automatic Control*, special issue, 1983).

State estimators rely on a model to relate the state variables to each other, to the observations and to the forcing. Commonly, some model parameters are initially unknown and must be identified, before or during state estimation. Section 8.9 considers combined state and parameter estimation.

Example 1.2.4 When a digital message is transmitted over a communication channel at a rate close to the maximum attainable, the channel dynamics smear out each signalling pulse and cause each received pulse to overlap several others. This inter-symbol interference must be corrected if the transmitted message is to be recovered (Clark, 1977). Fixed filters called equalisers can do the job if the channel dynamics are fixed. In a switched system such as the public telephone network the channel varies from connection to connection. It is also affected by temperature changes, and there is noise due to switching, poor contacts, crosstalk and induction from power apparatus. The equaliser must therefore adapt to the channel and, ideally, the noise characteristics. Many adaptive equaliser structures have been proposed, one being to estimate the message computationally as the state of an initially unknown and varying system consisting of the channel, its filters, coder and decoder, modulator and demodulator. The message estimator requires a model of the channel dynamics, which is updated continually (Lee and Cunningham, 1976; Luvison and Pirani, 1979). △

1.2.4 Modelling for Diagnosis of Faults and Inadequacies

A great benefit of identification, seldom acknowledged, is its ability to uncover shortcomings and anomalies. For instance, when an attempt is made to identify the dynamics of a system, the measurements are often found to be inadequate. Examples are a noisy thermocouple on a distillation column, incomplete economic statistics, too few rain gauges or too many ungauged inflows in hydrology, and physiological measurements too widely spaced because of the discomfort they cause. A deficiency like this may not be easy to put right, but the awareness of its importance may be worth the effort vainly spent on identification. Disclosure of unexpected or untypical behaviour of the system is equally valuable whether unpremeditated, as in the first example below, or the main reason for identifying a model, as in the second.

Example 1.2.5 A digital simulation of a pilot-scale gas-heated catalytic reactor was developed from results of tests on one reactor tube, physical chemistry and design information (Norton and Smith, 1972). The simulation was initially unable to match the observed steady-state temperature profile of the reagents. The trouble was traced to stagnation of the heating gas at one end of the reactor, pointing to a potential design improvement. The reactor model was also useful in a positive way in explaining an unexpected difference between the responses of the reagent temperature to perturbations of the inlet heating-gas temperature and flow rate. △

Example 1.2.6 A methionine tolerance test consists of giving a human subject an oral dose of methionine then sampling its concentration in the blood at five to ten instants over the next three hours or so. Abnormality in the variation of the concentration may be due to liver disease or diabetes. To aid interpretation and classification of the response, a model made up of two or three rate equations may be fitted to it (Brown *et al.*, 1979). △

1.2.5 Models for Simulation and Operator Training

Models make it possible to explore situations which in the actual system would be hazardous, difficult or expensive to set up. Aircraft and space-vehicle simulators are well-known examples. Comprehensiveness and accuracy are at a premium for this application, whereas cheapness and simplicity are less so. As well as operator training, simulation models are valuable for "what if?" analyses. Accuracy and completeness may be less crucial for the latter, when qualitative outcomes are being explored rather than precise numerical consequences. The discussion and thought stimulated by notorious world-growth models some years ago (Forrester, 1971; Meadows, 1973) justified their construction, imperfect as they may have been.

1.2.6 Models and Serendipity

We all sometimes stumble across something interesting when looking for something else entirely. This also happens in identification.

Example 1.2.7 In 1758 Messier was searching for Halley's comet, to validate Halley's orbit model. He found the Crab nebula and labelled it M1 in his catalogue. It was subsequently found to be a strong radio source (1948) and to have a pulsar at its centre (1968). In fact, it turned out to be more interesting than Halley's comet. △

Serendipity is hardly a motive for modelling, but it can be a weighty retrospective justification.

1.3 WHAT SORT OF MODEL?

One special family of dynamical models has identification methods far more fully developed than the rest: linear, lumped, time-invariant, finite-order models. The reason is that they are versatile, yet comparatively simple to identify, analyse and understand. We had better examine the properties of these models. In doing so, we shall incidentally say what we mean by "dynamical".

1.3.1 Dynamical Models; Model Order

The feature that distinguishes a *dynamical system* is that its output at any instant depends on its history, not just on the present input as in an *instantaneous system*. In other words, a dynamical system has memory. Often the memory can be attributed to some easily recognisable stored energy. If the present output can be expressed in terms of the input and output values an infinitesimal time ago, the input–output relation is a differential equation.

Example 1.3.1 The voltage $v(t)$ at time t across the capacitor in Fig. 1.3.1 is related to the source voltage $u(t)$ by $\dot{v}(t) = [u(t) - v(t)]/CR$. This equation is the limit, as time increment δt tends to zero, of $v(t) = v(t - \delta t) + \delta t[u(t - \delta t) - v(t - \delta t)]/CR$.

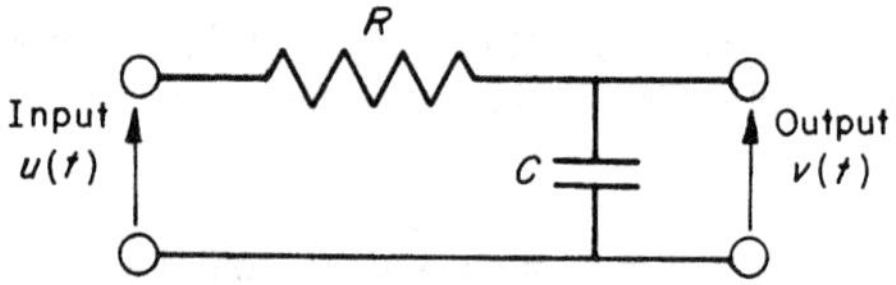

Fig. 1.3.1 A dynamical system.

An initial condition $v(t_0)$ and knowledge of $u(t)$ from t_0 onwards will give $v(t)$ from t_0 on. Conceptually, we find $v(t)$ from $u(t - \delta t)$ successively at $t = t_0 + \delta t$, $t_0 + 2\,\delta t$ and so on. In the limit, we integrate the differential equation to obtain

$$v(t) = v(t_0)\exp\left(\frac{t_0 - t}{CR}\right) + \frac{1}{CR}\int_{t_0}^{t} u(\tau)\exp\left(\frac{\tau - t}{CR}\right)d\tau$$

whatever the value of t_0, so $v(t_0)$ is enough to determine the effects of the history up to t_0 on the later behaviour. The capacitor charge $q(t_0)$ is $Cv(t_0)$, so the stored electric-field energy $q^2(t_0)/2C$ is determined by $v(t_0)$ and can be regarded as the memory of the system. △

The past of a system or model influences the future by way of a number of initial conditions or stored energies (one, in Example 1.3.1). The number is called the *system or model order*. Any model must describe how each of its energy storages contributes to the output, so the number of model parameters is at least equal to the model order. Some parameters may be known in advance, e.g. known to be zero, and the size of the model may well be reduced further by ignoring some energy storages because they give rise to dynamics which are too rapid, too slow or too small in amplitude to show.

Example 1.3.2 The resistor and wiring in Example 1.3.1 have a small stray inductance L, which modifies the input–output relation to $LC\ddot{v}(t) + RC\dot{v}(t) + v(t) = u(t)$. Two initial conditions are now needed before we can solve for $v(t)$, as there are two energy storages, electric-field energy in the capacitor and magnetic-field energy $LC^2\dot{v}^2(t)/2$ in the inductance. However, the magnetic-field energy can be ignored unless we are interested in circumstances giving very large rates of change of $v(t)$. △

The story is more complicated if there is significant pure delay in the system. Some delay is always present, since changes propagate through the system at finite velocity, but it may be negligible. A delay t_d right at the input merely delays the response by t_d, so it adds nothing to the difficulty of analysing the response. The same goes for a delay right at the output, and we have only to write $y(t + t_d)$ for $y(t)$ at the end of the analysis or $u(t - t_d)$ for $u(t)$ at the start in either case. Such delay is called *dead time*. With noticeable delay anywhere else, the response from any instant onwards depends not just on the conditions at that instant and the forcing, but also on the behaviour of the delayed variable throughout the delay interval, i.e. at an infinite number of instants. The number of initial conditions to be specified, and hence the system order, is infinite, making analysis more difficult (Problem 1.1).

We shall be paying most attention to models which relate the output at a succession of evenly spaced sampling instants to input and output at earlier sample instants. They also are hard to analyse when they contain a delay not at the input or output, even if it is an integer number of samples long, if the variables being sampled exist between samples as well. The difficulty does not arise, however, if the delay is in a part of the system which is entirely discrete in time, such as a digital filter or controller, since a complete specification of the delayed variable then only amounts to a finite number of sample values (Problem 1.2).

1.3.2 Lumped Models

When we write the variables describing a system as functions of time only, we imply that each is located at one point in space or has no spatial significance: the system is lumped, not distributed.

Example 1.3.3 Halley's comet extends over a considerable distance and alters in shape as it orbits. Its velocity is theoretically a function of three spatial dimensions and time. An astronomer is, in practice, content to know fairly precisely how its centre of gravity moves and, as a separate issue, approximately what happens to its shape. △

Example 1.3.4 Studies of water quality in rivers and lakes are concerned with diffusion and circulation of pollutants, nutrients and dissolved oxygen. Rather than modelling these quantities through partial differential equations as functions of two or three spatial dimensions as well as time, it may be permissible to represent rivers as cascades of well-mixed reaches, and lakes by two- or three-dimensional arrays of compartments. Exchanges of material between reaches or compartments are then described by a set of ordinary differential equations (Whitehead *et al.*, 1979). △

When a distributed variable is represented by one or more lumped variables, approximation error is incurred in the dynamics in addition to loss of resolution. Even so, the question is how to lump, not whether to lump, since digital computation requiring a lumped representation will be necessary at some stage in the analysis of the system unless the system and its boundary conditions are very simple.

1.3.3 Time-Invariant Models

A dynamical system is time-invariant if the sole effect of delaying its forcing and initial conditions is to delay its response by the same amount. In other words, the input–output relations do not vary with time and are relatively easy to analyse.

A time-varying model may be preferred to a more comprehensive but complicated time-invariant model, with the time variation showing the effects of the omitted part of the time-invariant model. Section 8.1 discusses this point further.

Example 1.3.5 (Dumont, 1982) Chip refiners in the wood-pulp and paper industry consist of two contra-rotating grooved plates which grind a mixture of wood chips and water. The wood-chip feed rate and motor power to the plates must be adjusted to control the energy input per unit mass of wood fibres. The motor load is adjusted by a hydraulic actuator which varies the gap between the grinding plates. Unfortunately the gain from plate gap to load power is non-stationary, because the plates wear, relatively slowly in normal operation but rapidly if the plates clash.

An "open-loop" estimate of the gain may be obtained from a wear index which records the plate age and number and severity of clashes. This is implicitly a model of the mechanism determining the gain. A more satisfactory solution is to update an empirical estimate of the gain at short intervals, as part of a closed identification and control loop. △

1.3.4 Linear Models

Consider a system with no initial stored energy. If its response to an input $u_1(t)$ is $y_1(t)$ and its response to $u_2(t)$ is $y_2(t)$, it is linear if its response to $\alpha u_1(t)+\beta u_2(t)$, with α and β any constants, is $\alpha y_1(t)+\beta y_2(t)$. A similar definition of linearity goes for systems with more than one input variable and response (output) variable. Linearity allows us to find the response to any input, however complicated, by breaking the input into simple components then adding the responses to each component. In identification, this implies that only the response to a suitable standard input need be identified, as in Chapters 2 and 3.

From Chapter 4 onwards, the model is required to be linear in its coefficients but not necessarily in its dynamics. That is, if we write the model as

$$f(y, y^{(1)}, \ldots, y^{(n)}, u, u^{(1)}, \ldots, u^{(m)}, \theta_1, \theta_2, \ldots, \theta_p) = 0 \qquad (1.3.1)$$

where $y^{(i)}$ means d^iy/dt^i and similarly for u, and θ_1 to θ_p are the coefficients to be estimated, the derivative of f with respect to each θ must be independent of all the θ's but the same is not necessary for the input- and output-dependent arguments of f, as it would be for linear dynamics.

Example 1.3.6 The circuit in Exercise 1.3.2 has linear dynamics but is nonlinear in parameters L, C and R. It is linear in LC and RC, though, and they can be regarded as its parameters if we do not insist on keeping L, C and R separate. △

1.3.5 Other Model Categories

Within the family of linear, lumped, time-invariant, finite-order models there are quite a few further distinctions to be made.

(i) *Input–Output versus State-Variable Models.* State-variable models have the structure shown in Fig. 1.3.2, with state variables as intermediaries between the inputs and outputs. The dynamics are expressed by a set of first-order ordinary differential equations, one per state variable. By this means the analysis of linear models of different orders is unified and brought within the

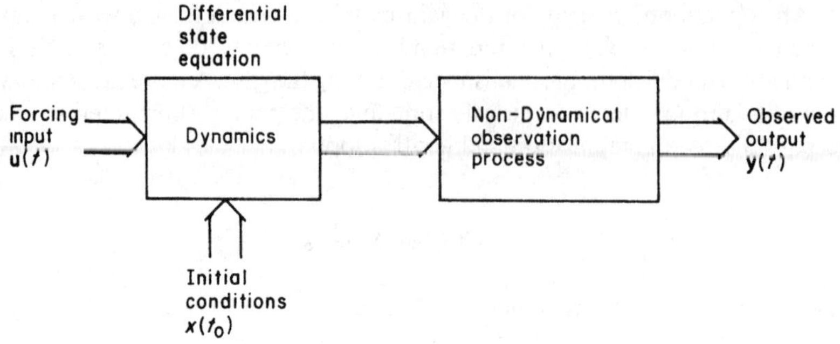

Fig. 1.3.2 State-variable model.

scope of linear algebra (Kailath, 1980). Models with multiple inputs and/or outputs fit into this framework as comfortably as single-input–single-output systems.

Any given input–output relation can be realised with any one of an infinity of equally valid choices of state variables. A suitable choice can either put the state equations into a convenient form, e.g. with certain coefficients zero (Problem 1.3), or make as many state variables physically meaningful as possible. This free choice, all within a standard form of model, is helpful in general but makes identification more complicated. The trouble is that the coefficients in a preferred state-variable model may not be identifiable from input–output behaviour alone, a point which must be checked for every candidate choice of state.

Example 1.3.7 The second-order input–output relation $LC\ddot{v} + RC\dot{v} + v = u$ of Exercise 1.3.2 can be rewritten as two first-order equations in state variables v and i, the capacitor voltage and source current:

$$\dot{i} = (-Ri - v + u)/L$$
$$\dot{v} = i/C$$

or equally well in terms of v and $v + Ri$:

$$\frac{d}{dt}(v + Ri) = \left(\frac{1}{RC} - \frac{R}{L}\right)(v + Ri) - \frac{v}{RC} + \frac{Ru}{L}$$
$$\dot{v} = \frac{1}{RC}(v + Ri) - \frac{v}{RC}$$

or v and $\dot{v}$, the latter denoted by w to avoid a cryptic equation $\dot{v} = \dot{v}$:

$$\dot{w} = (-RCw - v + u)/LC$$
$$\dot{v} = w$$

Any two independent linear combinations of v and i will do as state variables. Notice that the second and third alternatives have $1/RC$, $1/LC$ and R/L as parameters, easily related to the parameters of the second-order input–output equation and thus identifiable from the input–output behaviour. The first pair of state equations, however, has parameters R/L, $1/L$ and $1/C$, only the first of which can be identified from the input–output relation between u and v.

△

Since the choice of state-variable models for identification requires more background in linear algebra than the identification of input–output models, we touch on it only briefly among the more advanced topics in Chapter 8.

(ii) *Time-Domain versus Transform Models.* Linear, time-invariant models may be differential equations or impulse responses in the time domain, or transfer functions in the frequency or Laplace-transform domain (Gabel and Roberts, 1980; Ziemer *et al.*, 1983). The two are formally equivalent; we can move readily from one to the other, and the choice is a matter of practical convenience (Ljung and Glover, 1981). The great majority of recent developments in identification concern time-domain methods and so does most of this book.

(iii) *Deterministic versus Stochastic Models.* Identification methods span a range from making no provision for uncertainty in the measurements or the model to treating the model coefficients as random variables and modelling the errors and impairments in the measurements in some detail. The resulting models are respectively deterministic, i.e. certain, and stochastic, i.e. probabilistic with time as an independent variable (Helstrom, 1984; Melsa and Sage, 1973). As a compromise, some methods identify deterministic models but take care to allow for impaired measurements. Chapters 2–4 examine such methods.

(iv) *Single-Input–Single-Output versus Multi-Input–Multi-Output Models.* Identification methods for single-input–single-output (s.i.s.o.) models will be our main focus. They form the foundation of methods for multi-input–multi-output (m.i.m.o.) models. Section 8.7 looks briefly at m.i.m.o. models. An important feature of linearity in the dynamics is additivity of the output responses to separate inputs. Non-linear dynamics cause the response to any one input variable to be affected by the behaviour of the other inputs, and so it is necessary to identify the relations between all the input variables and the output simultaneously. The relations may be identified one at a time in a linear system, in principle, treating all but one of the inputs as sources of output disturbance (admittedly structured) while each input–output relation is identified.

(v) *Continuous-Time versus Discrete-Time Models.* All large systems and many small ones are identified from records taken at a succession of discrete instants, either because the data-logging or parameter estimation is digital or because the observations become available only periodically. Typical periodical observations are quarterly or monthly economic statistics, shift records from industrial processes and sampled signals from digital communication channels. The natural thing to do with discrete-time records is to identify a model which relates sample–time values but says nothing about what happens between samples. Such a model is convenient if it is intended for digital control design, state estimation or periodic prediction.

Information in continuous-time variables is lost when they are sampled unless the sampling rate is high enough (Gabel and Roberts, 1980, Chapter 5; Reid, 1983, Chapter 3; Ziemer *et al.*, 1983, Chapter 7). Without going into details, we note that the rate must be at least $2f$ to preserve a component at a frequency f, in the sense that accurate recovery of the component is theoretically possible by lowpass filtering of the sample sequence, if the sampler and filter are perfect. Allowing for imperfections and the gradual rather than abrupt decline of significant content with increasing frequency, a realistic sampling rate is 10 or so per period at the frequency at which the power starts to drop off rapidly, or at the cut-off frequency of a lowpass filter applied to the variable before sampling. Conversely, it is unwise to draw conclusions from a discrete-time model about behaviour at frequencies approaching half the sampling frequency.

Too-rapid sampling has its own drawbacks, fortunately serious only in rather specialised circumstances. It yields non-minimum-phase discrete-time models of some minimum-phase continuous-time systems (Åström *et al.*, 1984), making some adaptive control methods unfeasible. We shall not pay any further attention to this problem. Sampling will be assumed to be at a satisfactory rate and uniform in time whenever we consider discrete-time models.

(vi) *Parametric versus Non-Parametric Models.* One way to represent dynamical behaviour is by a function, say the output response $h(t)$ of a linear system to an impulse input, which is not parameterised. That is, the function is specified directly by the result it gives for each value of its argument. The alternative is to nominate first a family of functions, such as all sums of exponentials, then one or more parameter values to pick one member out, such as the number of exponentials to be included in our model, and finally coefficient values, like the initial value and coefficient of time in the exponent for each exponential.

The benefit of restricting the model to a parametric family is economy; relatively few parameters and coefficients are needed to describe it. Such economy is achieved by taking the trouble to find a suitable model structure, then going through a more complicated and less general identification procedure than for a non-parametric model. The impulse-response, step-response and frequency-response methods of Chapters 2 and 3 are non-parametric, while the methods of later chapters are most often applied to parametric models. Wellstead (1981) reviews non-parametric identification methods from a practical viewpoint.

In practice the distinction between parametric and non-parametric models is not sharp. For instance, the number of instants we evaluate $h(t)$ at, and the interval, are in effect parameters. A working definition of "parametric model" is "pretty restrictive model, identified in stages (structure, parameters, coefficients)". "Non-parametric model" might be interpreted as "fairly unrestrictive model, identified all in one go".

A potential source of confusion is that "parameter" is used in identification both in the sense employed here and to mean any number which is not a variable, e.g. the coefficient of each term in a non-parametric linear model. The mix-up is firmly established and there is a lack of alternative words with precise enough meanings, so we shall just try not to read too much into the word.

(vii) *Sectioned versus Unitary Models.* Sectioning a model can simplify identification by separating aspects of behaviour which can be identified one at a time. A natural and often intuitively obvious basis for sectioning is local–regional–global. For example, a distillation column might be modelled on two scales, local s.i.s.o. relations between feed flow rate or temperature and product flow rate or column temperature, for instance, forming parts of a model of the column as a whole. The column model might then form part of a model of a refinery. Control is conveniently split up in the same way, with local single-loop controllers co-ordinated by manipulation of their set-points, and overall control exerted through a relatively long-term scheduling process, in a control hierarchy.

Differences in time scale are also a basis for sectioning. Indeed, models are often split up by time scale with little conscious thought, treating slow components of an output as drift while identifying faster dynamics, and fast dynamics as instantaneous while identifying slower ones. When the spread of speeds is large, separate treatment of fast and slow dynamics is highly desirable, as otherwise the choice of sampling rate is difficult. A rate high enough for the fastest dynamics implies a large number of samples to cover the slowest. Resolution of the fast dynamics may be lost, or estimation of the slow dynamics spoilt by cumulative error.

Example 1.3.8 With $R^2C/L = 100$ in the circuit of Exercise 1.3.7, the response $v(t)$ to an impulse input $u(t)$ contains two decaying exponentials. The ratio between the exponents is 98:1. In an interval L/R, the faster exponential decays by a factor 2.718 and the slower by less than 1%. To determine the faster component accurately the output should be sampled at intervals not much over $0.4L/R$, the time to decay by 33%. An unparameterised impulse-response model would then need over 750 terms to cover the time up to when the slower component has decayed to 5% of its initial value.

A parametric second-order transfer-function model (Section 2.3) gives the same response with only three coefficients. However, an error of 0.1% in the decay of the slower component over one $0.4L/R$ interval gives an error of 112% in the calculated value of the impulse response at a lag of 750 samples, so the model coefficients would have to be found to an impossible degree of accuracy. The remedy is to use a longer sampling interval, around $40L/R$, in estimating the slower component (Problem 2.3.) △

Large systems with relatively few internal connections are easily split into sections if those connections are accessible for measurement. So are systems with response components spanning a large range of speeds but not too many of similar speed. It is a different story for large systems with complicated internal connections and systems with a fairly uniform spread of response speeds. Reduction of models with linear dynamics has received a good deal of attention, with an eye on simplified models for control design, and is the subject of Section 8.5. Methods for decomposing large systems for identification (Mahmoud and Singh, 1981) have not yet had much impact on the cut-and-try approach usual in identification. A fundamental obstacle to automation of the identification of large systems (and small ones, come to that) is the need, in the end, to decide part by part whether the model is credible. If model testing and validation has to be piecemeal, and it has, there is less incentive to avoid a piecemeal approach to identification.

We should note that a system may be large and strongly interacting without necessarily being difficult to section for identification and control design. The accessibility of variables for measurement is the determining factor. An example is a steelworks cold-rolling mill (Bryant and Higham, 1973), where the effects of control action at different stands interact strongly but through a small number of well-instrumented variables such as strip gauge, tension and speed.

(viii) *Markov-Chain Models.* We shall look only at models based directly or indirectly on differential equations, or, in discrete time, difference equations, with the system inputs and outputs as variables. Other types of model are valuable in some applications, notably Markov-chain models. These specify the probability of every possible transition from a state at one instant to a state at the next (Luenberger, 1979, Chapter 7; Wadsworth and Bryan, 1974, Chapter 9). The variables are the probabilities of being in each possible state at each instant. The transition probabilities are assumed constant and independent of previous history. At first sight a Markov-chain model looks very different from a difference-equation model, which gives the next output in terms of input values over a *range* of times. When the difference equation includes an additive random-variable "noise" term to account for unknown disturbances and measurement error, some similarities can be traced. We can rewrite the difference-equation model as a set of first-order difference equations, i.e. a state-variable model, with the next state expressed in terms of present, not past, state, input and noise. Moreover, the next state is given as a probability distribution, determined by the probability distributions of the noise and present state, although for simplicity we usually quote (and compute) only the mean value of the output and perhaps some measure of its variability about the mean. Thus we see that a stochastic difference-equation model also gives future state probabilities from present state probabilities. The remaining difference between Markov-chain and ordinary difference-equation models is one of emphasis. The difference-equation model gives the next state mainly as a deterministic function of present state and known input, with the uncertainty brought in via the noise. The Markov-chain model is entirely probabilistic, with no mechanism describing how the next state is determined, other than the transition probabilities.

Identification of a Markov-chain model presupposes either enough observations of each possible state transition to yield its probability, or a good knowledge of the causes which determine the probabilities.

1.4 HOW DO YOU CONSTRUCT A MODEL?

No straight answer can be given to the question of how to construct a model. The best way to construct a model depends on a host of practicalities, not all foreseeable, and all we can do is generalise by pointing out a few of the stages and some of the constraints on the choice of technique.

1.4.1 Stages in Identification

Chapters 2–8 introduce identification methods, Chapter 9 talks about experiment design and Chapter 10 discusses the analysis of results. Very seldom does an identification project consist of a single pass through the sequence (i) pick a model, (ii) design the experiment, (iii) do the experiment and (iv) analyse the results. Figure 1.4.1 sketches identification more realistically. Most items in Fig. 1.4.1 are just common sense, but two of these items deserve more attention than they tend to get in the literature, namely informal checks on the records and model validation. Both are discussed at length in Chapter 10.

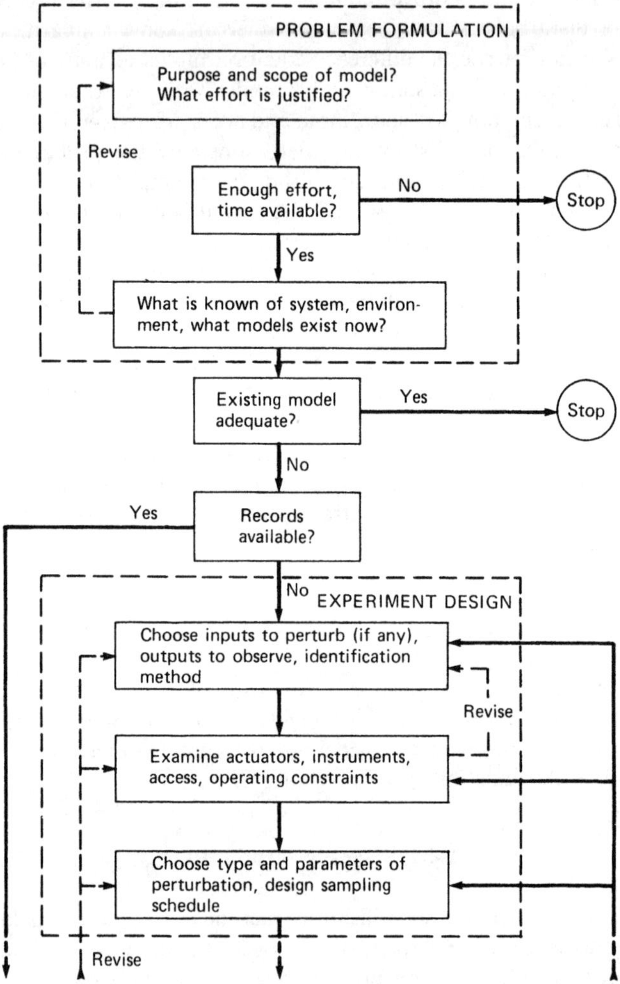

Fig. 1.4.1 Identification.

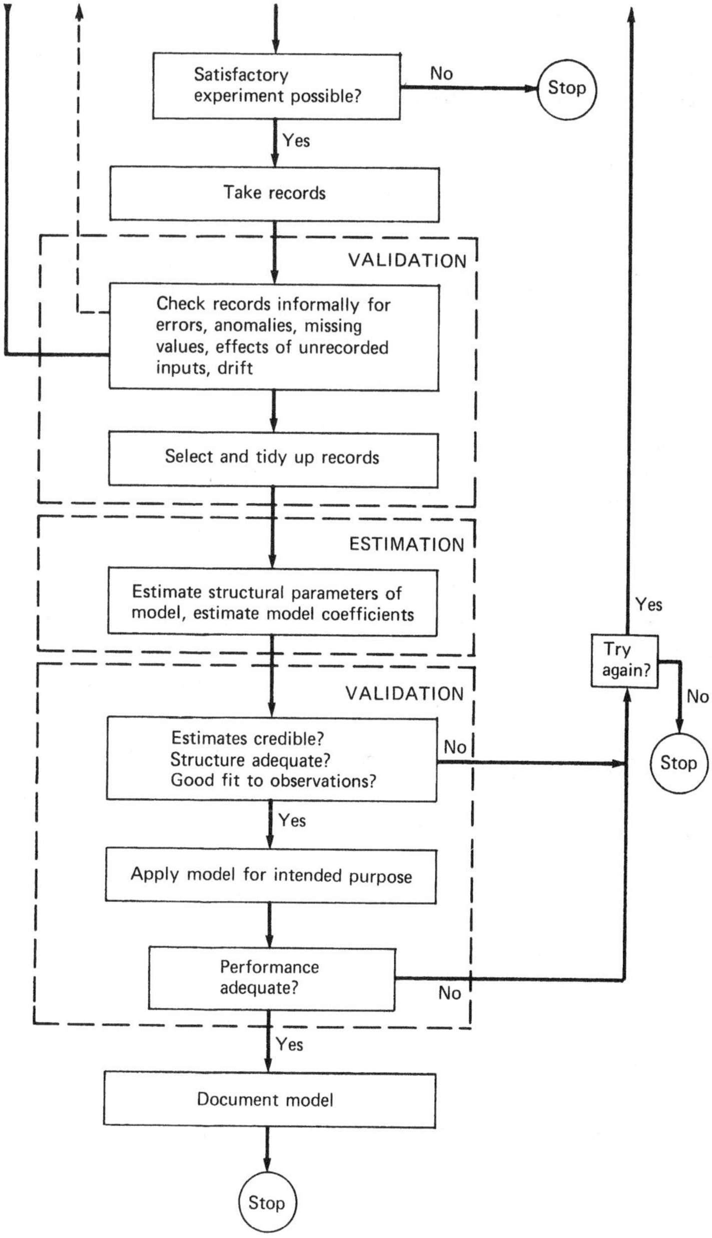

Fig. 1.4.1 (*continued*)

1.4.2 Constraints on Identification Methods and Results

Once the aim of the identification exercise is clear and the system to be modelled has been defined, the methods adopted and results obtained depend on

(i) *access to the system:* historical records, e.g. economic or hydrological, with no opportunity to influence them, or normal operating records with no control over the input but some chance to improve instrumentation, or responses to planned input perturbations in the presence of other disturbances and drift, or responses from bench tests in controlled conditions, or detailed examination of a system which can be dismantled and tested in sections, or, most likely of all, some combination of these; repeatability of tests and volume of records are other important factors;

(ii) *time available:* the response of a bandpass communication channel might be found in milliseconds or less at little cost, while a blast furnace with transient responses lasting a day or two has to be logged for weeks or months to get sufficient records (Norton, 1973), so its identification is costly (but so is ignorance of its dynamics);

(iii) *instruments and actuators:* the average power, instantaneous value, rate of change and smoothness of change of input perturbations are limited by the input actuators and the system; the size of an output response to a perturbation is normally stringently limited in process plant, to ensure usable product; the sampling rate may be limited by instruments or logger, or by the time taken to collect records, particularly in the life sciences or economics; test duration may be limited by instrument and actuator unreliability or by human endurance when manual collection of records is involved (Stebbing *et al.*, 1984); instrument noise or sampling error may be the predominant factor limiting the quality of the records;

(iv) *computing facilities:* lack of computing power is not the constraint it once was, but an important question is still whether the model can be updated sequentially as measurements are taken, or requires off-line iteration through the records; the back-of-an-envelope calculation of approximate model parameters, e.g. gains and time constants from Bode plots or step responses (Chapters 2 and 3), is still a great attraction of some classical identification methods;

(v) *availability of specialised equipment and methods:* transfer-function analysers and cross-correlators (Chapters 2 and 3) make identification relatively quick and easy in circumstances that suit them; specialised methods of estimating models have developed in many fields, such as reverberation testing of rooms (Parkin and Humphreys, 1958) or "curve-peeling" for separating exponentials in biomedical test results (Jacquez, 1972, Chapter 7), but we have not the space to cover them;

(vi) *precision and completeness required:* effort is not unlimited, even for academics, and the law of diminishing returns applies as much to identification as to anything else.

1.5 HOW TO READ THIS BOOK

Identification is not a spectator sport. The only way to find out what the various techniques really can and cannot do is to try them. There *are* pieces of fairly portable technique with a firm theoretical base, otherwise this book would hardly be justified, but they all have their weaknesses, and every application seems to have some non-standard feature to test them. The book recognises the numerical and empirical nature of identification by resorting to numerical examples as often as possible. They are intended to be followed through in detail and often raise significant points; they are not just illustrations. In many of them it would be worthwhile to alter some of the numbers or details and explore the consequences. Similarly, the end-of-chapter problems are not primarily drill exercises, but are intended to encourage scrutiny of further practical issues. The best accompaniments to the text, though, are a set of records from an actual dynamic system, and someone who knows their peculiarities and wants a model from them.

REFERENCES

Ashworth, M. J. (1982). "Feedback Design of Systems with Significant Uncertainty". Research Studies Press, Wiley, New York.

Asimov, I. (1983). "The Universe", 3rd ed. Penguin, London.

Åström, K. J., Hagander, P., and Sternby, J. (1984). Zeros of sampled systems. *Automatica* **20**, 31–38.

Brown, R. F., Godfrey, K. R., and Knell, A. (1979). Compartmental modelling based on methionine tolerance test data: a case study. *Med. Biol. Eng. Comput.* **17**, 223–229.

Bryant, G. F., and Higham, J. D. (1973). A method for realizable non-interactive control design for a five stand cold rolling mill. *Automatica* **9**, 453–466.

Clark, A. P. (1977). "Advanced Data-Transmission Systems". Pentech Press, London.

D'Azzo, J. J., and Houpis, C. H. (1981). "Linear Control System Analysis and Design", 2nd ed. McGraw-Hill Kogakusha, Tokyo.

Dumont, G. A. (1982). Self-tuning control of a chip refiner motor load. *Automatica* **18**, 307–314.

Forrester, J. W. (1971). "World Dynamics". Wright-Allen, Cambridge, Massachusetts.

Gabel, R. A., and Roberts, R. A. (1980). "Signals and Linear Systems", 2nd ed. Wiley, New York.

Helstrom, C. W. (1984). "Probability and Stochastic Processes for Engineers". Macmillan, New York.

IEEE (1983). Special issue on Application of Kalman Filtering. *IEEE Trans. Autom. Contr.* **AC-28**, 3.

Jacquez, J. A. (1972). "Compartmental Analysis in Biology and Medicine". Elsevier, Amsterdam.

Kailath, T. (1980). "Linear Systems". Prentice-Hall, Englewood Cliffs, New Jersey.

Kashyap, R. L., and Rao, A. R. (1976). "Dynamic Stochastic Models from Empirical Data". Academic Press, New York and London.

Kitanidis, P. K., and Bras, R. L. (1980). Real-time forecasting with a conceptual hydrologic model. *Water Resour. Res.* **16**, 1025–1044.

Lee, T. S., and Cunningham, D. R. (1976). Kalman filter equilization for QPSK communications. *IEEE Trans. Commun.* **COM-24**, 361–364.

Ljung, L., and Glover, K. (1981). Frequency domain versus time domain methods in system identification. *Automatica* **17**, 71–86.

Luenberger, D. G. (1979). "Introduction to Dynamic Systems". Wiley, New York.

Luvison, A., and Pirani, G. (1979). Design and performance of an adaptive Kalman receiver for synchronous data transmission. *IEEE Trans. Aerospace Electron. Sys.* **AES-15**, 635–648.

Mahmoud, M. S., and Singh, M. G. (1981). "Large Scale System Modelling". Pergamon, Oxford.

Meadows, D. L. (1973). "Dynamics of Growth in a Finite World". Wright-Allen, Cambridge, Massachusetts.

Melsa, J. L., and Sage, A. P. (1973). "An Introduction to Probability and Stochastic Processes". Prentice-Hall, Englewood Cliffs, New Jersey.

Murdin, P., and Allen, D. (1979). "Catalogue of the Universe". Cambridge Univ. Press, London and New York.

Norton, J. P. (1973). Practical problems in blast furnace identification, *Meas. Control* **6**, 29–34.

Norton, J. P., and Smith, W. (1972). Digital simulation of the dynamics of a fixed-bed catalytic reactor. *Meas. Control* **5**, 147–152.

Parkin, P. H., and Humphreys, H. R. (1958). "Acoustics, Noise and Buildings". Faber, London.

Reid, J. G. (1983). "Linear System Fundamentals". McGraw-Hill, New York.

Richards, R. J. (1979). "An Introduction to Dynamics and Control". Longman, London.

Stebbing, A. R. D., Norton, J. P., and Brinsley, M. D. (1984). Dynamics of growth control in a marine yeast subjected to perturbation. *J. Gen, Microbiol.* **130**, 1799–1808.

Wadsworth, G. P., and Bryan, J. G. (1974). "Applications of Probability and Random Variables", 2nd ed. McGraw-Hill, New York.

Wellstead, P. E. (1981). Non-parametric methods of system identification. *Automatica* **17**, 55–69.

Whitehead, P. G., Young, P. C., and Hornberger, G. (1979). A systems model of flow and water quality in the Bedford Ouse river system—1. Stream-flow modelling. *Water Res.* **13**, 1155–1169.

Ziemer, R. E., Tranter, W. H., and Fannin, D. R. (1983). "Signals and Systems: Continuous and Discrete". Macmillan, New York.

PROBLEMS

1.1 A system containing a pure delay τ is described by

$$\dot{y}(t) + ay(t-\tau) = bu(t)$$

Its output $y(t)$ is zero over the interval τ up to time zero. The input $u(t)$ is zero at all times except zero, when it is a very short, unit-area impulse. Find $y(\tau)$, $y(2\tau)$, $y(3\tau)$, and if you have the patience, $y(4\tau)$, by integration over $0 < t \leq \tau$, $\tau < t \leq 2\tau$ and so on. Compare these values of $y(t)$, and the effort it costs to get them, with those of the system described by

$$\dot{y}(t) + ay(t) = bu(t)$$

1.2 The output $y(t)$ from a digital controller is related to the input $u(t)$ by the discrete-time equation

$$\frac{y(iT) - y[(i-1)T]}{T} + ay[(i-k)T] = bu[(i-1)T], \qquad i = 1, 2, 3, \ldots$$

where T is the sampling interval, a and b are constants and k is a fixed positive integer. The output is zero up to and including time zero. A unit-pulse input, which is one at time zero and zero at all other times, is applied. For (i) $k = 1$, (ii) $k = 3$ find the resulting output, over a long enough period for its behaviour to become clear. Does the extra delay in (ii) make the output any more complicated?

1.3 Verify that for the system of Exercises 1.3.2 and 1.3.7, state variables

$$x_1 = (1 + \alpha)v/2 + L\alpha i/CR, \qquad x_2 = (1 - \alpha)v/2 - L\alpha i/CR$$

give decoupled state equations of the form

$$\dot{x}_1 = \lambda_1 x_1 + \alpha u/CR, \qquad \dot{x}_2 = \lambda_2 x_2 - \alpha u/CR$$

where $\alpha = (1 - 4L/CR^2)^{1/2}$ and λ_1, λ_2 are the poles (eigenvalues) of the system. What is the observation equation relating $v(t)$ to these state variables? By integrating the state equations find the response $v(t)$ to a unit-impulse input $u(t)$ for (i) $4L < CR^2$, (ii) $4L > CR^2$. What functions of R, L and C can be identified from this response? If the amplitude of the response is unknown but its waveform is otherwise known accurately, is there any change in what can be identified?

1.4 (Mainly for electrical engineers) How would you find LC and RC in the input–output o.d.e. of Exercises 1.3.2 and 1.3.7, if you could choose $u(t)$ freely but only record $v(t)$, making no other measurements? If the network were in a vandal-proof box on the bench, with the input and output terminals labelled, but you knew only that the box contained passive, bilateral components, could you identify the nature, configuration and values of the components? If so, you would be relying on electrical engineering background.

This problem illustrates the large difference between "black box" identification of input–output dynamics and identification of the internal structure of the system from external measurements. The latter requires more thought and more background knowledge about the system.

Chapter 2

Classical Methods of Identification: Impulse, Step and Sine-Wave Testing

2.1 TIME-DOMAIN DESCRIPTION OF RESPONSE

The identification methods of this chapter rely on the theory of s.i.s.o., linear, time-invariant dynamical systems as covered by electrical and control engineering degree courses and many others. Suitable textbooks, useful also for later chapters, include Gabel and Roberts (1980), Reid (1983) and Ziemer *et al.* (1983). A reminder of the theoretical background will be given for each method to minimise the need for background reading.

2.1.1 Impulse Response and Initial-Condition Response

Linearity allows us to find the response of a linear system to any forcing $u(t)$, $t \geq 0$, by

(i) calculating or measuring the response to some very simple standard waveform, say a step of size 1 or a short rectangular pulse of area 1, then

(ii) breaking $u(t)$ up into a collection of components, each a scaled and delayed version of the standard waveform, approximating $u(t)$ if necessary, and finally,

(iii) building up the total response by summing the responses to the components ("superposition").

Figure 2.1.1 illustrates this procedure with a short, rectangular, unit-area pulse as the standard waveform. The shorter the pulse, the better $u(t)$ can be approximated. If we make the pulse width tend to zero, keeping the area unity, the standard waveform becomes a *unit (area) impulse* or *Dirac δ function*. The unit impulse at time zero is defined by

$$\delta(t) = 0 \quad \text{for all} \quad t \neq 0, \qquad \int_{-\infty}^{\infty} \delta(t)\, dt = 1 \tag{2.1.1}$$

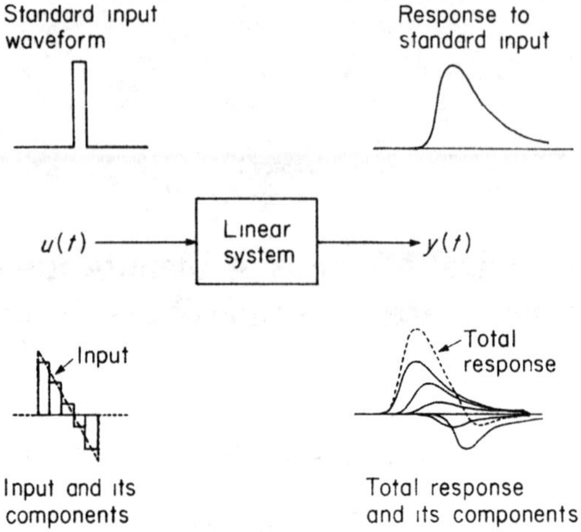

Fig. 2.1.1 Finding response of linear system by superposition.

The section of input covering a short time $\Delta\tau$ about time τ is roughly a rectangular pulse with area $u(\tau)\,\Delta\tau$. As $\Delta\tau$ tends to zero, this tends to $u(\tau)\,\Delta\tau$ times a unit impulse occurring at time τ, $\delta(t-\tau)$. If $h(t)$ is the *unit-impulse response* of the system to $\delta(t)$, the response to the impulse $\delta(t-\tau)u(\tau)\,\Delta\tau$ is $h(t-\tau)u(\tau)\,\Delta\tau$. Summing the contributions from infinitesimally short sections of input all the way from its start at $\tau = 0$ to the present time $\tau = t$ we find the total forced response as the *convolution or superposition integral*

$$y(t) = \int_0^t h(t-\tau)u(\tau)\,d\tau, \qquad t \geq 0 \tag{2.1.2}$$

Any non-zero initial conditions, due to initial stored energy in the system, also contribute to the response. The initial-condition response just adds to the forced response, because of linearity. In fact, it can be thought of as the response to an input before time zero, which was carefully designed to establish the specified conditions at time zero. The presence of an initial-condition response in an identification experiment means that to identify the system we must first know, or find out, the initial conditions, e.g. $y(0)$ and $\dot{y}(0)$ for a second-order system. As they generally include derivatives of a noisy signal, they may not be easy to measure. We therefore either arrange zero initial conditions by letting the system settle to quiescence before perturbing it, or observe $y(t)$ for long enough for the forced response to predominate as the initial-condition response dies away. The (unit-) impulse response is found by solving (2.1.2), e.g. by Laplace transforms. If a parametric model such as a differential equation is then fitted, the response to any specified initial conditions (i.c.) can also be found.

Example 2.1.1 A stable system is modelled by $\ddot{y}+a_1\dot{y}+a_2y=bu$. Taking Laplace transforms of this equation,

$$s[sY(s)-y(0_-)]-\dot{y}(0_-)+a_1[sY(s)-y(0_-)]+a_2Y(s)=bU(s)$$

so

$$Y(s)=\text{forced response}\frac{bU(s)}{s^2+a_1s+a_2}+\text{i.c. response}\frac{(s+a_1)y(0_-)+\dot{y}(0_-)}{s^2+a_1s+a_2}$$

Consider the case where $s^2+a_1s+a_2$ can be factorised into $(s-\alpha_1)(s-\alpha_2)$ with α_1 and α_2 being real, and $u(t)$ is a unit impulse, so $U(s)$ is 1. We split $Y(s)$ into partial fractions and invert them to find the total response

$$y(t)=\frac{1}{\alpha_1-\alpha_2}\{[b-\alpha_2y(0_-)+\dot{y}(0_-)]\exp(\alpha_1t)-[b-\alpha_1y(0_-)+\dot{y}(0_-)]\exp(\alpha_2t)\}$$

Provided α_1 and α_2 are not too close, we can estimate them from the impulse-response test by saying that $y(t)$ is dominated by the slower-to-decay exponential towards the end of the transient, fitting a single exponential to this tail, subtracting it from $y(t)$ and fitting a faster exponential to the remainder. Hence a_1 and a_2 are estimated. However, b can be found only if $y(0_-)$ and $\dot{y}(0_-)$ are known in advance. The response to any specified initial conditions can be determined without knowing b, and therefore without knowing the $y(0_-)$ and $\dot{y}(0_-)$ in the impulse-response test. △

2.1.2 Discrete-Time Forced Response: Unit-Pulse Response

The forced response of a discrete-time system is easily found by superposition. The obvious standard input is a unit pulse, i.e. one at time zero. The zero-initial-condition response to this input, the *unit-pulse response* (u.p.r.), consists of a sequence of pulses of size $h_0, h_1, h_2, \ldots$ at times $0, T, 2T, \ldots$, where T is the sampling period. No real system responds instantly to an input, so h_0 is zero, although it sometimes makes sense to ignore a delay much smaller than T. Superposition gives the response at sample instant t to an input sequence $u_0, u_1, u_2, \ldots$ at $0, T, 2T, \ldots$ as the *convolution sum*

$$y_t=\sum_{k=0}^{t}h_{t-k}u_k, \qquad h_0 \text{ probably zero}, \quad \text{integer } t\geq 0 \tag{2.1.3}$$

Example 2.1.2 At intervals of T seconds, a radar gives a sample of the position x of a target moving in a straight line. A microprocessor forms an estimate $g_t=(x_t-2x_{t-1}+x_{t-2})/T^2$ of the target's acceleration, which is added to the signal controlling the torque of the motor rotating the radar antenna, by way of a digital-analogue converter, hold and power amplifier.

The response g to a unit pulse $x_0 = 1$ at time zero is $1/T^2$ at time t_c, $-2/T^2$ at $T + t_c$ and $1/T^2$ at $2T + t_c$, where t_c is the delay in the microprocessor. Hence the initial response g_0 is zero, and the hold output is $1/T^2$ at time T, $-2/T^2$ at $2T$ and $1/T^2$ at $3T$. However, if $t_c \ll T$ and the motor torque responds rapidly to the control signal, the antenna angular position is more accurately calculated by ignoring the delay t_c. On the other hand, the inertia of the motor and antenna prevents angular position from responding rapidly to torque changes, so the response of sampled angular position to x_t first shows at sample $t+1$, and h_0 is taken as zero in the overall u.p.r. △

Commonly an estimated u.p.r. describes the sampled behaviour of a continuous-time system. We must be careful to recognise the limitations of a u.p.r. in those circumstances. The u.p.r. may miss significant fast dynamics by having too large a sampling period, or slow dynamics by being too restricted in duration. The dead time (pure delay) is determined only to within one sampling period.

Example 2.1.3 (i) Figure 2.1.2a shows two u.p.r.'s, one with $h_t = 1.1(0.6)^{t-1} - 0.1(0.95)^{t-1}$, $t \geq 1$ and the other with $h_t = 0.555^{t-1}$, $t \geq 1$. Over the first five samples they look very similar, yet the corresponding steady-state gains (final responses to a sampled unit step) are 0.75 and 2.247, respectively. The explanation is that the steady-state gain is the sum of all the u.p.r. values h_0 to h_∞, and the low-amplitude slowly decaying second component of the first u.p.r. ultimately cancels 73% of the sum due to the faster first component, even though its effect on h_1 to h_5 is small.
(ii) The two sections of u.p.r. in Fig. 2.1.2b differ by at most 5% or so. They are $h_t = 0.8^t - 0.1^t$, $t \geq 1$ and $h_t = 0.72(0.828)^{t-1}$, $t \geq 1$. They represent the sampled behaviour of zero-dead-time continuous-time systems with greatly differing impulse responses. That of the first system rises at a finite rate from zero, while that of the latter jumps abruptly to 0.87 (Problem 2.5). The former has fast dynamics which are initially important in the impulse response but hardly show at all in the u.p.r.
(iii) Consider the second u.p.r. in (i). It could derive from sampling a continuous-time system with negligible dead time, impulse response $\exp[(t/T)\ln 0.555]/0.555$, steady-state gain (area under the impulse response) 3.06 and impulse-response peak 1.802. At the other extreme, it might come from a continuous-time system with dead time just less than T, impulse response $\exp[(t/T - 1)\ln 0.555]$, $t \geq T$, steady-state gain 1.698 and impulse-response peak 1. △

The difficulties and dangers of fitting models to impulse responses are further illustrated, with many numerical examples, by Godfrey (1983).

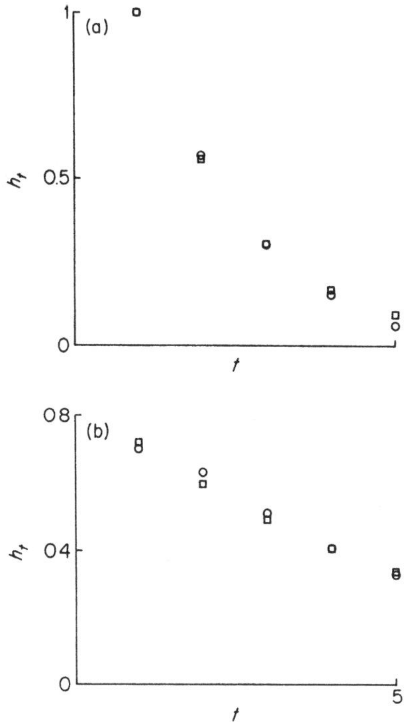

Fig. 2.1.2 (a) Unit-pulse responses, Example 2.1.3(i). $h_t = 0.555^{t-1}$, ○; $h_t = 1.1(0.6)^{t-1} - 0.1(0.95)^{t-1}$, □. (b) Unit-pulse responses, Example 2.1.3(ii). $h_t = 0.8^t - 0.1^t$, ○; $h_t = 0.72(0.828)^{t-1}$, □.

Another point to note when identifying the u.p.r. of a continuous-time system is that a sampled input is usually applied to a continuous-time system through a hold circuit, which reconstructs a continuous-time signal from the samples. The dynamics of the hold circuit will be included with those of the system in the identified model, and might noticeably affect them. The simplest and most widely used hold is the zero-order hold, which provides a constant output equal to the most recent input. If it is taken to respond to the area of the input pulse rather than its amplitude, as is realistic, its impulse response $h(t)$ is a pulse of height 1 and duration T, the sampling interval. Hence $h(t)$ and its frequency transfer function are

$$h(t) = \mu(t) - \mu(t - T), \qquad H(j\omega) = (1/j\omega)(1 - e^{-j\omega T}) \tag{2.1.4}$$

where $\mu(t)$ is the unit step at time zero. The effects of the hold are negligible at frequencies much below $1/\pi T$ Hz (Problem 2.7).

2.1.3 Step Response

A convenient feature of linear systems is that an ideal integrator at the input has the same effect as one at the output. The response to an integrated unit impulse, i.e. a unit step, is therefore the time integral of the impulse response. We can, in principle, measure the step response and find the impulse response by differentiation. How practical this is we discuss in Section 2.2.2.

We must be rather careful with discrete-time systems on this point. The u.p.r. can be found, in principle, by differencing the response to a *sampled* unit step but not the sampled response to an unsampled unit step.

Example 2.1.4 A continuous-time system has unit-impulse response $ae^{-t/\tau}$. Its unit-step response is therefore $a\tau(1-e^{-t/\tau})$, which when sampled gives $a\tau(1-e^{-kT/\tau})$ at sample k. The response at time kT to a sampled-unit-step input, on the other hand, is

$$a\tau(e^{-kT/\tau}+e^{-(k-1)T/\tau}+\cdots+e^{-T/\tau}+1)$$
$$=a\tau(1-e^{-(k+1)T/\tau})(1-e^{-T/\tau}) \qquad \triangle$$

2.2 DIRECT MEASUREMENT OF IMPULSE AND STEP RESPONSES

2.2.1 Measurement of Impulse Response

The simplest of all identification techniques is to find the impulse response by putting in an impulse and seeing what comes out. The model is acquired in a single response measurement if noise is not excessive. The influence of noise can be reduced if necessary by repeating the perturbation and averaging the responses. The averaging depends on the inconsistency of the noise. Summing N responses gives N times the consistent part but less than N times the inconsistent part, and thus improves the signal:noise ratio (s.n.r.). A more formal statistical justification is investigated in Problem 5.4. Measured impulse responses may also contain structured disturbances such as slow drift or periodic responses to unmeasured forcing inputs, not necessarily reduced by averaging. They must be filtered out instead. Freehand interpolation and subtraction of the disturbance on a response plot may be enough. Other options are bandpass filtering or even fitting an explicit model to the disturbance so that it can be subtracted.

A practicable test input has finite duration and amplitude, and can only approximate the δ function. The duration must be short compared with the fastest feature of interest in the impulse response, and the amplitude large enough to give an acceptable output s.n.r. The actual response is the ideal impulse response convolved with the input pulse waveform, as in (2.1.2). The effect is to blur the impulse response. For instance, a rectangular pulse of duration t_d and area 1 would give by (2.1.2)

$$y(t) = \frac{1}{t_d} \int_0^{t_d} h(t-\tau)\, d\tau \simeq h(t) \tag{2.2.1}$$

rather than the ideal $h(t)$.

A large-amplitude short pulse may be difficult to produce. Moreover, it leaves some doubt as to whether the response is linear and typical or is affected by large-signal non-linearity such as saturation. The pulse amplitude is limited by the input actuator range, and usually by the maximum size of perturbation regarded as acceptable.

Example 2.2.1 The silicon content of the pig iron produced by a blast furnace is a good indicator of how the furnace is running. It is influenced by the temperature of the hot air blast. The response of silicon content to a blast temperature perturbation lasts of the order of a day and affects several successive casts of iron (Unbehauen and Diekmann, 1983; Norton, 1973). The temperature can be altered quite rapidly, but a short pulse of higher or lower temperature would have to be very large for its effect to be discernible in the normal cast-to-cast silicon variation. Even if it could be produced, such a pulse would not be risked. The prime concern is to run the furnace smoothly. A large temperature pulse might cause irregular behaviour, difficult to predict or correct and untypical of normal operation. In an extreme case, hanging and slipping of the burden (ore, sinter and coke) could occur, perhaps damaging the refractory lining. △

Another potential difficulty is that the actuator may have dynamics comparable in time scale with, and inseparable from, the dynamics of the rest of the system, but we may not want to include them in the model.

2.2.2 Measurement of Step Response

A step is an indefinite succession of contiguous, equal, short, rectangular pulses and produces a much larger response and higher s.n.r. than does one short pulse of the same peak amplitude. Conversely, a given peak output amplitude is achievable with a smaller step input than pulse input, and therefore there is less risk of saturation within the system.

The effect of noise on a parameter estimate differs according to whether the estimate is obtained from a step response or from an impulse or unit-pulse response. For example, steady-state gain is easily found, in the absence of drift, from the initial and final values of the step response, even in considerable noise. By contrast, the value obtained by integrating the impulse response includes an unknown contribution from noise, and a value found by summing u.p.r. samples would be susceptible to dead-time uncertainty introduced by the sampling, as in Example 2.1.3(iii). Another example, favouring the impulse response this time, is estimation of the impulse-response peak. Differentiation of a noisy step response would exaggerate wideband noise because differentiation amounts to filtering with gain proportional to frequency. Unless the peak of the derivative were found from a parametric model fitted to the whole step response with high confidence, its value would be very uncertain.

We note finally that a step is the easiest of all inputs to produce with acceptable fidelity, and repeated steps in a square wave are equally easy.

2.3 TRANSFORM DESCRIPTION OF RESPONSE

Laplace and Fourier transforms are the basis of classical control design and much of the analysis of electrical systems. We recall their definitions

$$\begin{aligned}\mathscr{L}[f(t)] \equiv F(s) &\triangleq \int_0^\infty f(t)e^{-st}\,dt \\ \mathscr{F}[f(t)] \equiv F(j\omega) &\triangleq \int_0^\infty f(t)e^{-j\omega t}\,dt\end{aligned} \tag{2.3.1}$$

Their popularity is due to the way they simplify the input–output convolution relation (2.1.2) to a multiplication of Laplace transforms

$$Y(s) = H(s)U(s) \tag{2.3.2}$$

or the Fourier version with $j\omega$ for s. In identification we are interested in solving (2.3.2) for the transfer function $H(s)$, which is the Laplace transform of the impulse response $h(t)$. Usually (2.3.2) is much easier to solve than the integral equation (2.1.2), and if we want $h(t)$, we can find it by inverse Laplace transformation of $H(s)$.

Example 2.3.1 The output of a certain initially quiescent, linear, time-invariant system forced by $u(t) = e^{\alpha_1 t}$, $t \geq 0$, is

$$y(t) = \beta(e^{\alpha_1 t} - e^{\alpha_2 t})$$

so the convolution relation (2.1.2) is

$$\beta(e^{\alpha_1 t} - e^{\alpha_2 t}) = \int_0^t h(t-\tau)e^{\alpha_1 \tau}\,d\tau$$

We start a time-domain solution for $h(t)$ by differentiating:

$$\beta(\alpha_1 e^{\alpha_1 t} - \alpha_2 e^{\alpha_2 t}) = h(0)e^{\alpha_1 t} + \int_0^t e^{\alpha_1 \tau} \frac{dh(t-\tau)}{dt} d\tau$$

We then put $-dh(t-\tau)/d\tau$ for $dh(t-\tau)/dt$ and integrate by parts:

$$\left[e^{\alpha_1 \tau} \int -\frac{dh(t-\tau)}{dt} d\tau \right]_0^t + \int_0^t \alpha_1 e^{\alpha_1 \tau} h(t-\tau)\, d\tau$$
$$= -e^{\alpha_1 t}h(0) + h(t) + \alpha_1 \beta(e^{\alpha_1 t} - e^{\alpha_2 t})$$

We now collect terms and cancel, to find $h(t) = (\alpha_1 - \alpha_2)\beta \exp(\alpha_2 t)$. By contrast, Equation (2.3.2) is

$$(\alpha_1 - \alpha_2)\beta/(s - \alpha_1)(s - \alpha_2) = H(s)/(s - \alpha_1)$$

giving $H(s)$ and hence $h(t)$ with negligible effort. △

2.3.1 Identification of Laplace Transfer Function

Through (2.3.2) we can in principle identify $H(s)$ as $Y(s)/U(s)$ using any Laplace-transformable signal as input. We have to fit a Laplace transform to the waveform $y(t)$ by breaking it into components with known Laplace transforms, such as a constant, a ramp and exponentials. By doing so we are selecting a parameterised model, a more demanding business than merely recording an unparameterised impulse response. To add to the complication, $Y(s)$ contains components stemming from $U(s)$ as well as $H(s)$, as in Example 2.3.1. This weighs against any input more complicated than a step, but a more elaborate waveform may be forced on us by input-actuator limitations.

2.3.2 Discrete-Time Transfer Function

We now review briefly the description of sampled signals by z transforms and at the same time establish notation which will be useful later. In discrete-time systems the relation between a sampled signal and its transform description is very direct. If a signal $f(t)$ is sampled every T seconds, the result is fully described by listing the sample values and their times. As before, we denote sample $f(kT)$ by f_k, and we shall usually be concerned with signals which start at time zero. The list of sample values $f_0, f_1, f_2, \ldots$ is denoted by $\{f\}$, the curly brackets being read as "sequence". To complete the description we need an operator to shift a sample along to the correct place in the sequence. If a plain number x is taken to mean a sample at time zero of value x, and a T-second delaying of a sample is denoted by an operator z^{-1}, a sample f_k at time kT is $z^{-k}f_k$, conventionally written $f_k z^{-k}$. The entire sampled signal is then

$$f_0 + f_1 z^{-1} + f_2 z^{-2} + \cdots + f_k z^{-k} + \cdots \equiv F(z^{-1}) \tag{2.3.3}$$

This polynomial in z^{-1} is the *z-transform* of the sampled $f(t)$. Its length is finite or infinite according to when $f(t)$ ends, and for most simple waveforms $F(z^{-1})$ can be written concisely in closed form as the quotient of two polynomials in z^{-1} (that is, a rational polynomial function). For example, when a finite-duration exponential $f(t) = e^{\alpha t}$, $0 \le t \le NT$, is sampled, its z transform is

$$\begin{aligned} F(z^{-1}) &= 1 + e^{\alpha T} z^{-1} + e^{2\alpha T} z^{-2} + \cdots + e^{N\alpha T} z^{-N} \\ &\equiv \frac{1 - e^{(N+1)\alpha T} z^{-N-1}}{1 - e^{\alpha T} z^{-1}} \end{aligned} \tag{2.3.4}$$

A z-transform input–output equation comes straight from superposition of the responses to individual input samples. With u.p.r. and input respectively

$$\begin{aligned} H(z^{-1}) &\equiv h_0 + h_1 z^{-1} + h_2 z^{-2} + \cdots \\ U(z^{-1}) &\equiv u_0 + u_1 z^{-1} + u_2 z^{-2} + \cdots \end{aligned} \tag{2.3.5}$$

the output is altogether

$$\begin{aligned} Y(z^{-1}) &= (h_0 + h_1 z^{-1} + \cdots)u_0 + z^{-1}(h_0 + h_1 z^{-1} + \cdots)u_1 \\ &\quad + z^{-2}(h_0 + h_1 z^{-1} + \cdots)u_2 + \cdots \\ &= (h_0 + h_1 z^{-1} + \cdots)(u_0 + u_1 z^{-1} + \cdots) \equiv H(z^{-1})U(z^{-1}) \end{aligned} \tag{2.3.6}$$

Clearly (2.3.6) is the discrete-time counterpart of (2.3.2). Sample t can be picked out:

$$\begin{aligned} y_t z^{-t} &= h_t z^{-t} u_0 + z^{-1} h_{t-1} z^{-t+1} u_1 + \cdots + z^{-t} h_0 u_t \\ &= \sum_{k=0}^{t} h_{t-k} u_k z^{-t}, \qquad t \ge 0 \end{aligned} \tag{2.3.7}$$

We recognise this as the convolution sum (2.1.3), with a time marker z^{-t} attached. In discrete time the step from the operational input–output relation (2.3.6) to the explicit expression (2.3.7) for the output is trivial, compared with the step from the transform relation (2.3.2) to the convolution (2.1.2) in continuous time. Furthermore, the discrete-time convolution sum is easy to compute.

The convolution sum suggests an identification method in which y_0 to y_N and u_0 to u_N are recorded, then the $N+1$ simultaneous linear equations given by (2.1.3) are solved for h_0 to h_N. Assuming the input is zero before time zero, (2.1.3) gives

$$y_0 = h_0 u_0, \qquad y_1 = h_1 u_0 + h_0 u_1, \qquad y_2 = h_2 u_0 + h_1 u_1 + h_0 u_2, \ldots \tag{2.3.8}$$

so it looks as if we can solve successively for h_0, h_1, h_2, etc. The only thing that spoils the idea is the presence of output noise. Careful choice of u_0 to u_N might keep the noise-induced errors in h_0 to h_N tolerable (Problem 2.8), but a better solution is to record more input and output samples, and find estimates of h_0 to h_N which give a good overall fit to the observed output. Chapter 4 follows up this idea in detail.

Let us now think about identifying the z-transform transfer function written as a rational polynomial function of z^{-1} rather than a long, or even infinite, power series. A typical u.p.r. consists of n sampled exponentials, and can be written as

$$\begin{aligned} H(z^{-1}) &= \gamma_1(1 + \beta_1 z^{-1} + \beta_1^2 z^{-2} + \cdots \infty) + \gamma_2(1 + \beta_2 z^{-1} + \beta_2^2 z^{-2} + \cdots \infty) \\ &\quad + \cdots + \gamma_n(1 + \beta_n z^{-1} + \beta_n^2 z^{-2} + \cdots \infty) \\ &= \frac{\gamma_1}{1-\beta_1 z^{-1}} + \frac{\gamma_2}{1-\beta_2 z^{-1}} + \cdots + \frac{\gamma_n}{1-\beta_n z^{-1}} \\ &= \frac{\gamma_1(1-\beta_2 z^{-1})(1-\beta_3 z^{-1})(\cdots)(1-\beta_n z^{-1}) + \gamma_2(\cdots) + \cdots \gamma_n(\cdots)}{(1-\beta_1 z^{-1})(1-\beta_2 z^{-1})(\cdots)(1-\beta_n z^{-1})} \\ &= \frac{b_0 + b_1 z^{-1} + \cdots + b_{n-1} z^{-n+1}}{1 + a_1 z^{-1} + \cdots + a_n z^{-n}} \end{aligned} \tag{2.3.9}$$

Here the dead time has been taken as zero; more often it will be non-zero, with the effect of making b_0, and perhaps other leading numerator coefficients, zero. From (2.3.9) and (2.3.6),

$$(1 + a_1 z^{-1} + \cdots + a_n z^{-n})Y(z^{-1}) = (b_0 + b_1 z^{-1} + \cdots + b_{n-1} z^{-n+1})U(z^{-1}) \tag{2.3.10}$$

The coefficients of z^{-t} on each side give us a difference equation

$$y_t + a_1 y_{t-1} + \cdots + a_n y_{t-n} = b_0 u_t + b_1 u_{t-1} + \cdots + b_{n-1} u_{t-n+1} \tag{2.3.11}$$

from which

$$y_t = -a_1 y_{t-1} - a_2 y_{t-2} - \cdots - a_n y_{t-n} + b_0 u_t + b_1 u_{t-1} + \cdots + b_{n-1} u_{t-n+1} \tag{2.3.12}$$

Like the convolution sum, this equation for y_t is linear in its transfer-function coefficients, and we should expect that the a's and b's can be estimated in much the same way as the u.p.r. coefficients h_0 to h_N. Later chapters will show how far this is true. The most important advantage of (2.3.9)–(2.3.12) is that for most systems n can be quite small, typically 2 or 3, whereas the number of significant u.p.r. samples may be very much larger.

2.3.3 Frequency Transfer Function

Although we shall concentrate on the identification of discrete-time systems, we should not ignore the most widely used identification method of all, sine-wave testing to obtain the frequency transfer function $H(j\omega)$ of a continuous-time system.

The response of a system to a sine-wave input can be found algebraically by first writing down its response to $e^{j\omega t}$, applied from time $-\infty$. The response to $e^{-j\omega t}$ follows by putting $-j$ for j, and the response to $\sin \omega t$ or $\cos \omega t$ by expressing them in terms of $e^{j\omega t}$ and $e^{-j\omega t}$. The convolution integral gives the response to $e^{j\omega t}$ as

$$y(t) = \int_{-\infty}^{t} h(t-\tau)e^{j\omega\tau}\,d\tau \tag{2.3.13}$$

Substitution of τ' for $t-\tau$ then gives

$$y(t) = \int_{\infty}^{0} h(\tau')e^{j\omega(t-\tau')}(-d\tau') = \left(\int_{0}^{\infty} h(\tau')e^{-j\omega\tau'}\,d\tau'\right)e^{j\omega t} = H(j\omega)e^{j\omega t} \tag{2.3.14}$$

where $H(j\omega)$ is the bracketed integral, i.e. the one-sided Fourier transform of the impulse response. If $H(j\omega)$ is $A+jB$ with A and B real (and frequency-dependent), a sine-wave input

$$u(t) = \sin \omega t = (e^{j\omega t} - e^{-j\omega t})/2j \tag{2.3.15}$$

gives an output

$$\begin{aligned} y(t) &= [H(j\omega)e^{j\omega t} - H(-j\omega)e^{-j\omega t}]/2j \\ &= [(A+jB)(\cos \omega t + j\sin \omega t) - (A-jB)(\cos \omega t - j\sin \omega t)]/2j \\ &= A\sin \omega t + B\cos \omega t = (A^2+B^2)^{1/2}\sin(\omega t + \tan^{-1}(B/A)) \end{aligned} \tag{2.3.16}$$

In other words, the output is a sine-wave of the same frequency as $u(t)$, multiplied in amplitude by $(A^2+B^2)^{1/2}$, which is $|H(j\omega)|$, and advanced in phase by $\tan^{-1}(B/A)$, which is $\angle H(j\omega)$. Since the system is linear and time-invariant, a scaled or time-shifted input produces a similarly scaled or time-shifted output, so the gain $|H(j\omega)|$ and phase change $\angle H(j\omega)$ apply to *any* input sine-wave at frequency $\omega/2\pi$ Hz.

For sinusoidal or periodic inputs we need the complex value of $H(j\omega)$ only at the frequency of the input and its harmonics, if any. This is the situation, for instance, in steady-state analysis of power systems. More generally we want the *frequency transfer function* $H(j\omega)$ over the whole range of frequencies passed by the system, so we can predict the response to any input waveform which has a Fourier transform, using

$$Y(j\omega) = H(j\omega)U(j\omega) \tag{2.3.17}$$

The impulse response can be recovered from $H(j\omega)$, in theory by inverse Fourier transformation and in practice by fitting a parametric model with a known inverse to an experimental $H(j\omega)$, as in the next section.

At this point it is worth noting that the delay operator z^{-1} in discrete-time systems can be interpreted as a compressed notation for e^{-sT}, where T is the sampling period. That is, sample f_k at time kT is regarded as a δ function of area f_k, which has a Laplace transform $f_k e^{-skT}$, i.e. $f_k z^{-k}$. Two benefits are conferred by this view. First, we can put $j\omega$ for s and obtain the frequency transfer function of a discrete-time system from its z-transform transfer function, and the spectrum $F(e^{-j\omega T})$ of a signal from its z-transform. Second, we can use Laplace transforms to analyse the sampled input response of a system *between* the sampling instants, writing $U(e^{-sT})$ for $U(z^{-1})$ and $H(e^{-sT})$ for $H(z^{-1})$. The only thing that then marks out discrete-time linear systems from any others is the relative complication of their Laplace and frequency transfer functions.

Example 2.3.2 A sampled exponential $\{u\} = e^{\alpha t}, t = 0, T, 2T, \cdots \infty$ with $\alpha < 0$ has z transform $U(z^{-1}) = 1 + e^{\alpha T}z^{-1} + e^{2\alpha T}z^{-2} + \cdots \infty = 1/[1 - e^{\alpha T}z^{-1}]$, and Laplace transform $1/[1 - e^{(\alpha - s)T}]$. The response of a system with transfer function $1/(s - \gamma)$ to $\{u\}$ has Laplace transform

$$Y(s) = 1/\{[1 - e^{(\alpha - s)T}](s - \gamma)\} = \frac{1}{s - \gamma} + e^{\alpha T}\frac{e^{-sT}}{s - \gamma} + e^{2\alpha T}\frac{e^{-2sT}}{s - \gamma} + \cdots \infty$$

so

$$y(t) = e^{\gamma t}\mu(t) + e^{\alpha T}e^{\gamma(t - \tau)}\mu(t - T) + e^{2\alpha T}e^{\gamma(t - 2T)}\mu(t - 2T) + \cdots \infty$$

where $\mu(t)$ is the unit step at time zero. That is, each sample from $\{u\}$ excites a new response starting at the sampling instant and persisting for ever. The expression for $y(t)$ is good for all times, not just the sample instants.

The spectrum of $\{u\}$ is $1/[1 - e^{(\alpha - j\omega)T}]$. Since $e^{2\pi j}$ is 1, $e^{(\alpha - j\omega)T} = e^{(\alpha - j(\omega - (2\pi k/T))T}$ for any integer k, so the spectrum repeats itself at intervals of $2\pi/T$ in ω. △

2.3.4 Measurement of Frequency Transfer Function

Some input actuators cannot apply a sine-wave. Many valves in process plant open and close at fixed rates, for instance. Any hysteresis (backlash) in an input actuator will distort a sine-wave but not spoil a step. If we are lucky and a sine-wave can be applied, and the output is undistorted and the s.n.r. high, the gain and phase change at a number of frequencies can be measured by just looking at input and output, once the initial-condition response due to switching the input on has subsided and the response looks as if the sine-wave has been going for ever. More often the output contains noise, harmonics caused by non-linearity (i.e. components at integer multiples of the input frequency), constant bias and perhaps drift. Periodic disturbances such as diurnal variation in biological systems, seasonal factors in economic records or mains-frequency interference in electrical systems are also common. Gain and phase-change measurements can be made less susceptible to such impairments by extracting the fundamental Fourier component from the output and measuring its amplitude and phase. That is, we compute the averages

$$S = \frac{\omega}{2N\pi}\int_0^{(2N\pi)/\omega} y(t)\sin\omega t\,dt \qquad C = \frac{\omega}{2N\pi}\int_0^{(2N\pi)/\omega} y(t)\cos\omega t\,dt \tag{2.3.18}$$

over N cycles of the output. An input $V\sin\omega t$ produces an output $GV\sin(\omega t+\theta)+v(t)$ where G is the gain, θ is the phase change and $v(t)$ comprises all the impairments. Now

$$\begin{aligned} S &= \frac{GV\omega}{2N\pi}\int_0^{(2N\pi)/\omega} [\tfrac{1}{2}(\cos\theta-\cos(2\omega t+\theta))+v(t)\sin\omega t]\,dt \\ &= \frac{GV}{2}\cos\theta + \frac{GV\omega}{2N\pi}\int_0^{(2N\pi)/\omega} v(t)\sin\omega t\,dt \end{aligned} \tag{2.3.19}$$

and the last integral is zero if $v(t)$ is constant, a ramp or any sinusoid at an integer multiple of the input frequency (Problem 2.11). It is small if N is large and $v(t)$ is random noise independent of the input or a sinusoid at a frequency unrelated to that of the input. Similarly, C gives $\frac{1}{2}GV\sin\theta$ plus a term which can be made small, so altogether

$$G \simeq 2(S^2+C^2)^{1/2}/V, \qquad \theta \simeq \tan^{-1}(C/S) \tag{2.3.20}$$

Commercial transfer-function analysers work on this principle.

Frequency transfer functions are attractive for several reasons. They are familiar to electrical engineers through their role in a.c. circuit analysis, and to control engineers through classical stability analysis and control design. The Bode plots used in control engineering, plots of $\log G$ against $\log\omega$ and θ

against $\log \omega$ (D'Azzo and Houpis, 1981; Melsa and Schultz, 1969), are particularly useful in suggesting parameterised rational polynomial transfer-function models. Think of

$$H(j\omega) = \frac{K(j\omega\tau_1 + 1)(j\omega\tau_2 + 1)\cdots(j\omega\tau_{n-1} + 1)}{(j\omega T_1 + 1)(\cdots)(j\omega T_n + 1)} = G\angle\theta \quad (2.3.21)$$

with the τ's and T's real. As we increase the frequency from near zero, the low-frequency gain is roughly K, then each factor $j\omega\tau_i + 1$ contributes significant gain and phase advance from about $\omega = 1/\tau_i$ up, and each $1/(j\omega T_i + 1)$ contributes attenuation and phase lag from about $\omega = 1/T_i$ up. If the decibel gain $20\log_{10} G$ is plotted, each $j\omega\tau_i + 1$ is asymptotically proportional to ω, and contributes 20 dB more gain per decade increase in ω. Each $1/(j\omega T_i + 1)$ gives 20 dB less. With ω also plotted logarithmically the asymptote is a straight line, approached within 0.17 dB (2%) at $\omega = 5/\tau_i$ or $5/T_i$. If we fit straight-line sections with slopes integer multiples of ± 20 dB/decade to the measured gain Bode plot, ω at each junction between two successive sections determines a $1/\tau_i$ or $1/T_i$, according as the bend is upwards or downwards. The phase plot provides a rough check, since $j\omega\tau_i + 1$ gives a 45° lead at $\omega = 1/\tau_i$, $1/(j\omega T_i + 1)$ a 45° lag at $\omega = 1/T_i$, and the other factors little lag or lead at frequencies well below their values of $1/\tau$ or $1/T$ but almost $\pm 90°$ well above those frequencies.

Dead time is visible on the phase plot as a constant rate of increase of phase lag with ω, since a delay t_d gives rise to $e^{-j\omega t_d}$ in the transfer function. Complex conjugate roots of the numerator or denominator of $H(j\omega)$ are less straightforward to estimate, but Exercise 2.3.3 will give an example. Factors $j\omega$ in the denominator contribute 90° lag and -20 dB/decade gain over all ω, and are easiest to detect from the low-frequency behaviour which they dominate.

Example 2.3.3 The open-loop dynamics of a motor-speed control system, sketched in Fig. 2.3.1, are to be identified by sine-wave testing. The sinusoid is added to the d.c. speed reference voltage, and the gain and phase change measured by the tachogenerator voltage normally fed back. The amplidyne is

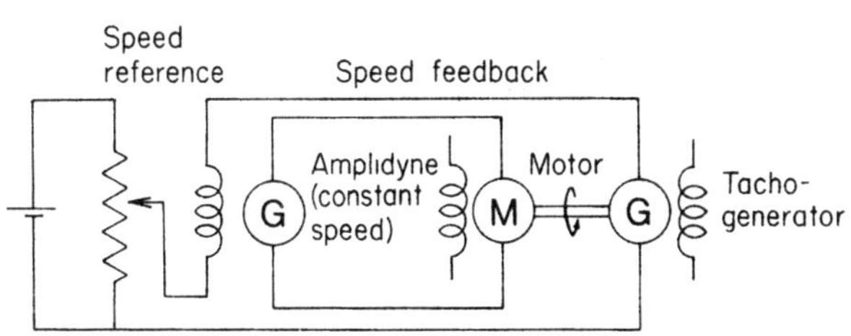

Fig. 2.3.1 Speed-control system.

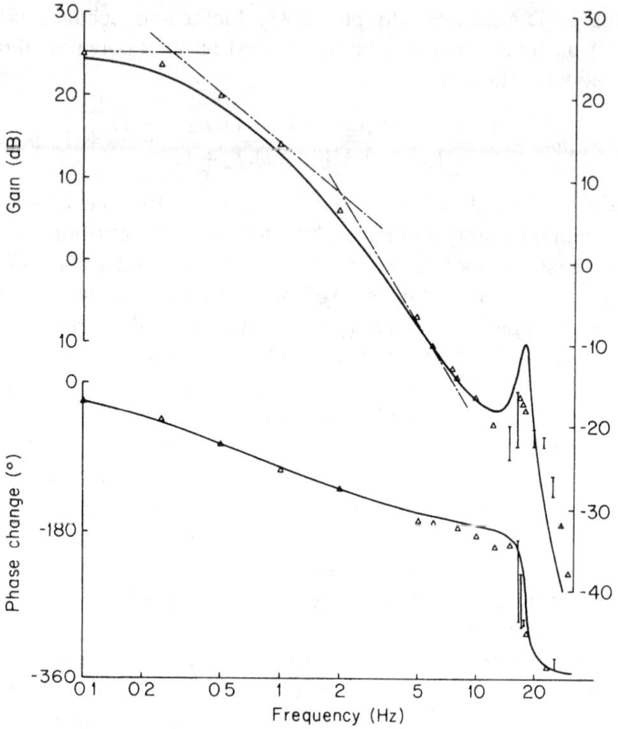

Fig. 2.3.2 Bode gain and phase-change plots for speed-control system. Vertical bars indicate uncertain measurements.

a high-gain d.c. generator acting as an amplifier. The dynamics might be quite complicated, as they include the amplidyne field time constant, the armature-circuit time constant of the amplidyne and motor, and the mechanical time constant of the motor–armature:tachogenerator–armature:motor–load combination. The output, of r.m.s. amplitude between about 0.25 V and a few volts, has superimposed on it about 25 V d.c., a near-sinusoidal 200 Hz commutator ripple of 5–10 V r.m.s. and a few volts r.m.s. of wideband commutation noise. Given the opportunity, one would test the various machines separately, but testing the overall dynamics is quicker and needs less instrumentation.

The Bode plots of the test results are given in Fig. 2.3.2. The downward breaks of the straight-line approximations to the gain plot, to -20 dB/decade slope at 1.8 rad/s and -40 dB/decade at 12.6 rad/s, indicate denominator factors $j\omega T+1$ with $T_1=0.556$ and $T_2=0.080$. The phase changes at those frequencies are close to $-45°$ and $-90°$ $-145°$, confirming the values of T_1

and T_2 and leaving no extra lag to suggest significant dead time. The low-frequency gain gives $K = 18$ in the transfer function. The high-frequency behaviour is more thought-provoking. A very sharp resonance peak at about 17 Hz suggests a transfer-function factor $1/[(j\omega/\omega_n)^2 + 2\zeta j\omega/\omega_n + 1]$ with ζ, the damping ratio (D'Azzo and Houpis, 1981), very small. The full-line curves in Fig. 2.3.2 given by $\omega_n = 113$ and $\zeta = 0{\cdot}05$ are a good fit in phase but only a moderate fit in gain. It looks as if even this ζ may be too high. At this stage some physical insight is essential to make sense of the results. A likely source of very lightly damped resonance is torsional oscillation between the armatures of motor and tachogenerator. The motor–tachogenerator combination is, in fact, a laboratory motor–generator set with two identical large armatures. Further testing with the tachogenerator electrically loaded confirmed this explanation; a 10 W load provided enough damping to reduce the resonance peak height to about 1.5 dB.

Two lessons may be drawn from this example. First, surprisingly smooth and apparently accurate results are obtained by averaging $y(t) \sin \omega t$ and $y(t) \cos \omega t$ (over 10 cycles up to 5 Hz, 100 cycles up to 10 Hz, then 1000 cycles) even in the presence of extreme impairment of the output. Second, a convincing model can only be found by interplay between test results and background knowledge of the system. △

Frequency-response identification is not short of disadvantages to balance its virtues. It requires a succession of tests at different frequencies, taking time and necessitating trial and error to arrive at a suitable range and spacing of frequencies. At each frequency the initial-condition response due to sudden transition from no input to a sine-wave must die out before the steady-state forced response is observed. For that reason sweeping of the frequency may not be acceptable. A frequency-domain model is not convenient for some applications. For instance, the intersymbol interference and echo behaviour of a data-communication channel is better modelled in the time domain, by impulse responses or u.p.r.'s. Finally, frequency transfer functions of discrete-time systems are complicated, as Problem 2.10 discovers.

REFERENCES

D'Azzo, J. J., and Houpis, C. H. (1981). "Linear Control System Analysis and Design", 2nd ed. McGraw-Hill, New York.

Gabel, R. A., and Roberts, R. A. (1980). "Signals and Linear Systems", 2nd ed. Wiley, New York.

Godfrey, K. R. (1983). "Compartmental Models and Their Application". Academic Press, New York and London.

Melsa, J. L., and Schultz, D. G. (1969). "Linear Control Systems". McGraw-Hill, New York.

Norton, J. P. (1973). Practical problems in blast furnace identification. *Meas. Control* **6**, 29–34.

Reid, J. G. (1983). "Linear System Fundamentals". McGraw-Hill, New York.
Unbehauen, H., and Diekmann, K. (1983). Application of m.i.m.o. identification to a blast furnace. In "Identification and System Parameter Estimation 1982" (G. A. Bekey and G. N. Saridis, eds.), pp. 180–185. Pergamon, Oxford.
Ziemer, R. E., Tranter, W. H., and Fannin, D. R. (1983). "Signals and Systems: Continuous and Discrete". Macmillan, New York.

PROBLEMS

2.1 A system with impulse response $h(t) = \exp(-10t)$ is quiescent at time zero. Find and sketch (i) its response to an input $u(t) = t$, $0 \leq t < 0.2$; (ii) the sampled response to this input, the samples being at time intervals of 0.05 from time zero; (iii) the sampled response to the sampled version of this input, i.e.

$$u(t) = 0.05\delta(t - 0.05) + 0.1\delta(t - 0.1) + 0.15\delta(t - 0.15)$$

(iv) the continuous-time response to the input in (iii).

2.2 Investigate the effects of a non-ideal input in an impulse-response test by plotting the response of the system with impulse response $h(t) = \exp(-t) - \exp(-5t)$ to a rectangular pulse input of unit area and duration (i) 0.1, (ii) 0.2, (iii) 0.5. Compare each response with $h(t)$.

2.3 A system has a u.p.r. consisting of two sampled exponential components, one fast and one slow. In each sampling interval, the fast one falls to α_1 times its value at the start of the interval. The corresponding figure for the slow one is α_2, and α_1 is about α_2^r with r a large integer. As in Example 1.3.8, the difference in speed causes difficulty in identifying α_1 and α_2 adequately. They could be identified in two separate experiments with sampling intervals differing by a factor r. Should the results for α_1 and α_2^r be combined in a z-transform transfer function of the form

$$H(z^{-1}) = b_1/(1 - \alpha_1 z^{-1}) + b_2/(1 - \alpha_2^r z^{-1})?$$

If not, how should they be combined?

2.4 Find and sketch the sampled unit-step responses of the two systems of Example 2.1.3(i).

2.5 By treating the first u.p.r. in Example 2.1.3(ii) as two sampled exponentials and the second u.p.r. as a single sampled exponential, verify that the continuous-time impulse responses of the two systems just after time zero behave as stated in that example.

2.6. Two first-order continuous-time systems have the same u.p.r.'s when their outputs are sampled, but one system has a dead time much less than one sampling interval, and the other a dead time just less than one sampling interval, as considered in Example 2.1.3(iii). Show that the ratio of their steady-state gains equals the common ratio between successive samples in their u.p.r.'s, and so does the ratio of their peak impulse-response values.

2.7 Show that the gain and phase change of the zero-order hold described in Section 2.1.2 are $2\sin(\omega T/2)/\omega$ and $-\omega T/2$, where ω is the angular frequency. Explain the asymptotic gain as ω tends to zero. At what frequency does the gain differ by 1 dB from its zero-frequency value?

2.8 For the identification method tentatively suggested in Section 2.3.2, solving (2.3.8) for the u.p.r. ordinates, consider the effects of an estimation error δh_i in h_i on the estimates of later ordinates h_{i+1}, etc. What features of the input sequence would cause the error to increase as it propagated? Would a diverging sampled exponential be a good test input? Would a converging exponential?

2.9 Find and sketch the amplitude and phase spectra of the sampled exponential $u(iT) = \exp(\alpha iT)$, $i = 0, 1, 2, \ldots$, considered in Example 2.3.2, with α real and negative. [Rather than grinding out algebraic expressions, think of $\exp(\alpha - j\omega)T$ as a vector of length $\exp(\alpha T)$ at an angle $-\omega T$ radians to the positive real axis, and do some geometry.]

2.10 A microprocessor takes samples $f_0, f_1, \ldots$ of a signal $f(t)$ at intervals T and forms the three-term moving average $g_t = (f_t + f_{t-1} + f_{t-2})/3$. Find the transfer function $G(j\omega)/F(j\omega)$. Sketch how the gain and phase change vary with frequency.

2.11 Referring to (2.3.19), verify that constant, ramp or harmonic-frequency components in the output have no effect on the frequency transfer function measured by the Fourier analysis method of Section 2.3.4.

2.12 Two adjacent break frequencies on a Bode gain plot are separated by a factor of 3. The gain contribution at the lower break frequency of the transfer-function factor which gives rise to the upper break is therefore less than 0.5 dB, negligible for most practical purposes. What is its phase contribution? What do you conclude about the relative convenience of gain and phase Bode plots for identification?

2.13 Roughly, what dead time would be the smallest reliably detectable in test results of the apparent quality of those in Example 2.3.3?

2.14 (For control engineers) What limitations, if any, would the high-frequency resonance found in Example 2.3.3 place on the steady-state-error capability of the closed-loop control system? Would your answer change if the resonance peak were at -3 dB rather than -15 dB or so?

Chapter 3

Identification Based on Correlation Functions

3.1 TIME AVERAGING TO REDUCE EFFECTS OF NOISE

3.1.1 Time-Average Relations between Signals

A basic problem in identification is to distinguish the effects of the input from noise in the observed output. Averaging the responses was recommended in the last chapter for impulse and step tests, and in a frequency response test, Example 2.3.3, we achieved impressive rejection of noise and other impairments by averaging the product of the output and $\sin \omega t$ or $\cos \omega t$, signals derived from the input. The idea of averaging can be extended to other forms of input by employing the *cross-correlation function* (c.c.f.) $r_{uy}(\tau, t)$ between input $u(t)$ and output $y(t)$, defined as

$$r_{uy}(\tau, t) = E[u(t)y(t+\tau)] \tag{3.1.1}$$

The notation $E[\cdot]$ signifies "the expected value of $\cdot$", so $r_{uy}(\tau, t)$ is the average of $u(t)y(t+\tau)$ over all possible values, regarding $u(t)y(t+\tau)$ for given t and τ as a random variable (Helstrom, 1984; Melsa and Sage, 1973). For our immediate purposes, we can read it as "the average of $\cdot$" and interpret it as the *time* average

$$r_{uy}(\tau) = \lim_{T\to\infty} \frac{1}{2T} \int_{-T}^{T} u(t)y(t+\tau)\,dt \tag{3.1.2}$$

This average is a function of lag τ but not of t. The two averages do not coincide for all signals, since $E[u(t)y(t+\tau)]$ might well vary with t. That is, averaging over time in one long experiment need not give the same result as averaging at one time over a large number of experiments, even in the limit as experiment length and number of experiments tend to infinity. A random variable for which they do coincide, as we shall be assuming, is called *ergodic*. The reason for letting the start time $-T$ in (3.1.2) tend to $-\infty$ rather than fixing it at zero is that we shall be interested in the steady-state forced response to an input which started long ago, not the transient response to a signal which starts at time zero. The c.c.f. defined by (3.1.1) measures how closely $y(t+\tau)$ is related to $u(t)$; the value α which minimises $E[(\alpha u(t) - y(t+\tau))^2]$ is easily shown to be $r_{uy}(\tau)/E[u^2(t)]$ (Problem 3.1).

A discrete-time counterpart of (3.1.2) is

$$r_{uy}(k) = \lim_{N\to\infty} \frac{1}{2N+1} \sum_{i=-N}^{N} u_i y_{i+k} \quad k \text{ integer} \tag{3.1.3}$$

which can be approximated by an average computed from finite records.

The *autocorrelation function* (a.c.f.) of a single signal is defined analogously. For instance, the continuous-time a.c.f. of an ergodic $u(t)$ is

$$r_{uu}(\tau) = E[u(t)u(t+\tau)] = \lim_{T\to\infty} \frac{1}{2T} \int_{-T}^{T} u(t)u(t+\tau)\,dt \tag{3.1.4}$$

and the discrete-time a.c.f. is

$$r_{uu}(k) = \lim_{N\to\infty} \frac{1}{2N+1} \sum_{i=-N}^{N} u_i u_{i+k} \tag{3.1.5}$$

The a.c.f. covers negative as well as positive lags τ or k, and is an even function of lag. From (3.1.5), for instance,

$$\begin{aligned} r_{uu}(k) &= \lim_{N\to\infty} \frac{1}{2N+1} \sum_{i=-N+k}^{N+k} u_{i-k} u_i \\ &= \lim_{N\to\infty} \frac{1}{2N+1} \sum_{i=-N+k}^{N+k} u_i u_{i-k} = r_{uu}(-k) \end{aligned} \tag{3.1.6}$$

since the starting point of the summation is immaterial. Although the c.c.f. also exists for negative lags, it is not an even function, and values of r_{uy} at negative lags are seldom of interest, as y does not then depend on u.

The a.c.f. r_{uu} and c.c.f. r_{uy} are important in identification because they are related through the impulse response or u.p.r. of the system with u as input and y as output. A good estimate of the impulse response or u.p.r. is often obtainable from the relation, since r_{uy}, and if necessary r_{uu}, can be measured by a time average on which noise has little effect.

3.1.2 Input–Output Relation in Terms of Correlation Functions

The discrete-time relation is of most use to us. The output at sample instant $i+k$ due to an input that started an indefinitely long time ago, so that the initial-condition response has long since vanished, is

$$y_{i+k} = \sum_{j=0}^{\infty} h_j u_{i+k-j} \tag{3.1.7}$$

so the c.c.f. between $\{u\}$ and $\{y\}$ is

$$\begin{aligned} r_{uy}(k) &= \lim_{N\to\infty} \frac{1}{2N+1} \sum_{i=-N}^{N} u_i \sum_{j=0}^{\infty} h_j u_{i+k-j} \\ &= \sum_{j=0}^{\infty} h_j \lim_{N\to\infty} \frac{1}{2N+1} \sum_{i=-N}^{N} u_i u_{i+k-j} = \sum_{j=0}^{\infty} h_j r_{uu}(k-j) \end{aligned} \tag{3.1.8}$$

This is the *Wiener–Hopf equation*. The continuous-time version, derived in much the same way, is

$$r_{uy}(\tau) = \int_0^{t_s} h(t) r_{uu}(\tau - t)\, dt \tag{3.1.9}$$

where t_s is the settling time beyond which $h(t)$ is negligible. The equation originally arose in optimal filter design. We want to solve (3.1.8) for the u.p.r. $\{h\}$, which can be truncated at h_s, say, in an asymptotically stable system, cutting off the negligible part of the decaying tail. The unknowns h_0 to h_s enter linearly, so we can compute $r_{uy}(k)$ for $s + 1$ values of k, insert them into (3.1.8) with the corresponding $r_{uu}(k)$ to $r_{uu}(k - s)$, and solve by matrix inversion. We go to the trouble of computing $\{r_{uy}\}$ and $\{r_{uu}\}$ to reduce the influence of noise. If the observed output $\{y\}$ is composed of clean output $\{y^c\}$ plus noise $\{v\}$, we compute

$$r_{uy}(k) \simeq \frac{1}{N - M + 1 - k} \sum_{i=M}^{N-k} u_i (y^c_{i+k} + v_{i+k}) \simeq r^c_{uy}(k) + r_{uv}(k) \tag{3.1.10}$$

So long as $\{v\}$ is unrelated to $\{u\}$ and zero-mean, the long-term average of $u_i v_{i+k}$ is very likely to be close to zero. We formalise this by saying $\{u\}$ and $\{v\}$ are *mutually uncorrelated* if $r_{uv}(k)$ is zero for all lags k. A more restrictive way to ensure that $r_{uv}(k)$ is zero is to assume that u_i and v_{i+k} are statistically independent, so that $E[u_i v_{i+k}]$ is $E[u_i]E[v_{i+k}]$, and one of them is zero-mean, i.e. $E[u_i]$ or $E[v_{i+k}]$ is zero. The precise assumption is unimportant, as we cannot in any case verify that $\{r_{uv}\}$ is negligible during the experiment since $\{v\}$ is unobservable. A heuristic "engineering" assumption is that we can be pretty sure $\{r_{uv}\}$ is negligible if we take care to avoid treating any input-dependent or constant components of the output as noise. We can be surer still that the computed $\{r_{uy}\}$ in (3.1.8) is less affected by noise than $\{y\}$ in the input–output convolution (2.3.8) which offered an alternative way to find the u.p.r.

Solution for $\{h\}$ is very easy if we employ an input with an uncomplicated a.c.f. The best of all is a *white* input which has $r_{uu}(k-j)$ zero except when lag $k-j$ is zero, yielding $\{h\}$ directly from

$$r_{uy}(k) = \sum_{j=0}^{\infty} h_j r_{uu}(k-j) = h_k \sigma_n^2, \qquad k = 0, 1, \ldots, s \tag{3.1.11}$$

where σ_n^2 is $r_{uu}(0)$, the m.s. value of u.

Before we examine specific white or near-white input signals for identification, it is worth taking a look at the frequency-domain significance of correlation functions and of white signals.

3.1.3 Power Spectral Density: White Noise

As we are considering signals which go on for ever, we cannot discuss their frequency-domain characteristics without first ensuring that their Fourier transforms exist. The transform of $u(t)$ may not exist unless $u(t)$ is absolutely integrable, i.e. $\int_{-\infty}^{\infty} |u(t)| dt$ is finite (Gabel and Roberts, 1980; Bracewell, 1978). To make sure it is, we restrict $u(t)$ to a finite but long duration from $-T$ to T, and correspondingly redefine $r_{uy}(\tau)$ as

$$r_{uy}(\tau) = \frac{1}{2T} \int_{-T}^{T} u(t) y(t+\tau)\, dt \tag{3.1.12}$$

and similarly for $r_{uu}(\tau)$. Discrete-time signals are treated the same way, with the a.c.f. and c.c.f. regarded as finite-time averages over as long an interval as we please.

The z transforms of $\{u\}$ and $\{y\}$, both extending from sample M to sample N, are

$$\begin{aligned} U(z^{-1}) &= u_M z^{-M} + u_{M+1} z^{-M-1} + \cdots + u_N z^{-N} \\ Y(z^{-1}) &= y_M z^{-M} + \cdots + y_N z^{-N} \end{aligned} \tag{3.1.13}$$

and it is easy to see that

$$[\text{coefficient of } z^{-k} \text{ in } U(z)Y(z^{-1})] = u_M y_{M+k} + u_{M+1} y_{M+k+1} + \cdots + u_{N-k} y_N \quad \text{for } 0 \le k \le N-M \tag{3.1.14}$$

If we divide through by $N-M+1-k$, this approximates $r_{uy}(k)$, and as

$N - M$ is increased, the z-transform of the c.c.f. gets closer and closer to $U(z)Y(z^{-1})/(N - M + 1)$. If the sampling interval is T', say, the frequency-domain behaviour of the c.c.f. is therefore found by writing z as $\exp(j\omega T')$ in $U(z)Y(z^{-1})/(N - M + 1)$. The a.c.f. is dealt with similarly, using $U(z)U(z^{-1})/(N - M + 1)$, and is more informative since

$$U(z)U(z^{-1}) = |U(z^{-1})|^2 = |U(e^{-j\omega T'})|^2 \tag{3.1.15}$$

That is, the Fourier transform of the a.c.f. gives the square of the amplitude of the signal spectrum. By analogy with a sinusoid, the square of the amplitude is proportional to the signal power at the frequency in question, or more accurately the power per unit frequency, since the signal is represented as a continuum of frequency components. The Fourier transform of the a.c.f. is therefore called the *power (auto-) spectral density*. The c.c.f. gives rise to the *cross-spectral power density*, less readily interpreted.

For completeness let us Fourier transform the c.c.f. and a.c.f. of continuous-time signals and find a similar interpretation. With $r_{uy}(\tau)$ as in (3.1.12),

$$\begin{aligned} R_{UY}(j\omega) = \mathscr{F}[r_{uy}(\tau)] &= \int_{-\infty}^{\infty} r_{uy}(\tau)e^{-j\omega\tau}\,d\tau \\ &= \int_{-\infty}^{\infty} \frac{1}{2T} \int_{-T}^{T} u(t)e^{j\omega t}y(t+\tau)e^{-j\omega(t+\tau)}\,dt\,d\tau \\ &= \frac{1}{2T} \int_{-T}^{T} u(t)e^{j\omega t} \int_{-\infty}^{\infty} y(t+\tau)e^{-j\omega(t+\tau)}\,d\tau\,dt \end{aligned} \tag{3.1.16}$$

In the inner integral t stays constant, so $d\tau$ equals $d(t + \tau)$ and the integral gives the Fourier transform $Y(j\omega)$ of the output. The transform exists provided the response to any short section of $u(t)$ is absolutely integrable, which is so in any system of which all poles have negative real parts. As $u(t)$ extends only from $-T$ to T, the outer integral is the Fourier transform integral with j in place of $-j$, so

$$R_{UY}(j\omega) = (1/2T)U(-j\omega)Y(j\omega) \tag{3.1.17}$$

In exactly the same way, the transform of the a.c.f. of $u(t)$ is

$$R_{UU}(j\omega) = (1/2T)U(-j\omega)U(j\omega) = (1/2T)|U(j\omega)|^2 \tag{3.1.18}$$

A frequency-response identification method can be based on

$$H(j\omega) = Y(j\omega)/U(j\omega) = R_{UY}(j\omega)/R_{UU}(j\omega) \tag{3.1.19}$$

and some commercial frequency-response analysers work that way, but the time-domain alternative is more convenient, as it avoids the practical problems of aliasing, leakage and windowing in numerical Fourier transformation (Ziemer *et al.* 1983).

Although we have written the transform of the input a.c.f. in terms of the input spectrum in deriving the power spectrum, we would in practice usually find the p.s.d. from the a.c.f. directly.

Example 3.1.1 A sampled signal has a.c.f. $r_{uu}(k) = a^{|k|}$ with $-1 < a < 1$. Its p.s.d. is found from $\{r_{uu}\}$ written as the z transform $1 + a(z + z^{-1}) + a^2(z^2 + z^{-2}) + \cdots \infty$ with $\exp j\omega T'$ for z. Since $|a| < 1$, the infinite series can be summed to get

$$R_{UU}(j\omega) = \frac{1}{1 - a\exp j\omega T'} + \frac{1}{1 - a\exp - j\omega T'} - 1 = \frac{1 - a^2}{1 + a^2 - 2a\cos\omega T'}$$

or written as

$$\begin{aligned} &2(1 + a\cos\omega T' + a^2\cos 2\omega T' + \cdots \infty) - 1 \\ &\quad = Re[2(1 + a\exp j\omega T' + a^2\exp 2j\omega T' + \cdots \infty) - 1] \end{aligned}$$

and summed.

The p.s.d. $R_{UU}(j\omega)$ is periodic in ω and oscillates between $(1 - a)/(1 + a)$ at $\omega = \pm\pi/T', \pm 3\pi/T', \ldots$ and $(1 + a)/(1 - a)$ at $\omega = 0, \pm 2\pi/T', \pm 4\pi/T', \ldots$.

△

One special case merits close attention. A sequence $\{w\}$ with a.c.f. zero except at lag zero has a constant p.s.d., since the z-transform of $\{r_{ww}\}$ is just σ_w^2, the m.s. value of w, and putting $\exp j\omega T'$ for z does not change it. In continuous time, a δ-function $r_{ww}(\tau)$ transforms to a flat, infinite-bandwidth p.s.d. A signal with a flat power spectrum is called *white noise*, by loose analogy with white light. The total power of such an infinite-bandwidth signal would be infinite if the power in any finite bandwidth were non-zero, so pure white noise is pure fiction; we are actually concerned with finite-bandwidth, finite-power signals with flat power spectra. A signal with a δ-function or single-impulse a.c.f. has no consistent time structure. Its future values do not depend on its present value. Because of its lack of structure, white noise represents an ideal against which to measure the output errors of a model, since no model can do more than embody all the structure of the output. White noise is also a convenient raw material for modelling structured signals. The idea is to describe a structured noise or input signal as the result of linear filtering of white noise.

Example 3.1.2 A test signal $u(t)$ is generated by filtering white noise $w(t)$ of constant p.s.d. p over the bandwidth of interest. The filter transfer function is $H(j\omega) = 1/(j\omega T_f + 1)$. The p.s.d. of $u(t)$ is to be found.

Since $U(j\omega)$ is $H(j\omega)W(j\omega)$, $|U(j\omega)|^2$ is $|H(j\omega)|^2|W(j\omega)|^2$ so the p.s.d. of $u(t)$ is $|H(j\omega)|^2 p$, i.e. $p/(\omega^2 T_f^2 + 1)$. The larger T_f, the narrower the spectral spread of $u(t)$. △

It is important to recognise the limitations of white noise as a basis for modelling structured signals. Real-life inputs and noise more often than not contain features which cannot be represented as filtered white noise. Sustained deterministic components or sporadic disturbances are often present. Sometimes the model can be extended to include them, but sometimes we are reduced to hoping they do not matter too much, or selecting records where they are not too prominent.

3.2 CORRELATION-BASED IDENTIFICATION WITH SPECIAL PERTURBATION

3.2.1 White-Noise Test

A white discrete-time perturbation signal $\{u\}$ can readily be generated by sampling a physical source of wideband flat-spectrum noise such as thermal noise (Helstrom, 1984), looking up a random-number table or computing a deterministic long-period number sequence indistinguishable from noise samples, as in the pseudo-random number generators provided by high-level languages. It is wise always to compute the sample a.c.f. and reject any unrepresentative test sequence. The standard deviation of a sample a.c.f. computed from N samples of a genuinely white, zero-mean $\{u\}$ is $E[u^2]/\sqrt{N}$ at any non-zero lag (Problem 3.4). As $E[u^2]$ is the autocorrelation at lag zero, a trial $\{u\}$ might reasonably be rejected if the sample a.c.f. at any small non-zero lag exceeded $\pm 2/\sqrt{N}$ times the sample autocorrelation at lag zero, roughly the 95% confidence limits for a Gaussian variate. A more refined test is hardly justified, as the criterion for rejection is subjective anyway.

Identification of the u.p.r. from (3.1.8) using a white input $\{u\}$ is known as white-noise testing. A white test signal is called "noise" to emphasise its unstructured nature, even when it is completely known. White-noise testing has some drawbacks. Very large input values may occur, depending on the source, and be clipped by the digital–analogue converter or input actuator, altering the a.c.f. A signal from a genuinely random noise source is not reproducible but can, of course, be recorded to allow direct comparison of experiments. More seriously, a finite stretch of a white sequence has a finite risk of a far-from-ideal sample a.c.f., invalidating the use of (3.1.11) on the sample a.c.f. and c.c.f.

The reliability of results from finite records can be assessed more easily if

$\{u\}$ is a deterministic pseudo-noise signal, with a.c.f. behaviour over a finite interval precisely known. A family of pseudo-noise signals with several convenient features is considered next.

3.2.2 Pseudo-Random Binary Sequence Test

A pseudo-random binary sequence (p.r.b.s.) (Peterson, 1961; Golomb, 1967; Godfrey, 1969) is

(i) deterministic but pseudo-random in the sense that its a.c.f. is close to zero, compared with the value at lag zero, over a range of non-zero lags;

(ii) binary, a great advantage as it maximises the power for a given maximum amplitude, simplifies digital generation and storage of the sequence, suits most actuators and makes c.c.f. computation easy as explained later;

(iii) periodic, so its a.c.f. is periodic and is found accurately by averaging over a single period, since if we take $2p$ periods,

$$r_{uu}(k) = \lim_{p\to\infty} \frac{1}{2p} \sum_{j=-p}^{p-1} \frac{1}{P} \sum_{i=jP}^{(j+1)P-1} u_i u_{i+k}$$

$$= \frac{1}{P} \sum_{i=0}^{P-1} u_i u_{i+k} = \frac{1}{P} \sum_{i=0}^{P-1} u_i u_{i+k+lP} = r_{uu}(k+lP) \qquad (3.2.1)$$

where l is an integer and P the period;

(iv) synchronous, i.e. the samples are produced regularly at one per bit interval t_b.

The best-known p.r.b.s. is the *maximal-length sequence* or *m*-sequence. An *m*-sequence has period $2^n - 1$ during which every *n*-bit binary number except *n* zeros starts exactly once. For example, one period of the *m*-sequence with $n = 3$ is 1011100, so the three-bit numbers 101, 011, 111, 110, 100, 001, 010 start successively in that period (and run into the first two bits of the next). The a.c.f. characteristics of *m*-sequences are best seen in an example.

Example 3.2.1 We calculate the a.c.f. of the seven-bit *m*-sequence by averaging $u_i u_{i+k}$ over seven successive samples, starting at $i = 1$ and taking u_{i+k} from the second period 1011100 as required. We find

$$r_{uu}(0) = (1+0+1+1+1+0+0)/7 = 4/7$$
$$r_{uu}(1) = (0+0+1+1+0+0+0)/7 = 2/7$$
$$r_{uu}(2) = (1+0+1+0+0+0+0)/7 = 2/7$$

and so on: $r_{uu}(3) = r_{uu}(4) = r_{uu}(5) = r_{uu}(6) = \frac{2}{7}$, $r_{uu}(7) = \frac{4}{7} = r_{uu}(0)$, $r_{uu}(8) = r_{uu}(1), \ldots$. The a.c.f. is far from that of white noise but can be brought closer by a d.c. shift in $\{u\}$. If we subtract c from every bit, forming $\{u'\}$,

$$r_{u'u'}(k) = \sum_{i=1}^{7} \frac{(u_i - c)(u_{i+k} - c)}{7} = r_{uu}(k) - 2c\bar{u} + c^2$$

where $\bar{u}$ is the mean of $\{u\}$, $\frac{4}{7}$. The choice $c = (4 \pm \sqrt{2})/7$ makes $r_{u'u'}(k)$ zero at all lags except multiples of 7. The resulting binary levels 0.227 and -0.773 or 0.631 and -0.369 may be less convenient than symmetrical levels. Symmetrical levels give non-zero autocorrelations at all lags, but the measured $\{r_{uy}\}$ can be adjusted to allow for them, as described shortly. △

Figure 3.2.1a shows the a.c.f. of a general m-sequence with binary levels $\pm b$. The sequence is normally applied via a zero-order hold to the input of a continuous-time system, and the output is sampled. The u.p.r. is then a description of the sample-time dynamics of the zero-order hold and system combined. An alternative is to view the output of the zero-order hold as a continuous-time input, observe the system output continuously and form

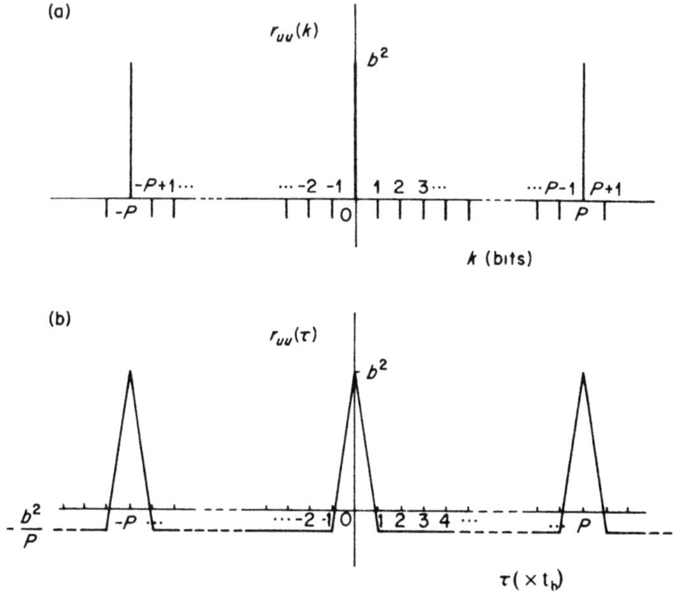

Fig. 3.2.1 (a) Autocorrelation function of m-sequence. The negative values of r_{uu} are all $-b^2/P$. (b) autocorrelation function of output of zero-order hold driven by m-sequence.

$r_{uy}(\tau)$ by analogue means, for any lag τ of interest. Hence $h(\tau)$ at particular values of τ is found, rather than the u.p.r. Figure 3.2.1b shows the a.c.f. of the zero-order hold output. It deviates from the ideal more than the discrete-time a.c.f. of the m-sequence, by virtue of the two-bit-interval width of the spikes.

The input–output c.c.f. corresponding to Fig. 3.2.1a, with output noise $\{v\}$ present, is

$$r_{uy}(k) = \sum_{j=0}^{s} h_j r_{uu}(k-j) + r_{uv}(k)$$

$$= b^2\left(h_k - \sum_{\substack{j=0\\ j\neq k}}^{s} \frac{h_j}{P}\right) + r_{uv}(k)$$

$$\simeq b^2 \frac{(P+1)h_k - g}{P}$$

where g is the steady-state gain $\sum_{j=0}^{s} h_j$ of the system, i.e. the final value of its unit-sampled-step response. It can be measured in a step-test, inferred from steady input and output levels, calculated as $P\sum_{k=0}^{s} r_{uy}(k)/((P-s)b^2)$ or estimated from steady-state performance specifications.

Continuous- or discrete-time cross-correlation is particularly easy when the binary input has symmetrical levels or one level zero, since multiplication by the input only requires sign reversal or switching on and off of the lagged output.

Experiment design for p.r.b.s. tests is straightforward. The bit interval should be short compared with the shortest feature of interest in the impulse response or u.p.r., to avoid blurring it or missing it, respectively. The period should be longer than the settling time of the impulse response or u.p.r., to avoid superimposing the effects of h_j and h_{j+P} in (3.1.8) or similarly in (3.1.9). The swing between binary levels should be as large as permitted to maximise the output s.n.r., and the experiment as long as possible to minimise the contribution of noise to the input–output c.c.f. Several short periods of p.r.b.s. are preferable to one long one, to prevent one or two short breakdowns from ruining the experiment; the good a.c.f. properties apply only to complete periods.

Industrial applications of p.r.b.s. testing are described by Godfrey (1970) and Cumming (1972), who discuss their results in detail.

Example 3.2.2 The open-loop speed-control system of Example 2.3.3 was perturbed by a 63-bit m-sequence with bit interval 25 ms and mean-square value similar to that of the input in Example 2.3.3.

Two different cross-correlators were used. One produced the continuous-time cross-correlation one lag at a time by analogue integration of the output, with sign reversed whenever the input changed binary level. Irregular drift of about 0.2 Hz bandwidth in the output made values at successive lags inconsistent and unrepeatable, even with careful biasing of the output to zero long-term d.c. level and averaging over several periods. Susceptibility to drift is common in open-loop systems normally controlled by feedback, since they have high gain at low frequencies.

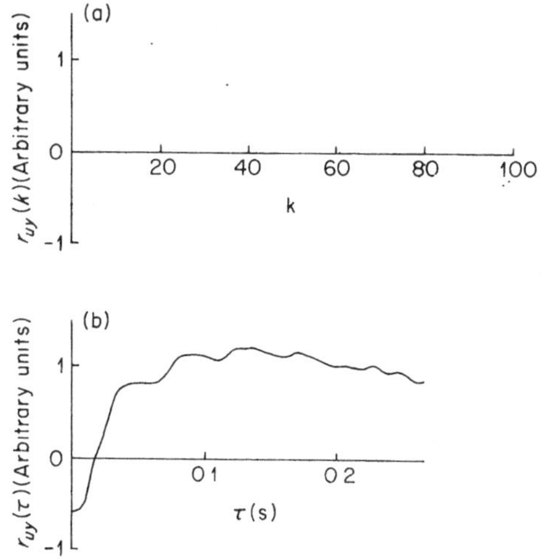

Fig. 3.2.2 Results of open-loop p.r.b.s. test of speed-control system (a) c.c.f. computed from 2048 samples showing dominant time constants (sampling interval 10 ms), and (b) c.c.f. computed from 16384 samples showing resonance as ringing.

Drift-correction schemes for *m*-sequence testing are well investigated (Brown, 1969) and fairly straightforward, but only compensate for drift adequately represented by a polynomial or short Fourier series. Unavoidable erratic drift is better modelled as non-deterministic, as in some methods in Chapter 7, where the disturbances as well as the dynamics are explicitly identified.

The second correlator sampled the input and output at a rate well above the bit rate and computed the c.c.f. over a range of lags from the same samples. As Fig. 3.2.2 shows, acceptable results were obtained, now repeatable. They are consistent with the dominant time constants 0.08 and 0.56 s found in Example 2.3.3, and reveal a similar high-frequency resonance. Averaging over 15 or more p.r.b.s. periods was necessary to obtain sufficiently repeatable and plausibly smooth u.p.r.'s. △

Pseudo-random binary sequence results have the disadvantage that non-linearity or drift may have effects indistinguishable from those of noise, in contrast to sine-wave responses. Inspection of the response waveform before any processing is highly advisable in either case. A major advantage of p.r.b.s. testing is its relative speed, even with considerable averaging, compared with a succession of sine-wave tests, providing the cross-correlations at all lags are found from the same records. Fast frequency-sweeping sine-wave-based transfer-function analysers exist, however (Doebelin, 1980). In them, the output is passed through a narrowband filter to extract the fundamental without long averaging, and the variable-frequency output is heterodyned to a fixed frequency so that a fixed filter can be used. Determining the fastest permissible sweep rate may not be very easy when, as usual, the required frequency resolution is initially unknown.

M-sequences are easy to generate compared with the other main family of periodic p.r.b.s., quadratic-residue codes (Everett, 1966; Godfrey, 1969). The *m*-sequence of period $2^n - 1$ appears in the right-most stage of an *n*-stage shift register fed at the left-hand end by the modulo-2 sum (i.e. the remainder of the sum when divided by 2) of the outputs of the right-most stage and one or more others. Table 3.2.1 gives the stages summed for *n* up to 11.

Table 3.2.1
Feedback connections to generate m-sequences

Number of stages n	Sequence period (bits)	Input to stage 1 is mod-2 sum of stages
3	7	2, 3
4	15	3, 4
5	31	3, 5
6	63	5, 6
7	127	4, 7
8	255	2, 3, 4, 8
9	511	5, 9
10	1023	7, 10
11	2047	9, 11

Modulo-2 addition of two bits A and B obeys the accompanying exclusive-OR truth table

A	B	$(A + B) \bmod 2$
0	0	0
0	1	1
1	0	1
1	1	0

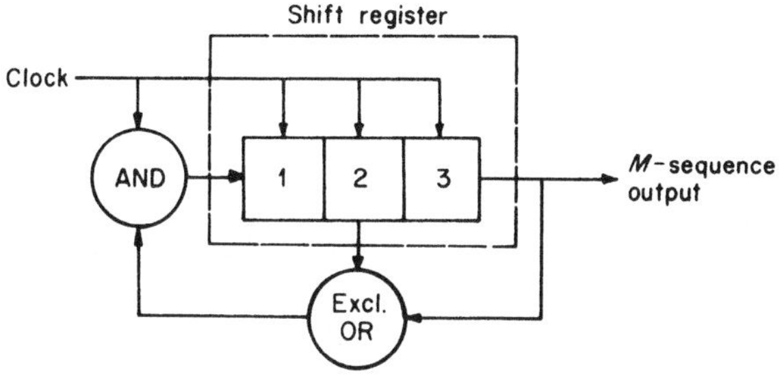

Fig. 3.2.3 Shift-register generation of 7-bit m-sequence.

Example 3.2.3 The contents of the 3-stage shift register in Fig. 3.2.3 are initially set to 001. The first clock pulse shifts them rightward one stage and transfers 1, the modulo-2 sum of bits 2 and 3, into stage 1, making the contents 100. Succeeding clock pulses produce contents 010, 101, 110, 111, 011, 001; then the sequence repeats. Stage 3 gives the 7 bit m-sequence 1001011. Stages 1 and 2 give the sequence delayed by 5 and 6 bits (not 2 and 1 bits!), and other delayed versions are easily obtainable; for instance, the modulo-2 sum of stage 1 and 3 contents is 1100101, the sequence delayed by 1 bit. △

FURTHER READING

Bendat and Piersol (1980) and Jenkins and Watts (1968) cover spectral and correlation methods of identification in depth. Further examples of p.r.b.s. identification are given by Billings (1981) and Hogg (1981).

REFERENCES

Bendat, J. S., and Piersol, A. G. (1980). "Engineering Applications of Correlation and Spectral Analysis". Wiley, New York.

Billings, S. A. (1981). Modelling and identification of three-phase electric-arc furnace. In "Modelling of Dynamical Systems" (H. Nicholson, ed), Vol. 2. Peter Peregrinus, London.

Bracewell, R. N. (1978). "The Fourier Integral and Its Applications". McGraw-Hill, New York.

Brown, R. F. (1969). Review and comparison of drift-correction schemes for periodic cross-correlation. *Electron. Lett.* **5**, 179–181.

Cumming, I. G. (1972). On-line identification of a steel mill. *Automatica* **8**, 531–541.

Doebelin, E. O. (1980). "System Modeling and Response". Wiley, New York.

Everett, D. (1966). Periodic digital sequences with pseudonoise properties. *GEC J.* **33**, 115–126.

Gabel, R. A., and Roberts, R. A. (1980). "Signals and Linear Systems", 2nd ed. Wiley, New York.

Godfrey, K. R. (1969). The theory of the correlation method of dynamic analysis and its application to industrial processes and nuclear power plant. *Meas. Control* **2**, T65–T72.

Godfrey, K. R. (1970). The application of pseudorandom sequences to industrial processes and nuclear power plant. *2nd IFAC Symp. on Identification and System Parameter Estimation, Prague, Czechoslovakia.* Pap. 7.1.

Golomb, S. W. (1967). "Shift Register Sequences". Holden-Day, San Francisco, California.

Helstrom, C. W. (1984). "Probability and Stochastic Processes for Engineers". Macmillan, New York.

Hogg, B. W. (1981). Representation and control of turbogenerators in electric power systems. *In* "Modelling of Dynamical Systems" (H. Nicholson, ed.), Vol. 2. Peter Peregrinus, London.

Jenkins, G. M., and Watts, D. G. (1968). "Spectral Analysis and Its Applications". Holden-Day, San Francisco, California.

Melsa, J. W., and Sage, A. P. (1973). "An Introduction to Probability and Stochastic Processes". Prentice-Hall, Englewood Cliffs, New Jersey.

Peterson, W. W. (1961). "Error Correcting Codes". *MIT Technical Press*, Cambridge, Massachusetts.

Ziemer, R. E., Tranter, W. H., and Fannin, D. R. (1983). "Signals and Systems: Continuous and Discrete". Macmillan, New York.

PROBLEMS

3.1 Show that, in the notation of Section 3.1.1, the gain α which makes $\alpha u(t)$ as close as possible to $y(t+\tau)$, in the sense that $E[(\alpha u(t) - y(t+\tau))^2]$ is minimised, is $r_{uy}(\tau)/r_{uu}(0)$.

3.2 A discrete-time system with unit-pulse response $H(z^{-1}) = h_0 + h_1 z^{-1} + h_2 z^{-2} + \cdots$ is driven by a white-noise input $\{w\}$ and has output $\{y\}$. By writing $r_{yy}(0)$ in terms of the autocorrelation function of $\{w\}$, show that the power gain of the system, i.e. m.s. output/m.s. input, is $h_0^2 + h_1^2 + h_2^2 + \cdots$ for this input.

3.3 A linear system with input $u(t)$ has an output $z(t)$ consisting of noise-free output $y(t)$ plus noise uncorrelated with $y(t)$. If the transfer function $H(j\omega)$ of the system is to be identified from correlation functions and the effect of noise is to be as small as possible, would $R_{UZ}(j\omega)/R_{UU}(j\omega)$ or $R_{ZZ}(-j\omega)/R_{UZ}(-j\omega)$ be preferable as the estimator of $H(j\omega)$?

3.4 A sample autocorrelation is computed as $\hat{r}_{uu}(k) = \sum_{i=1}^{N} u_i u_{i+k}/N$. Show that if successive samples from $\{u\}$ are independent, zero-mean and of constant m.s. value σ_u^2, $\hat{r}_{uu}(k)$ has a mean zero and r.m.s. value $\sigma_u^2/\sqrt{N}$. [Note that no assumption need be made about the amplitude probability distribution of $\{u\}$.]

3.5 A discrete-time deterministic test signal has a period of M samples and sampling interval T. Find (i) the interval in lag at which its autocorrelation function repeats itself; (ii) the interval in frequency at which its discrete power spectrum repeats itself; (iii) how many values on its discrete power spectrum can be specified independently when the signal is being designed in the frequency domain.

3.6 Find modulo-2 sums of stage contents and/or gate output in Fig. 3.2.3 to give the 7-bit m-sequence delayed by 2, 3 and 4 bits.

3.7 An inverse-repeat sequence of period $2P$ is the result of changing the sign of alternate bits, say the even-numbered bits, in 2 periods of a P-bit m-sequence. Find the a.c.f. of such a sequence. Compare the contribution of a constant-plus-linear-drift output component to the input–output c.c.f., when the input is an inverse-repeat sequence, with that given by an m-sequence input.

3.8 Show that the cross-correlation function between an m-sequence and the inverse-repeat sequence derived from it as in Problem 3.7 is zero at all lags. Verify that this makes it possible to identify the two u.p.r.'s of a two-input, one-output linear system simultaneously, using the Wiener–Hopf equation.

3.9 The quadratic-residue sequence of period 7 is $1\ 1\ -1\ 1\ -1\ -1\ \pm 1$. Take whichever sign you like for the last bit and check whether the a.c.f. of this sequence is the same as that of the 7-bit m-sequence 1001011. How is this m-sequence related to the q.r. sequence? [Note that not all q.r. sequences have the same period as an m-sequence; a q.r. sequence with period $4k-1$ exists whenever $4k-1$ is a prime (Godfrey, 1969).]

3.10 In an identification experiment, the first bit in each period of the 7-bit m-sequence 1001011 used as an input is inadvertently changed to 0. The mistake is later discovered. What alterations to the u.p.r. estimates obtained by way of the Wiener–Hopf equation are necessary to correct the error?

3.11 A discrete-time model of a linear, time-invariant s.i.s.o. system is to be found by a p.r.b.s. test. The system is known to have a continuous-time impulse response approximating to $h(t) = 100(\exp(-0.1t) - \exp(-t))$. Find acceptable values of the p.r.b.s. bit interval, sequence length $2^N - 1$ bits (N integer) and amplitude, if the mean-square value of the sampled output is not to exceed 25.

[*Hint for last part:* find the discrete-time power gain of the system for an uncorrelated input, on the reasonable assumption that the p.r.b.s. approximates white noise.]

Chapter 4

Least-Squares Model Fitting

4.1 FINDING THE "BEST-FIT" MODEL

In the last two chapters we reduced the influence of noise on the estimate of step response, impulse response or transfer function by time-averaging. The justification was essentially statistical. We relied on the zero-mean noise-dependent terms affecting the estimates becoming negligible if averaged over a long enough interval. For discrete-time models, we had to take many more observations than would be needed in the absence of noise, i.e. many more observations than unknowns in the model.

In this chapter we take a different approach to the problem of estimating a model from a large set of observations. We find the model, of specified structure, which fits the observations best according to a deterministic measure of error between model output and observed output, totalled over all the observations. Initially we appeal to statistical theory as little as possible (not at all, in fact, for most of this chapter). We examine the resulting estimators in a probabilistic setting in later chapters.

4.1.1 Least Squares

We shall find the values of the coefficients in a given model which minimise the sum of the squared errors between the model output and the observations of the output: *least-squares* estimates of the coefficients. We might well consider other measures of fit than output error squared, but this measure has two big advantages. First, large errors are heavily penalized: an error twice as large is four times as bad. This usually accords with common sense, but there are exceptions. For instance, when a few observations are very poor, or even totally spurious misreadings, the best thing may be to ignore them altogether, and the worst thing to take a lot of notice of them. The other advantage is mathematical tractability. The formula giving the least-squares estimates is obtained by quite simple matrix algebra, and the estimates are computed as the solution to a set of linear equations. Moreover, the properties of the estimates are relatively easy to analyse. Gauss, who devised least-squares estimation at roughly the same time as Legendre, wrote: "...of all these principles ours is the most simple; by the others we would be led into the most complicated calculations" (Gauss, 1809).

4.1.2 Ordinary Least Squares

The model we use relates an observed variable y_t, the *regressand*, to p explanatory variables, the *regressors* u_{1t} to u_{pt}, all known in advance or observed. In dynamical models the sample-indexing variable t is time, but the method is not restricted to dynamical models, and t need not represent time. For instance, an econometric model might relate expenditure y_t to such indicators as income, age and family size. In that case t would index observed individuals or groups.

The model has one unknown coefficient θ_i per explanatory variable. If the u's for one sample and the θ's are collected into p-vectors

$$\mathbf{u}_t = [u_{1t} \quad u_{2t} \quad \cdots \quad u_{pt}]^{\mathrm{T}}, \qquad \boldsymbol{\theta} = [\theta_1 \quad \theta_2 \quad \cdots \quad \theta_p]^{\mathrm{T}} \tag{4.1.1}$$

then the model is

$$y_t = f(\mathbf{u}_t, \boldsymbol{\theta}) + e_t, \qquad t = 1, 2, 3, \ldots, N \tag{4.1.2}$$

where e_t accounts for observation error (measurement noise) and modelling error, since even without observation error few models are perfect. We aim to find the value $\hat{\boldsymbol{\theta}}$ of $\boldsymbol{\theta}$ which minimises

$$S \triangleq \sum_{t=1}^{N} e_t^2 = \sum_{t=1}^{N} (y_t - f(\mathbf{u}_t, \boldsymbol{\theta}))^2 \tag{4.1.3}$$

for the practically useful case where $f(\cdot\,,\cdot)$ is *linear in the unknown coefficients* making up $\boldsymbol{\theta}$. That is,

$$y_t = \mathbf{u}_t^{\mathrm{T}}\boldsymbol{\theta} + e_t, \qquad t = 1, 2, 3, \ldots, N \tag{4.1.4}$$

It is important to realise that the model need not be linear in the physical variables giving rise to the u's. For example, we might model the smooth trajectory of a radar target in one dimension by

$$y_t = \theta_1 + \theta_2 t + \theta_3 t^2 + e_t \tag{4.1.5}$$

The model is clearly non-linear in t, but linear in θ_1, θ_2 and θ_3. Notice that θ_1 covers any constant component of y, so $\{e\}$ can be assumed zero-mean. The explanatory "variable" whose coefficient is θ_1 is 1 for all samples.

To make the algebra tidy, collect all the samples y_1 to y_N into an N-vector $\mathbf{y}$, all the $\mathbf{u}_t$ vectors into an $N \times p$ matrix U and e_1 to e_N into $\mathbf{e}$, giving

$$\mathbf{y} = U\boldsymbol{\theta} + \mathbf{e} \tag{4.1.6}$$

and

$$S = \mathbf{e}^{\mathrm{T}}\mathbf{e} = (\mathbf{y}^{\mathrm{T}} - \boldsymbol{\theta}^{\mathrm{T}} U^{\mathrm{T}})(\mathbf{y} - U\boldsymbol{\theta}) \tag{4.1.7}$$

The value $\hat{\boldsymbol{\theta}}$ which minimises S makes the gradient of S with respect to $\boldsymbol{\theta}$ zero:

$$\frac{\partial S}{\partial \boldsymbol{\theta}} = \left[\frac{\partial S}{\partial \theta_1}\,\frac{\partial S}{\partial \theta_2} \cdots \frac{\partial S}{\partial \theta_p}\right]^{\mathrm{T}} = \mathbf{0} \tag{4.1.8}$$

To evaluate $\partial S/\partial \boldsymbol{\theta}$ we need two standard results for derivatives of vector-matrix expressions, namely

$$\frac{\partial(\mathbf{a}^{\mathrm{T}}\boldsymbol{\theta})}{\partial \boldsymbol{\theta}} = \left(\text{vector with element } i \quad \frac{\partial}{\partial \theta_i} \sum_{j=1}^{p} a_j \theta_j \right) = \mathbf{a} \tag{4.1.9}$$

and

$$\begin{aligned}\frac{\partial(\boldsymbol{\theta}^{\mathrm{T}} A \boldsymbol{\theta})}{\partial \theta} &= \left(\text{vector with element } i \quad \frac{\partial}{\partial \theta_i} \sum_{j=1}^{p} \sum_{k=1}^{p} a_{jk} \theta_j \theta_k \right) \\ &= \left(\text{vector with element } i \quad \sum_{k=1}^{p} a_{ik}\theta_k + \sum_{j=1}^{p} a_{ji}\theta_j \right) \\ &= (A + A^{\mathrm{T}})\boldsymbol{\theta}\end{aligned} \tag{4.1.10}$$

We multiply out the expression for S in (4.1.7), note that $\boldsymbol{\theta}^{\mathrm{T}} U^{\mathrm{T}} \mathbf{y}$ is identical to $y^T U\boldsymbol{\theta}$ since it is a scalar, and putting $U^{\mathrm{T}}\mathbf{y}$ for $\mathbf{a}$ and $U^{\mathrm{T}}U$ for A in (4.1.9) and (4.1.10) obtain

$$\frac{\partial S}{\partial \boldsymbol{\theta}} = -2U^{\mathrm{T}}\mathbf{y} + 2U^{\mathrm{T}}U\boldsymbol{\theta} \tag{4.1.11}$$

The $\boldsymbol{\theta}$ that makes the gradient of S zero is therefore

$$\hat{\boldsymbol{\theta}} = [U^{\mathrm{T}}U]^{-1}U^{\mathrm{T}}\mathbf{y} \tag{4.1.12}$$

To check that $\hat{\boldsymbol{\theta}}$ gives a minimum of S, not a maximum or saddle point, we must see whether any small change $\delta\boldsymbol{\theta}$ about $\hat{\boldsymbol{\theta}}$ increases S. With $\partial S/\partial \boldsymbol{\theta}$ zero,

$$\delta S = 2\,\delta\boldsymbol{\theta}^{\mathrm{T}} U^{\mathrm{T}} U\,\delta\boldsymbol{\theta} = 2(U\,\delta\boldsymbol{\theta})^{\mathrm{T}} U\,\delta\boldsymbol{\theta} = 2\sum_{t=1}^{N}\{(\text{element } t \text{ of } U\,\delta\boldsymbol{\theta})^2\} \tag{4.1.13}$$

The condition for δS to be positive whatever the set of small changes $\delta\boldsymbol{\theta}$ is therefore that $U\,\delta\boldsymbol{\theta}$ should not be zero. In other words, the columns of U must not be linearly dependent, or to put it another way, none of the regressors may be totally redundant by being a linear combination of the others at every sample. A corollary states that $U^{\mathrm{T}}U$ is positive-definite, ensuring that $\delta\boldsymbol{\theta}^{\mathrm{T}}U^{\mathrm{T}}U\,\delta\boldsymbol{\theta}$ is positive and also guaranteeing that the inverse of $U^{\mathrm{T}}U$ exists, since $U^{\mathrm{T}}U\,\delta\boldsymbol{\theta}$ cannot be zero for any real, non-zero $\delta\boldsymbol{\theta}$. It does not guarantee that the inverse is easy to compute accurately, though, as we shall see in a moment.

The $\hat{\boldsymbol{\theta}}$ given by (4.12) is called the *ordinary least-squares* (*o.l.s.*) *estimate* of $\boldsymbol{\theta}$. We shall find that it has some out-of-the-ordinary properties.

Example 4.1.1 The positive x of a radar target moving in a straight line is observed at intervals of 0.2 s over 1 s. Its position is to be predicted. A simple way is to assume constant acceleration over the observation and prediction interval, estimate the initial position x_0, velocity v_0 and acceleration a and predict future position using the model $x(t) = x_0 + v_0 t + at^2/2$. Putting the time origin at the first observation, the radar gives

$t(s)$	0	0.2	0.4	0.6	0.8	1
$x(m)$	3	59	98	151	218	264

Here $\boldsymbol{\theta}$ is $[x_0 \quad v_0 \quad a]^{\mathrm{T}}$ and $\mathbf{y}$ is $[3 \quad 59 \quad \cdots \quad 264]^{\mathrm{T}}$. Matrix U has all 1's in column 1, the sample times $0, 0.2, \ldots, 1$ in column 2 and $t^2/2$ values $0, 0.02, \ldots, 0.5$ in column 3, so

$$U^{\mathrm{T}}\mathbf{y} = \begin{bmatrix} 793 \\ 580 \\ 238 \end{bmatrix}, \qquad U^{\mathrm{T}}U = \begin{bmatrix} 6 & 3 & 1.1 \\ 3 & 2.2 & 0.9 \\ 1.1 & 0.9 & 0.3916 \end{bmatrix}$$

and to three figures

$$[U^{\mathrm{T}}U]^{-1} = \begin{bmatrix} 0.821 & -2.95 & 4.46 \\ -2.95 & 18.2 & -33.5 \\ 4.46 & -33.5 & 67.0 \end{bmatrix}$$

so

$$\hat{\boldsymbol{\theta}}^{\mathrm{T}} = [\hat{x}_0 \quad \hat{v}_0 \quad \hat{a}] = [4.79 \quad 234 \quad 55.4] \qquad \triangle$$

Several practical points show up in this example. First, the dimensions of the *normal matrix* $U^{\mathrm{T}}U$ are fixed by the relatively small number of coefficients being estimated, three here, however many observations there may be. Second, $U^{\mathrm{T}}U$ is symmetric and, as we saw earlier, positive-definite, and there are special efficient methods of solving sets of linear equations with such coefficient matrices. Some are described in Section 4.2. However, if N is large, a lot of computation is required just to form the *normal equations*

$$U^{T}U\hat{\boldsymbol{\theta}} = U^{T}\mathbf{y} \tag{4.1.14}$$

A third point is that the normal matrix may be near-singular. Computing its inverse would then be ill-conditioned, involving at some stage small differences of large quantities. In Example 4.1.1 the computation is not very ill-conditioned, but $|U^{T}U|$ and the cofactors of some elements of $U^{T}U$ are an order or so smaller than any element of $U^{T}U$. In more serious cases like Example 4.1.2, ill-conditioning may prevent satisfactory solution of the normal equations. When this happens, it signals that at least one regressor is not pulling its weight, as it is close to being linearly dependent on the other regressors, which would cause U to lose rank and $U^{T}U$ to become singular. Poor numerical conditioning therefore indicates a badly constructed model, with near-redundancy among its explanatory variables. Such near-redundancy can be induced by a bad choice of co-ordinates for the observations, obscuring the information in one or more regressors, as Example 4.1.2 will show.

Example 4.1.2 The same radar observations are obtained as in Example 4.1.1, at intervals of 0.2 s but starting at 10 s rather than time zero. The same model as in Example 4.1.1, quadratic in t, is fitted. With the new values of t, 10(0.2)11, keeping eight significant figures to try to maintain accuracy gives

$$U^{T}\mathbf{y} = \begin{bmatrix} 793 \\ 8510 \\ 45687.96 \end{bmatrix}, \qquad U^{T}U = \begin{bmatrix} 6 & 63 & 331.1 \\ 63 & 662.2 & 3483.9 \\ 331.1 & 3483.9 & 18348.392 \end{bmatrix}$$

and

$$[U^{T}U]^{-1} = \begin{bmatrix} 20963.816 & -3995.7237 & 380.39474 \\ -3995.7237 & 761.74342 & -72.532895 \\ 380.39474 & -72.532895 & 6.9078947 \end{bmatrix}$$

so, rounding finally to three figures,

$$\hat{\boldsymbol{\theta}}^{T} = [157 \quad -52.4 \quad 5.71]$$

The position at 11 s given by this $\hat{\boldsymbol{\theta}}$ is -73.9 m, clearly at odds with the given observations, so something is amiss. The observations are actually

$$\begin{aligned} x(t) &= 5 + 250(t-10) + 5(t-10)^2 + e(t) \\ &= -1995 + 150t + 5t^2 + e(t) \end{aligned}$$

and the observation noise $e(t)$ has samples with r.m.s. value 6.12, so $\hat{\boldsymbol{\theta}}$ is wildly inaccurate. The normal matrix $U^{T}U$ is now very ill-conditioned, and the calculated $|U^{\mathrm{T}}U|$, 0.608, is about eight orders smaller than the individual product terms in its calculation. The cause is the new choice of time origin, which has made the regressors very nearly linearly dependent. Over the observation range $10 \leq t \leq 11$, denoting $t - 10.5$ by τ and recognising that τ is small,

$$t^2 = (\tau + 10.5)^2 \simeq 21\tau + 110.25 = 21t - 110.25$$

so in U, column 3 is almost equal to $10.5 \times$ column $2 - 55.125 \times$ column 1, making $U^{\mathrm{T}}U$ almost singular.

If 10 significant figures are kept, the ill-conditioning has less effect but $\hat{\boldsymbol{\theta}}^{\mathrm{T}}$ is $[417 \quad -310 \quad 53.4]$, still far from the correct values. Now the reason is that the coefficients of 1 and t are very sensitive to any noise-induced error in the acceleration and initial velocity implied by the observations. Specifically, with the observations generated by

$$x(t) = x_0 + v_0(t - T) + (a/2)(t - T)^2 + e(t)$$

the model is

$$\begin{aligned} x(t) &= [x_0 - v_0 T + (aT^2/2)] + (v_0 - aT)t + (at^2/2) + e(t) \\ &= x(-T) + v(-T)t + (at^2/2) + e(t) \end{aligned}$$

If T is large (10 here), a small error in a or v_0 corresponds to large errors in coefficients $x(-T)$ and $v(-T)$ of the present model. The 10-figure calculations give a $\hat{\boldsymbol{\theta}}$ which, although it looks poor, implies $x_0 = -15.9$, $v_0 = 224$, $a = 53{\cdot}4$, so the indirect estimates of v_0 and a are little worse than in Example 4.1.1. The error in x_0 is larger because of the ill-conditioning but much smaller than might have been guessed from the poor $\hat{\boldsymbol{\theta}}$. Nevertheless, it leads to a mean error of 26.3 between model output and observed position, enough to detect the model deficiency easily. △

Two lessons can be drawn from Example 4.1.2. First, the model and the observation co-ordinates should be chosen to avoid approaching linear dependence between the regressors. Second, the sensitivity of the model coefficient estimates to noise may be much higher than that of the goodness of fit. The importance of avoiding high sensitivity varies according to whether interest centres on the accuracy of the coefficients or their ability to fit the observations.

4.1.3 Orthogonality

A property of great value in interpreting least squares, orthogonality, becomes apparent when the model output

$$\hat{\mathbf{y}} = U\hat{\boldsymbol{\theta}} \tag{4.1.15}$$

is compared with the observed output $\mathbf{y}$. The vector of errors between $\hat{\mathbf{y}}$ and $\mathbf{y}$ is

$$\hat{\mathbf{e}} = \mathbf{y} - \hat{\mathbf{y}} \tag{4.1.16}$$

(Incidentally, it may seem odd to define error so that overestimating the output gives a negative error. It *is* odd, but it is an almost universal convention and we are stuck with it.) Looking at the sum of the products of corresponding model-output and output-error samples, we find that

$$\begin{aligned}\hat{\mathbf{y}}^{\mathrm{T}}\hat{\mathbf{e}} &= \hat{\boldsymbol{\theta}}^{\mathrm{T}} U^{\mathrm{T}}(\mathbf{y} - U\hat{\boldsymbol{\theta}}) = \mathbf{y}^{\mathrm{T}} U[U^{\mathrm{T}}U]^{-1}U^{\mathrm{T}}(\mathbf{y} - U[U^{\mathrm{T}}U]^{-1}U^{\mathrm{T}}\mathbf{y})\\ &= \mathbf{y}^{\mathrm{T}}(U[U^{\mathrm{T}}U]^{-1}U^{\mathrm{T}} - U[U^{\mathrm{T}}U]^{-1}U^{\mathrm{T}})\mathbf{y} = 0\end{aligned} \tag{4.1.17}$$

Two vectors whose inner product is zero, like the model-output and output-error vectors here, are said to be orthogonal. This follows from Pythagoras' theorem and the definition of Euclidean length of a vector as the square root of its inner product with itself, that is, the sum of squares of its elements. Denoting the length of a vector by $\|\cdot\|$, if $\hat{\mathbf{y}}^{\mathrm{T}}\hat{\mathbf{e}}$ is zero we have

$$\|\mathbf{y}\|^2 = \mathbf{y}^{\mathrm{T}}\mathbf{y} = \hat{\mathbf{y}}^{\mathrm{T}}\hat{\mathbf{y}} + 2\hat{\mathbf{y}}^{\mathrm{T}}\hat{\mathbf{e}} + \hat{\mathbf{e}}^{\mathrm{T}}\hat{\mathbf{e}} = \|\hat{\mathbf{y}}\|^2 + \|\hat{\mathbf{e}}\|^2 \tag{4.1.18}$$

i.e. $\hat{\mathbf{y}}$, $\hat{\mathbf{e}}$ and $\mathbf{y}$ form a right-angled triangle. That makes sense, as in o.l.s. estimation $\hat{\boldsymbol{\theta}}$ *is* chosen to make $\hat{\mathbf{e}}$ as short as possible, subject to $\hat{\mathbf{y}}$ being of the form $U\hat{\boldsymbol{\theta}}$, a linear combination of the column vectors $\mathbf{u}_1^{\mathrm{c}}$ to $\mathbf{u}_p^{\mathrm{c}}$ of U:

$$\hat{\mathbf{y}} = \hat{\theta}_1\mathbf{u}_1^{\mathrm{c}} + \hat{\theta}_2\mathbf{u}_2^{\mathrm{c}} + \cdots + \hat{\theta}_p\mathbf{u}_p^{\mathrm{c}} \tag{4.1.19}$$

In other words, $\hat{\mathbf{y}}$ has to lie in the hyperplane spanned by $\mathbf{u}_1^{\mathrm{c}}$ to $\mathbf{u}_p^{\mathrm{c}}$. Figure 4.1.1 shows the situation for $p = 2$. Plainly, the shortest error vector is obtained by

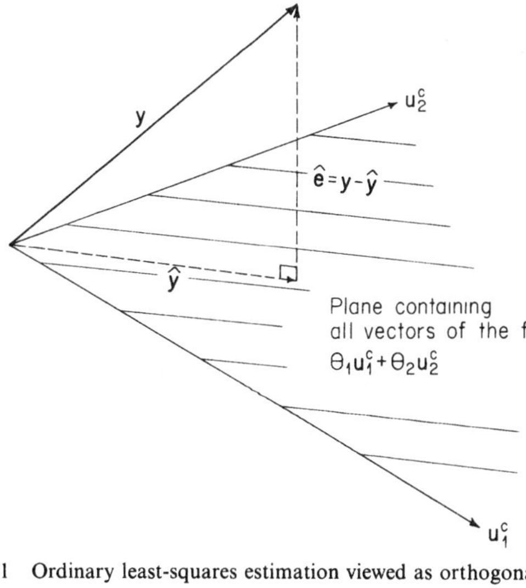

Fig. 4.1.1 Ordinary least-squares estimation viewed as orthogonal projection.

making $\hat{\mathbf{y}}$ the orthogonal projection of $\mathbf{y}$ on to the hyperplane of the $\mathbf{u}^c$'s. The picture also makes it clear that $\hat{\mathbf{e}}$ is orthogonal to each individual regressor vector $\mathbf{u}^c$ in the hyperplane. Algebraically,

$$U^T\hat{\mathbf{e}} = U^T(\mathbf{y} - U[U^TU]^{-1}U^T\mathbf{y}) = U^T\mathbf{y} - U^T\mathbf{y} = \mathbf{0} \tag{4.1.20}$$

Example 4.1.3 The o.l.s. model obtained in Example 4.1.1 gives, rounding 10-figure computations,

t	0	0.2	0.4	0.6	0.8	1
$\hat{y}$	4.786	52.79	103.0	155.4	210.1	266.9
$\hat{e}$	−1.786	6.214	−5.000	−4.429	7.929	−2.929

and $\hat{\mathbf{y}}^T\hat{\mathbf{e}}$ is -0.0012, near enough to zero. The regressors 1, t and $t^2/2$ give $\mathbf{u}^{cT}\hat{\mathbf{e}} = -8 \times 10^{-6}$, -4.6×10^{-6} and -1.7×10^{-6}. Notice that whenever a constant term is included in the regression equation, the sample mean of the model-output error will be zero. △

With the view that $\hat{\mathbf{y}}$ is the orthogonal projection of $\mathbf{y}$ on to the hyperplane defined by U, it is natural to speak of a *projection matrix* $P(U)$ projecting $\mathbf{y}$ on to U:

$$\hat{\mathbf{y}} = U\hat{\boldsymbol{\theta}} = U[U^TU]^{-1}U^T\mathbf{y} = P(U)\mathbf{y} \tag{4.1.21}$$

We see that o.l.s. is a linear estimator, in that $\hat{\boldsymbol{\theta}}$ and $\hat{\mathbf{y}}$ are both linear in the observations $\mathbf{y}$.

Another useful interpretation of o.l.s. is in terms of correlation functions. The discrete-time Wiener–Hopf equation approach to identifying the unit-pulse response (Section 3.1) was based on

$$r_{uy}(k) = \sum_{i=0}^{\infty} h_i r_{uu}(k-i), \qquad k = 1, 2, \ldots, L \tag{4.1.22}$$

where $r_{uy}(k)$ is the input–output cross-correlation at lag k, $r_{uu}(k-i)$ the input autocorrelation at lag $k-i$ and h_i the unit-pulse response at lag i. The sum in (4.1.23) is in practice from $i=1$ to $i=s$, the effective settling time beyond which the u.p.r. is negligible. Collecting $r_{uy}(1)$ to $r_{uy}(L)$ into a vector $\mathbf{r}_{uy}$, the significantly non-zero u.p.r. values h_1 to h_s into $\mathbf{h}$, and the r_{uu} values into an $L \times s$ matrix

$$R_{uu}(L, s) = \begin{bmatrix} r_{uu}(0) & r_{uu}(-1) & \cdots & r_{uu}(1-s) \\ r_{uu}(1) & r_{uu}(0) & \cdots & \\ \vdots & & & \\ r_{uu}(L-1) & r_{uu}(L-2) & \cdots & r_{uu}(L-s) \end{bmatrix} \tag{4.1.23}$$

(4.1.22) becomes

$$\mathbf{r}_{uy} = R_{uu}(L, s)\mathbf{h} \tag{4.1.24}$$

We can take L equal to s and compute the u.p.r. estimate

$$\hat{\mathbf{h}} = R_{uu}^{-1}(s, s)\mathbf{r}_{uy} \tag{4.1.25}$$

Here R_{uu} is symmetric, as autocorrelation values at equal positive and negative lags are, by definition, equal. It is also positive-definite and invertible unless some exact linear relation holds between any s successive input samples, an easy situation to avoid.

In practice the correlation values in $\mathbf{r}_{uy}$ and R_{uu} would be replaced by finite-sample estimates

$$\hat{r}_{uy}(k) = \frac{1}{N} \sum_{j=l}^{N+l-1} u_{j-k} y_j, \qquad k = 1, 2, \ldots, s \tag{4.1.26}$$

$$\hat{r}_{uu}(k) = \frac{1}{N} \sum_{j=m}^{N+m-1} u_{j-k} u_j, \qquad k = 0, 1, \ldots, s-1 \tag{4.1.27}$$

where l and m are any convenient starting times from which the necessary samples of u and y are available. We could, for instance, use

$$\hat{\mathbf{r}}_{uy} = (1/N)U^{\mathrm{T}}\mathbf{y}, \qquad \hat{R}_{uu} = (1/N)U^{\mathrm{T}}U \tag{4.1.28}$$

with

$$U \equiv \begin{bmatrix} u_0 & u_{-1} & \cdots & u_{1-s} \\ u_1 & u_0 & \cdots & \\ \vdots & & & \\ u_{N-1} & u_{N-2} & \cdots & u_{N-s} \end{bmatrix}, \qquad \mathbf{y} \equiv \begin{bmatrix} y_1 \\ y_2 \\ \vdots \\ y_N \end{bmatrix} \tag{4.1.29}$$

The u.p.r. estimate $\hat{\mathbf{h}}$ found by inverting the Wiener–Hopf equation would then be

$$\hat{\mathbf{h}} = [U^{\mathrm{T}}U]^{-1}U^{\mathrm{T}}\mathbf{y} \tag{4.1.30}$$

which is identical to the o.l.s. estimate based on the same observations. The model in both cases is

$$y_t = \sum_{i=1}^{s} h_i u_{t-i} + e_t \tag{4.1.31}$$

The sole difference, insignificant for long records, is that the two methods might start some calculations at slightly different points in the streams of samples.

A correlation interpretation of the orthogonality of $\hat{\mathbf{y}}$ and $\hat{\mathbf{e}}$ is that, if $\hat{\mathbf{y}}$ and $\hat{\mathbf{e}}$ are each composed of an ergodic sequence of samples, $\hat{\mathbf{y}}^T\hat{\mathbf{e}}$ is N times the sample discrete-time correlation between $\{\hat{y}\}$ and $\{\hat{e}\}$ at lag zero. In that case one can say that the output error is uncorrelated (in the finite sample) with the corresponding model output. Intuition agrees that there should be no correlation between the explained part $\hat{y}$ of the output and the unexplained part $\hat{e}$ if $\hat{e}$ cannot be reduced further by adjusting the explanatory terms.

4.1.4 Weighted Least Squares

Ordinary least squares estimation weights each output error in the same way; the significance of an error depends only on its size, not on its position in the succession of N samples. There are occasions when it makes sense to weight errors at some points more heavily than others. For example, in tracking a radar target one might be more worried by errors in recent position than by older ones. Another possibility is that some observations might be distrusted more than others, so that one would wish to take less notice of the corresponding errors. Such eventualities are the concern of this section.

The algebra deriving the o.l.s. estimate is scarcely altered if $\mathbf{e}^T W\mathbf{e}$ is used instead of $\mathbf{e}^T\mathbf{e}$, as the error measure to be minimised, with W a symmetric matrix showing the desired weighting of individual error terms contributing to the total

$$S_w \triangleq \mathbf{e}^T W\mathbf{e} = \sum_{i=1}^{N}\sum_{j=1}^{N} w_{ij}e_ie_j \tag{4.1.32}$$

A diagonal W can be used, each w_{ii} element weighting an individual squared error. We could even specify non-zero off-diagonal terms to penalize products of errors, but at the moment it is not clear what circumstances would require this. It will prove necessary when we consider estimation in autocorrelated noise, in Section 5.3.5.

The estimate $\hat{\boldsymbol{\theta}}_w$ of $\boldsymbol{\theta}$ minimising S_w is easily found by making

$$\partial S_w/\partial\boldsymbol{\theta} = 2(U^T WU\boldsymbol{\theta} - U^T W\mathbf{y}) = \mathbf{0} \tag{4.1.33}$$

giving as the weighted-least-squares (w.l.s.) estimate

$$\hat{\boldsymbol{\theta}}_w = [U^T WU]^{-1}U^T W\mathbf{y} \tag{4.1.34}$$

The expression for $\hat{\boldsymbol{\theta}}_w$ is little more complicated than the o.l.s. estimate, and $U^T WU$ is still symmetric. With W positive-definite to ensure that S_w is positive, $U^T WU$ is also positive-definite and therefore invertible. Generally, the presence of W might imply a large increase in computation, as W is $N \times N$. So long as W is diagonal, however, the increase is small. One need only multiply y_i and $\mathbf{u}_i$ in each regression equation by $w_{ii}^{1/2}$ then calculate as if for o.l.s. or, equally simply, multiply each column $\mathbf{u}_i$ of U^{T} by w_{ii}, producing Z^{T}, say, then calculate

$$\hat{\theta}_{\mathrm{w}} = [Z^{\mathrm{T}}U]^{-1}Z^{\mathrm{T}}\mathbf{y} \tag{4.1.35}$$

Both these ideas are capable of extension to non-diagonal W. They are followed up in Section 5.3 in connection with the Markov estimate and instrumental variables, respectively.

Example 4.1.4 Once more we use the radar observations from Example 4.1.1. This time we have prior knowledge that the third, fourth and fifth observations are subject to larger noise than the others. (The sequence $\{e\}$ is, in fact, 2 -3.8 7.8 5.8 -9.8 -4.) We decide to weight the squares of these errors $\frac{1}{4}$ as strongly as the others, making $w_{33} = w_{44} = w_{55} = 1$ and $w_{11} = w_{22} = w_{66} = 4$ in a diagonal W. Note that the absolute scaling of W is immaterial. $U^{\mathrm{T}}W$ is

$$\begin{bmatrix} 4 & 4 & 1 & 1 & 1 & 4 \\ 0 & 0.8 & 0.4 & 0.6 & 0.8 & 4 \\ 0 & 0.08 & 0.08 & 0.18 & 0.32 & 2 \end{bmatrix}$$

giving

$$\hat{\theta}_{\mathrm{w}} = \begin{bmatrix} 15 & 6.6 & 2.66 \\ 6.6 & 5.32 & 2.412 \\ 2.66 & 2.412 & 1.1428 \end{bmatrix}^{-1} \begin{bmatrix} 1771 \\ 1407.4 \\ 637.5 \end{bmatrix} = \begin{bmatrix} 4.59 \\ 250 \\ 19.0 \end{bmatrix}$$

The weighting has greatly improved the estimates of the initial velocity $\hat{v}_0 \equiv \hat{\theta}_2$ and acceleration $\hat{a} \equiv \hat{\theta}_3$. Even rough prior information on noise behaviour is seen to be valuable, particularly in estimating parameters which depend, as here, on derivatives of an observed variable. △

4.2 COMPUTATIONAL METHODS FOR ORDINARY LEAST-SQUARES

We have discovered two good reasons for seeking better methods of solving the normal equations than general-purpose matrix-inversion routines: the fact that the normal matrix $U^{\mathrm{T}}U$ has special properties which should be exploited, namely symmetry and positive-definiteness, and the possibility that $U^{\mathrm{T}}U$ is ill-conditioned, causing inaccuracy in its inversion. All the methods we shall examine entail splitting $U^{\mathrm{T}}U$ into matrix factors.

4.2.1 Choleski Factorisation

The simplest factorisation method is Choleski factorisation into a real triangular matrix and its transpose. Denoting $U^{\mathrm{T}}U$ by A for convenience, Choleski factorisation finds a lower-triangular *matrix square root* L of A satisfying

$$A = LL^{\mathrm{T}} \tag{4.2.1}$$

Matrix square roots are not unique, for if B is a square root of A and P is any orthogonal matrix of the appropriate dimension, then as PP^{T} is I,

$$A = BB^{\mathrm{T}} = BIB^{\mathrm{T}} = BPP^{\mathrm{T}}B^{\mathrm{T}} = BP(BP)^{\mathrm{T}} \tag{4.2.2}$$

so BP is also a square root of A. A lower-triangular square root may be made unique by requiring all its principal-diagonal elements to be positive. Once A is in the form LL^{T}, it is easy to invert, as A^{-1} is $L^{-\mathrm{T}}L^{-1}$ and a lower-triangular matrix is invertible, a row at a time, by successive substitution.

Choleski decomposition of A is nothing more exotic than *completing the square*, familiar from school algebra. The quadratic $\mathbf{z}^{\mathrm{T}}A\mathbf{z}$ is rewritten as a sum of squares of linear combinations of the z's:

$$\mathbf{z}^T A\mathbf{z} = (\mathbf{l}_1^T\mathbf{z})^2 + (\mathbf{l}_2^T\mathbf{z})^2 + \cdots + (\mathbf{l}_p^T\mathbf{z})^2 = (L^T\mathbf{z})^T(L^T\mathbf{z}) = \mathbf{z}^T LL^T\mathbf{z} \tag{4.2.3}$$

where

$$L^{\mathrm{T}} = \begin{bmatrix} \mathbf{l}_1^{\mathrm{T}} \\ \mathbf{l}_2^{\mathrm{T}} \\ \vdots \\ \mathbf{l}_p^{\mathrm{T}} \end{bmatrix} \tag{4.2.4}$$

and $\mathbf{l}_1^{\mathrm{T}}\mathbf{z}$ contains, in general, all p z's, $\mathbf{l}_2^{\mathrm{T}}\mathbf{z}$ contains the last $p-1$ of them, and so on, so that L^{T} is upper-triangular and L lower-triangular. As (4.2.3) is true whatever the value of $\mathbf{z}$, LL^{T} equals A as required.

Example 4.2.1 The normal matrix A in Example 4.1.1 gives

$$\begin{aligned}
\mathbf{z}^{\mathrm{T}}A\mathbf{z} &= 6z_1^2 + 6z_1z_2 + 2.2z_1z_3 + 2.2z_2^2 + 1.8z_2z_3 + 0.3916z_3^2 \\
&= 6(z_1^2 + z_1z_2 + 0.36\dot{}z_1z_3) + 2.2z_2^2 + 1.8z_2z_3 + 0.3916z_3^2 \\
&= (\sqrt{6}(z_1 + 0.5z_2 + 0.183\dot{}z_3))^2 + (2.2 - 1.5)z_2^2 \\
&\quad + (1.8 - 1.1)z_2z_3 + (0.3916 - 0.2016\dot{})z_3^2 \\
&= (\sqrt{6}(z_1 + 0.5z_2 + 0.183\dot{}z_3))^2 + (\sqrt{0.7}\,(z_2 + 0.5z_3))^2 + (0.1222z_3)^2
\end{aligned}$$

so

$$L^{\mathrm{T}}\mathbf{z} = \begin{bmatrix} \sqrt{6} & 0.5\sqrt{6} & 0.18\dot{3}\sqrt{6} \\ 0 & \sqrt{0.7} & 0.5\sqrt{0.7} \\ 0 & 0 & 0.1222 \end{bmatrix} \mathbf{z}$$

Now $L^{-\mathrm{T}}$ can be found by solving $L^{\mathrm{T}}\mathbf{z} = \boldsymbol{\zeta}$, starting with the last row and working upwards:

$$z_3 \simeq \zeta_3/0.1222 \simeq 8.183\zeta_3$$

$$z_2 = \zeta_2/\sqrt{0.7} - 0.5z_3 \simeq 1.195\zeta_2 - 4.092\zeta_3$$

$$z_1 = \zeta_1/\sqrt{6.6} - 0.5z_2 - 0.18\dot{3}z_3 \simeq 0.4082\zeta_1 - 0.5976\zeta_2 + 0.5455\zeta_3$$

so

$$A^{-1} = L^{-\mathrm{T}}L^{-1} \simeq \begin{bmatrix} 0.4082 & -0.5976 & 0.5455 \\ 0 & 1.195 & -4.092 \\ 0 & 0 & 8.183 \end{bmatrix} \begin{bmatrix} 0.4082 & 0 & 0 \\ -0.5976 & 1.195 & 0 \\ 0.5455 & -4.092 & 8.183 \end{bmatrix}$$

giving

$$A^{-1} = \begin{bmatrix} 0.821 & -2.95 & 4.46 \\ -2.95 & 18.2 & -33.5 \\ 4.46 & -33.5 & 67.0 \end{bmatrix}$$

rounded to three figures (from ten-figure calculations). This agrees with the inverse found in Example 4.1.1. △

An alternative and easily programmed way to find L is by direct identification of columns of LL^{T} with the same columns of A, one at a time. The first column of L is found by noting that l_{11} times column 1 of L equals column 1 of A, and l_{11}^2 equals a_{11}. Column 2 of L is then the only unknown in

$$l_{21}(\text{column 1 of } L) + l_{22}(\text{column 2 of } L) = \text{column 2 of } A \quad (4.2.5)$$

and l_{22} is obtainable from

$$l_{21}^2 + l_{22}^2 = a_{22} \quad (4.2.6)$$

The process is continued until all the columns of L are found.

Both methods of finding L will break down if at any stage i the expression for l_{ii}^2 is negative. The positive-definiteness of A ensures that this will not happen, as we can see by considering the situation after $i-1$ successful stages of completing the square. By then, $\mathbf{z}^{\mathrm{T}}A\mathbf{z}$ is in the form

$$\zeta_1^2 + \zeta_2^2 + \cdots + \zeta_{i-1}^2 + a_{ii}^{(i-1)}z_i^2 + q^{(i-1)} \quad (4.2.7)$$

where $q^{(i-1)}$ is the quadratic remainder in z_{i+1} to z_p, and $a_{ii}^{(i-1)}$ is the square of the coefficient of z_i in the next square to be computed, ζ_i^2. For ζ_i to be found, $a_{ii}^{(i-1)}$ must be positive. Now $\mathbf{z}^T A\mathbf{z}$ is positive for any real, non-zero $\mathbf{z}$, including that in which z_i is 1, z_{i+1} to z_p zero, and z_1 to z_{i-1} chosen to make ζ_1 to ζ_{i-1} all zero (certainly feasible since the coefficients of z_1 to z_{i-1} in ζ_1 to ζ_{i-1} form an upper-triangular matrix with non-zero principal-diagonal elements, which is consequently non-singular). For this $\mathbf{z}$, $\mathbf{z}^T A\mathbf{z}$ is just $a_{ii}^{(i-1)}$, so $a_{ii}^{(i-1)}$ is certainly positive.

A refinement removing the need to compute the square roots $a_{ii}^{(i-1)1/2}$ is to replace LL^T by LDL^T, where D is diagonal with $d_{ii} = a_{ii}^{(i-1)}$ and L is still lower-triangular but with 1's all along its principal diagonal. An upper-triangular matrix version is called *U–D factorisation*, and is the basis of a least-squares algorithm widely used in estimating the state of dynamical systems (Bierman, 1977). (State estimation is discussed briefly in Section 7.3.2.)

The aim of matrix-square-root methods for finding the o.l.s. estimate is to reduce inaccuracy due to ill-conditioning of the normal matrix. They work by reducing the range of number magnitudes, and hence the seriousness of rounding errors. Computation with matrix square roots gives accuracy comparable with that obtained by keeping twice as many significant figures without resort to matrix square roots.

4.2.2 Golub–Householder Technique

A more recent alternative way around numerical difficulties in solving the normal equations is the Golub–Householder technique (Golub, 1965). A linear transformation is applied to U and $\mathbf{y}$ (i.e. they are premultiplied by a matrix) so as to modify the regression equation, without changing $\boldsymbol{\theta}$, and make it easier to solve. Premultiplying the regression equations (4.1.6) by an $N \times N$ matrix Q gives

$$\mathbf{y}^* \triangleq Q\mathbf{y} = QU\boldsymbol{\theta} + Q\mathbf{e} \equiv U^*\boldsymbol{\theta} + \mathbf{e}^* \tag{4.2.8}$$

with modified observations, regressors and errors, but the same $\boldsymbol{\theta}$. The sum of squares of the output errors given by the o.l.s. estimate $\hat{\boldsymbol{\theta}}$ and the modified model is

$$S^* \triangleq (\mathbf{y}^* - U^*\hat{\boldsymbol{\theta}})^T(\mathbf{y}^* - U^*\hat{\boldsymbol{\theta}}) = (\mathbf{y} - U\hat{\boldsymbol{\theta}})^T Q^T Q(\mathbf{y} - U\hat{\boldsymbol{\theta}}) \tag{4.2.9}$$

This sum can be made equal to S, the sum of squared errors using $\hat{\boldsymbol{\theta}}$ in the original equation, by choosing an orthogonal matrix as Q, so that $Q^T Q$ is I. Furthermore, there is enough freedom in choosing Q for us to insist that U^* should be in the especially convenient form

$$U^* = QU \equiv \begin{bmatrix} V \\ 0 \end{bmatrix} \tag{4.2.10}$$

where V is an upper-triangular $p \times p$ matrix. Notice that, since

$$U^{*\mathrm{T}}U^* = V^{\mathrm{T}}V = U^{\mathrm{T}}Q^{\mathrm{T}}QU = U^{\mathrm{T}}U \tag{4.2.11}$$

and V is upper-triangular, V is a Choleski factor of $U^{\mathrm{T}}U$.

If we now denote the top p elements of $\mathbf{y}^*$ by $\mathbf{y}_1^*$ and the other $N-p$ by $\mathbf{y}_2^*$, we find from (4.2.9) and (4.2.10) that

$$S = S^* = (\mathbf{y}_1^* - V\hat{\boldsymbol{\theta}})^{\mathrm{T}}(\mathbf{y}_1^* - V\hat{\boldsymbol{\theta}}) + \mathbf{y}_2^{*\mathrm{T}}\mathbf{y}_2^* \tag{4.2.12}$$

In this sum of N squared terms, only the first p depend on $\hat{\boldsymbol{\theta}}$, so S is minimised by making them zero. Hence

$$\hat{\boldsymbol{\theta}} = V^{-1}\mathbf{y}_1^* \tag{4.2.13}$$

and

$$S = \mathbf{y}_2^{*\mathrm{T}}\mathbf{y}_2^* \tag{4.2.14}$$

The $\hat{\boldsymbol{\theta}}$ thus found is the product of the inverse of an upper-triangular matrix, easy to compute by back-substitution, and the first p samples from $\mathbf{y}^*$, a linearly filtered version of $\mathbf{y}$. If we can now find a reasonably simple method of constructing an orthogonal Q to give QU in the desired form (4.2.10) we shall have an attractive way to calculate the o.l.s. estimate of $\boldsymbol{\theta}$ without explicitly forming and inverting the normal matrix.

*4.2.3 Householder Transformation of Regressor Matrix

The required method is provided by Householder transformations, in which U is premultiplied successively by orthogonal matrices $P^{(1)}, P^{(2)}, \ldots, P^{(p)}$, each of the form

$$P^{(i)} = I - 2\mathbf{w}^{(i)}\mathbf{w}^{(i)\mathrm{T}} \tag{4.2.15}$$

Each $\mathbf{w}^{(i)}$ has to do two things: make $P^{(i)}$ orthogonal so that in the long run $P^{(p)}P^{(p-1)}\ldots P^{(1)}$ is orthogonal, and perform a stage of the upper-triangularisation of QU by making a column zero. Writing out $P^{(i)\mathrm{T}}P^{(i)}$ shows that $\mathbf{w}^{(i)\mathrm{T}}\mathbf{w}^{(i)}$ must be 1 to make $P^{(i)}$ orthogonal. The upper-triangularisation of U can be carried out column by column if each $\mathbf{w}^{(i)}$ is chosen to leave unaltered the first $i-1$ columns of the matrix

$$U^{(i)} = P^{(i-1)}P^{(i-2)}\ldots P^{(1)}U \tag{4.2.16}$$

when $U^{(i)}$ is premultiplied by $P^{(i)}$, while making elements $i+1$ to N of column i zero. Not only is this possible, but $\mathbf{w}^{(i)}$ can also be chosen to leave the first $i-1$ rows of $U^{(i)}$ unchanged, as follows. Denoting column i of $U^{(i)}$ and $U^{(i+1)}$ by $\mathbf{u}_i^{(i)}$ and $\mathbf{u}_i^{(i+1)}$,

$$\mathbf{u}_i^{(i+1)} = P^{(i)}\mathbf{u}_i^{(i)} = \mathbf{u}_i^{(i)} - 2\mathbf{w}^{(i)}\mathbf{w}^{(i)\mathrm{T}}\mathbf{u}_i^{(i)} \tag{4.2.17}$$

so

$$\mathbf{w}^{(i)} = (\mathbf{u}_i^{(i)} - \mathbf{u}_i^{(i+1)})/2\mathbf{w}^{(i)\mathrm{T}}\mathbf{u}_i^{(i)} \tag{4.2.18}$$

which says that $\mathbf{w}^{(i)}$ is a scalar times the change from $\mathbf{u}_i^{(i)}$ to $\mathbf{u}_i^{(i+1)}$. To make $P^{(i)}$ orthogonal, $\mathbf{w}^{(i)\mathrm{T}}\mathbf{w}^{(i)}$, that is $\|\mathbf{w}^{(i)}\|^2$, must be 1, so $\mathbf{w}^{(i)}$ is in fact $\pm(\mathbf{u}_i^{(i)} - \mathbf{u}_i^{(i+1)})/\|\mathbf{u}_i^{(i)} - \mathbf{u}_i^{(i+1)}\|$. The sign is immaterial, and we shall take it as positive. Now $\mathbf{w}^{(i)}$ can be fixed, except for $\mathbf{w}_i^{(i)}$, by specifying that the first $i-1$ elements of $\mathbf{u}_i^{(i+1)}$ be unchanged from those of $\mathbf{u}_i^{(i)}$ and the last $N-i$ elements be zero, giving

$$w_j^{(i)} = 0, \quad 1 \leq j \leq i-1; \qquad w_j^{(i)} = u_{ji}^{(i)}/\alpha^{(i)}, \quad i+1 \leq j \leq N \tag{4.2.19}$$

where

$$\alpha^{(i)} = \|\mathbf{u}_i^{(i)} - \mathbf{u}_i^{(i+1)}\| = \left\{(u_{ii}^{(i)} - u_{ii}^{(i+1)})^2 + \sum_{j=i+1}^{N} u_{ji}^{(i)2}\right\}^{1/2} \tag{4.2.20}$$

The only unknown element $u_{ii}^{(i+1)}$ of $\mathbf{u}_i^{(i+1)}$ is determined by noting that since $P^{(i)}$ is orthogonal

$$\begin{aligned}\|\mathbf{u}_i^{(i+1)}\|^2 &= \|P^{(i)}\mathbf{u}_i^{(i)}\|^2 = \mathbf{u}_i^{(i)\mathrm{T}}P^{(i)\mathrm{T}}P^{(i)}\mathbf{u}_i^{(i)} \\ &= \mathbf{u}_i^{(i)\mathrm{T}}\mathbf{u}_i^{(i)} = \|\mathbf{u}_i^{(i)}\|^2\end{aligned} \tag{4.2.21}$$

The first $i-1$ elements of $\mathbf{u}_i^{(i+1)}$ are the same as in $\mathbf{u}_i^{(i)}$ and the last $N-i$ are zero, so (4.2.21) requires that

$$u_{ii}^{(i+1)2} = \sum_{j=1}^{N} u_{ji}^{(i)2} \triangleq \sigma^2 \tag{4.2.22}$$

The sign of $u_{ii}^{(i+1)}$ is chosen to maximise $|w_i^{(i)}|$ so as to avoid unnecessary loss of numerical accuracy:

$$u_{ii}^{(i+1)} = -\sigma \operatorname{sgn} u_{ii}^{(i)} \tag{4.2.23}$$

where σ is positive, then

$$w_i^{(i)} = (u_i^{(i)} + \sigma \operatorname{sgn} u_{ii}^{(i)})/\alpha^{(i)} \tag{4.2.24}$$

and, substituting (4.2.23) into (4.2.20),

$$\alpha^{(i)} = \left\{\sigma^2 + 2\sigma|u_{ii}^{(i)}| + \sum_{j=i}^{p} u_{ji}^{(i)2}\right\}^{1/2} = \{2\sigma(\sigma + |u_{ii}^{(i)}|)\}^{1/2} \tag{4.2.25}$$

Some practical points to notice are that the square root in (4.2.25) need not be computed, as $P^{(i)}$ involves only second-degree terms in elements of $\mathbf{w}^{(i)}$; that $\mathbf{w}^{(i)}\mathbf{w}^{(i)\mathrm{T}}$ need not be computed and used explicitly, as $U^{(i+1)}$ is composed only of $U^{(i)}$, $\mathbf{w}^{(i)\mathrm{T}}U^{(i)}$ and $\mathbf{w}^{(i)}$; and that σ will be zero only if U is singular.

The Golub–Householder technique just described has excellent numerical properties in cases where other methods are vulnerable to ill-conditioning.

Example 4.2.2 Let us carry out the o.l.s. computation for the problem of Example 4.1.1 by the Golub–Householder method. Starting with

$$U^{(1)} \equiv U = \begin{bmatrix} 1 & 0 & 0 \\ 1 & 0.2 & 0.02 \\ 1 & 0.4 & 0.08 \\ 1 & 0.6 & 0.18 \\ 1 & 0.8 & 0.32 \\ 1 & 1 & 0.5 \end{bmatrix}$$

we compute in turn

$$\sigma = \{\text{sum of squares of elements } i \text{ to } p \text{ of column } i \text{ of } U^{(i)}\}^{1/2}$$

with $i = 1$, so

$$\begin{aligned} \sigma &= \sqrt{6} \simeq 2.4495 \\ u_{11}^{(2)} &= -\sigma \operatorname{sgn} u_{11}^{(1)} = -2.4495 \\ \alpha^{(1)} &= \{2\sigma(|u_{11}^{(1)}| + \sigma)\}^{1/2} = 4.1108 \\ w_j^{(1)} &= u_{j1}^{(1)}/\alpha^{(1)} = 0.24326 \qquad \text{for } j = 2, \ldots, 6 \\ w_j^{(1)} &= (u_{11}^{(1)} + \sigma \operatorname{sgn} u_{11}^{(1)})/\alpha^{(1)} = 0.83912 \end{aligned}$$

Then

$$\mathbf{w}^{(1)\mathrm{T}} U^{(1)} = [2.0554 \quad 0.72978 \quad 0.26759]$$

and

$$U^{(2)} = U^{(1)} - 2\mathbf{w}^{(1)} \cdot \mathbf{w}^{(1)\mathrm{T}} U^{(1)} = \begin{bmatrix} -2.4495 & -1.2247 & -0.44907 \\ 0 & -0.15505 & -0.11019 \\ 0 & 0.04495 & -0.05019 \\ 0 & 0.24495 & 0.04981 \\ 0 & 0.44495 & 0.18981 \\ 0 & 0.64495 & 0.36981 \end{bmatrix}$$

and similarly

$$\mathbf{y}^{(2)} = \mathbf{y}^{(1)} - 2\mathbf{w}^{(1)} \cdot \mathbf{w}^{(1)\mathrm{T}} \mathbf{y}^{(1)}$$

$$= \begin{bmatrix} 3 \\ 59 \\ 98 \\ 151 \\ 218 \\ 264 \end{bmatrix} - 2\mathbf{w}^{(1)} \cdot 194.69 = \begin{bmatrix} -323.74 \\ -35.722 \\ 3.2785 \\ 56.278 \\ 123.28 \\ 169.28 \end{bmatrix}$$

For $i = 2$, we compute

$$\sigma = 0.83666, \qquad u_{22}^{(3)} = 0.83666, \qquad \alpha^{(2)} = 1.2882$$

and

$$\mathbf{w}^{(2)\mathrm{T}} = [0 \quad -0.76984 \quad 0.034893 \quad 0.19015 \quad 0.34540 \quad 0.50066]$$

giving

$$\mathbf{w}^{(2)\mathrm{T}} U^{(2)} = [0 \quad 0.64410 \quad 0.34326]$$

Hence

$$U^{(3)} = \begin{bmatrix} -2.4495 & -1.2247 & -0.44907 \\ 0 & 0.83666 & 0.38916 \\ 0 & 0 & -0.07282 \\ 0 & 0 & -0.17315 \\ 0 & 0 & -0.03423 \\ 0 & 0 & 0.04451 \end{bmatrix} \quad \text{and} \quad \mathbf{y}^{(3)} = \begin{bmatrix} -323.74 \\ 219.32 \\ -8.2814 \\ -6.7169 \\ 8.8475 \\ 3.4120 \end{bmatrix}$$

Finally, for $i = 3$,

$$\sigma = 0.12220 = u_{33}^{(4)}, \qquad \alpha^{(3)} = 0.21906$$

$$\mathbf{w}^{(3)\mathrm{T}} = [0 \quad 0 \quad -0.89630 \quad -0.36852 \quad -0.21598 \quad 0.11915]$$

and

$$\mathbf{w}^{(3)\mathrm{T}} U^{(3)} = [0 \quad 0 \quad 0.10953]$$

Matrix $U^{(4)}$ has the same first two rows as $U^{(3)}$, only $u_{33}^{(4)}$ non-zero in row 3, and rows 4 to 6 all zero. As $U^{(4)}$ is U^*, with V as its first 3×3 sub-matrix, we have

$$V = \begin{bmatrix} -2.4495 & -1.2247 & -0.44907 \\ 0 & 0.83666 & 0.41833 \\ 0 & 0 & 0.12220 \end{bmatrix}$$

and

$$\mathbf{y}^* = \mathbf{y}^{(4)} = \begin{bmatrix} -323.74 \\ 219.32 \\ 6.7647 \\ -0.53066 \\ 12.473 \\ 1.4119 \end{bmatrix} \begin{matrix} \left.\begin{matrix} \\ \\ \\ \end{matrix}\right\} \mathbf{y}_1^* \\ \left.\begin{matrix} \\ \\ \\ \end{matrix}\right\} \mathbf{y}_2^* \end{matrix}$$

Solving $V\hat{\boldsymbol{\theta}} = \mathbf{y}^*$, we find $\hat{\theta}_3 = 55.4$, $\hat{\theta}_2 = 234$, $\hat{\theta}_1 = 4{\cdot}79$ by back-substitution. The sum of squares of output errors S is given by $\mathbf{y}_2^{*\mathrm{T}}\mathbf{y}_2^*$ as 157.8. △

4.2.4 Singular-Value Decomposition

A factorisation method for o.l.s. which indicates exactly where the cause of any ill-conditioning lies is singular-value decomposition (Forsythe *et al.*, 1977). It has much in common with the Golub–Householder method. The technique is based on decomposing the regressor matrix U into

$$U = PRQ^{\mathrm{T}} \tag{4.2.26}$$

where P and Q are orthogonal matrices, respectively $N \times N$ and $p \times p$, and R is an $N \times p$ matrix zero but for non-negative elements r_{ii}, $1 \leq i \leq p$. These elements are square roots of the eigenvalues of the normal matrix $U^{\mathrm{T}}U$. To see why, notice that

$$U^{\mathrm{T}}UQ = QR^{\mathrm{T}}P^{\mathrm{T}}PRQ^{\mathrm{T}}Q = QR^{\mathrm{T}}R \tag{4.2.27}$$

and $R^{\mathrm{T}}R$ is a $p \times p$ diagonal matrix with r_{ii}^2 as element (i, i), so each column $\mathbf{q}_i$ of Q satisfies

$$U^{\mathrm{T}}U\mathbf{q}_i = r_{ii}^2\mathbf{q}_i \tag{4.2.28}$$

In other words, r_{ii}^2 is an eigenvalue of $U^{\mathrm{T}}U$, with $\mathbf{q}_i$ as the corresponding eigenvector. Clearly a zero value for r_{ii} would indicate exact linear dependence between the columns of $U^{\mathrm{T}}U$, reflecting similar dependence between the regressors.

Again we transform the regression equation linearly, to make the normal equations easier to solve. Premultiplying by P^{T},

$$\mathbf{y}^* = P^{\mathrm{T}}\mathbf{y} = P^{\mathrm{T}}U\boldsymbol{\theta} + P^{\mathrm{T}}\mathbf{e} = P^{\mathrm{T}}PRQ^{\mathrm{T}}\boldsymbol{\theta} + P^{\mathrm{T}}\mathbf{e} = R\boldsymbol{\theta}^* + \mathbf{e}^* \tag{4.2.29}$$

where $\boldsymbol{\theta}^*$ denotes $Q^{\mathrm{T}}\boldsymbol{\theta}$ and $\mathbf{e}^*$ denotes $P^{\mathrm{T}}\mathbf{e}$. Because P^{T} is orthogonal, the transformation does not alter the sum of squared errors:

$$S^* \triangleq (\mathbf{y}^* - R\boldsymbol{\theta}^*)^{\mathrm{T}}(\mathbf{y}^* - R\boldsymbol{\theta}^*) = (\mathbf{y} - U\boldsymbol{\theta})^{\mathrm{T}}PP^{\mathrm{T}}(\mathbf{y} - U\boldsymbol{\theta}) = S \tag{4.2.30}$$

The first p elements of $R\boldsymbol{\theta}^*$ are $r_{ii}\theta_1^*$ to $r_{pp}\theta_p^*$ and the rest are zero, so S minimised by

$$\hat{\theta}_i = y_i^*/r_{ii}, \qquad i = 1, 2, \ldots, p \tag{4.2.31}$$

After the $\hat{\boldsymbol{\theta}}^*$ which minimises S^*, and therefore S, is found by this trivial calculation, $\hat{\boldsymbol{\theta}}$ is readily obtained as $Q\hat{\boldsymbol{\theta}}^*$, since the inverse of Q^{T}, an orthogonal matrix, is Q. Furthermore, near-linear dependence between the regressors will show up as a very small value for one or more of the r_{ii}'s. In an extreme case, some r_{ii} might be negligible or even brought to zero by roundoff error, so S would scarcely be affected by adding a large number α to $\hat{\theta}_i^*$. As $Q\hat{\boldsymbol{\theta}}^*$ is $\hat{\boldsymbol{\theta}}$, the corresponding change to $\hat{\boldsymbol{\theta}}$ is the addition to $\hat{\boldsymbol{\theta}}$ of

$$\alpha Q(\text{vector zero but for 1 in position } i) = \alpha(\text{column } i \text{ of } Q) = \alpha\mathbf{q}_i \tag{4.2.32}$$

If addition of $\alpha\mathbf{q}_i$ to $\hat{\boldsymbol{\theta}}$ has little effect on S, $U\mathbf{q}_i$ must be negligible. In other words, the linear combination of regressors with the elements of $\mathbf{q}_i$ as coefficients contributes nothing to the performance of the model. To summarise, the attraction of the singular-value decomposition is that it reveals any ill-conditioning clearly, and provides an easy way to prevent it from causing numerical difficulties. By setting θ_i^* to zero if r_{ii} is below some small specified value, a corresponding linear combination $\mathbf{q}_i^T\hat{\boldsymbol{\theta}}$ with little influence on S is set to zero. If it were not, it could be so large as to obscure the meaningful part of $\hat{\boldsymbol{\theta}}$.

The mechanics of obtaining P and Q are a little complicated (Forsythe *et al.*, 1977) and will not be detailed here. Briefly, Householder transformations are used to introduce zeros below the leading diagonal of U by premultiplication, and above the first superdiagonal by postmultiplication, then a version of the QR algorithm is employed to reduce the bidiagonal matrix thus obtained to diagonal form, iteratively.

4.3 NON-LINEAR LEAST-SQUARES ESTIMATION

In many situations where the model is not linear in all its coefficients, it still makes sense to find the coefficient values which minimise the sum of the squared errors between model output and observed output. An example is multi-exponential modelling of an impulse response

$$y(t) = a_1 \exp(\lambda_1 t) + a_2 \exp(\lambda_2 t) + \cdots + a_n \exp(\lambda_n t) + e(t) \qquad (4.3.1)$$

where the a's enter linearly but the λ's do not. The algebra which gave the o.l.s. estimate no longer applies, but we can still exploit the fact that the function being minimised is a sum of squared errors; we are not faced with a general unconstrained algebraic minimisation problem.

4.3.1 Generalised Normal Equations

Let the model be

$$\mathbf{y} = \mathbf{f}(U, \boldsymbol{\theta}) + \mathbf{e} \qquad (4.3.2)$$

where $\mathbf{y}$ is the vector composed of all the output samples, U is the matrix of explanatory-variable samples, and $\mathbf{f}$ comprises the functions modelling $\mathbf{y}$. The least-squares estimate of $\boldsymbol{\theta}$ has to make the gradient of S zero, i.e.

$$\partial S/\partial\boldsymbol{\theta} = -2[J_\theta f]^T(\mathbf{y} - \mathbf{f}) = \mathbf{0} \qquad (4.3.3)$$

where $J_\theta f$ is the Jacobian matrix of $\mathbf{f}$ with respect to $\boldsymbol{\theta}$, i.e. the matrix with $\partial f_i/\partial\theta_j$ as element (i, j). We have in (4.3.3) a generalised version of the normal equations. For a model linear in $\boldsymbol{\theta}$, $\mathbf{f}$ is $U\boldsymbol{\theta}$ and $J_\theta f$ is U. The extra difficulty with a non-linear model is that $J_\theta f$ is a function of $\boldsymbol{\theta}$, so the normal equations are no longer linear in $\boldsymbol{\theta}$, and have to be solved iteratively. If some of the coefficients in $\boldsymbol{\theta}$ enter $\mathbf{f}$ linearly, as in the multi-exponential model, they can be found at each iteration by o.l.s. after the non-linearly entering coefficients have been fixed.

Example 4.3.1 For many biomedical applications, such as the study of blood concentration of a drug after a dose, a low-order multi-exponential response model is employed, either as an end in itself or as a half-way stage to a set of rate equations making up a compartmental model. One such model, with initial and final output zero, is

$$y(t) = c(\exp(\lambda_1 t) - \exp(\lambda_2 t)) + e(t), \qquad t \geq 0$$

The vector $\boldsymbol{\theta}$ is $[c \quad \lambda_1 \quad \lambda_2]^{\mathrm{T}}$, and if $y(t)$ is sampled at times $t_1, t_2, \ldots, t_N$, we have $\exp(\lambda_1 t_i) - \exp(\lambda_2 t_i)$ as $\partial f_i/\partial\theta_1$, $ct_i \exp(\lambda_1 t_i)$ as $\partial f_i/\partial\theta_2$ and $-ct_i \exp(\lambda_2 t_i)$ as $\partial f_i/\partial\theta$. Collecting samples $\exp(\lambda_1 t_i)$ into a vector $\boldsymbol{\eta}_1$, $\exp(\lambda_2 t_i)$ into $\boldsymbol{\eta}_2$, $t_i \exp(\lambda_1 t_i)$ into $\boldsymbol{\zeta}_1$ and $t_i \exp(\lambda_2 t_i)$ into $\boldsymbol{\zeta}_2$, the normal equations become

$$-2[\boldsymbol{\eta}_1 - \boldsymbol{\eta}_2 \quad c\boldsymbol{\zeta}_1 \quad -c\boldsymbol{\zeta}_2]^{\mathrm{T}}(\mathbf{y} - c(\boldsymbol{\eta}_1 - \boldsymbol{\eta}_2)) = \mathbf{0}$$

From the first of these equations, c can be estimated as

$$\hat{c} = (\boldsymbol{\eta}_1 - \boldsymbol{\eta}_2)^{\mathrm{T}}\mathbf{y}/(\boldsymbol{\eta}_1 - \boldsymbol{\eta}_2)^{\mathrm{T}}(\boldsymbol{\eta}_1 - \boldsymbol{\eta}_2)$$

once λ_1 and λ_2 have been estimated. At each iteration $\hat{\lambda}_1$ and $\hat{\lambda}_2$ are adjusted to try to satisfy the other two normal equations, with $\hat{c}$ substituted for c, either numerically from the previous iteration or algebraically giving

$$\boldsymbol{\zeta}_j^{\mathrm{T}}\mathbf{y}(\boldsymbol{\eta}_1 - \boldsymbol{\eta}_2)^{\mathrm{T}}(\boldsymbol{\eta}_1 - \boldsymbol{\eta}_2) = (\boldsymbol{\eta}_1 - \boldsymbol{\eta}_2)^{\mathrm{T}}\mathbf{y}\boldsymbol{\zeta}_j^{\mathrm{T}}(\boldsymbol{\eta}_1 - \boldsymbol{\eta}_2), \qquad j = 1, 2 \qquad \triangle$$

4.3.2 Gauss–Newton Algorithm

Solving the normal equations amounts to adjusting $\hat{\boldsymbol{\theta}}$ until each of a collection of functions of $\hat{\boldsymbol{\theta}}$ becomes zero. One approach is Newton's method. In a scalar problem adjusting $\hat{\theta}$ to bring $g(\hat{\theta})$, say, to zero, iteration k of Newton's method replaces $\hat{\theta}^{(k-1)}$ by

$$\hat{\theta}^{(k)} = \hat{\theta}^{(k-1)} - (\partial g/\partial\theta|_{\theta=\hat{\theta}^{(k-1)}})^{-1} g(\hat{\theta}^{(k-1)}) \tag{4.3.4a}$$

The rationale is that for small changes about $\hat{\theta}^{(k-1)}$, $g(\hat{\theta})$ varies almost linearly with $\hat{\theta}$, so this adjustment will bring $g(\hat{\theta}^{(k)})$ close to zero. The multivariable version, similarly motivated, is

$$\hat{\boldsymbol{\theta}}^{(k)} = \hat{\boldsymbol{\theta}}^{(k-1)} - [J_\theta \mathbf{g}|_{\boldsymbol{\theta}=\hat{\boldsymbol{\theta}}^{(k-1)}}]^{-1}\mathbf{g}(\hat{\boldsymbol{\theta}}^{(k-1)}) \tag{4.3.4b}$$

Looking back at the normal equations,

$$g_i(\boldsymbol{\theta}) = -2(\text{row } i \text{ of } [J_\theta f]^{\mathrm{T}})(\mathbf{y}-\mathbf{f}) \tag{4.3.5}$$

so a typical element of the Jacobian matrix in (4.3.4b) is

$$\frac{\partial g_i}{\partial \theta_j} = 2\sum_{l=1}^{N}\left\{\frac{\partial f_l}{\partial \theta_i}\cdot\frac{\partial f_l}{\partial \theta_j} - \frac{\partial^2 f_l}{\partial \theta_i \partial \theta_j}(y_l - f_l)\right\} \tag{4.3.6}$$

Most of the labour in evaluating $J_\theta g$ is due to the $Np(p+1)/2$ second derivatives. However, if our current estimate $\hat{\boldsymbol{\theta}}^{(k-1)}$ already gives quite small output errors $y_l - f_l$, we may perhaps ignore all the terms in (4.3.6) involving second derivatives. If so,

$$[J_\theta g] \simeq 2\sum_{l=1}^{N}\frac{\partial f_l}{\partial \boldsymbol{\theta}}\cdot\frac{\partial f_l^{\mathrm{T}}}{\partial \boldsymbol{\theta}} = 2[J_\theta f]^{\mathrm{T}}[J_\theta f] \tag{4.3.7}$$

so, denoting $J_\theta f$ at $\hat{\boldsymbol{\theta}}^{(k-1)}$ by $J^{(k-1)}$ and similarly for $\mathbf{f}$, we have the Gauss–Newton algorithm

$$\hat{\boldsymbol{\theta}}^{(k)} = \hat{\boldsymbol{\theta}}^{(k-1)} + [J^{(k-1)\mathrm{T}}J^{(k-1)}]^{-1}J^{(k-1)\mathrm{T}}(\mathbf{y} - \mathbf{f}^{(k-1)}) \tag{4.3.8}$$

Example 4.3.2 The two-exponential impulse response model of Example 4.3.1 is to be fitted to observations

t	0	0.5	0.75	1.25	1.75	2.25
y	0	90	115	85	55	40

given by a methionine tolerance test (Brown *et al.*, 1979). Starting guesses -0.7 for $\hat{\lambda}_1^{(0)}$ and -2 for $\hat{\lambda}_2^{(0)}$ are made after inspection of a plot of the observations. They allow calculation of

$$\boldsymbol{\eta}_1^{(0)\mathrm{T}} \simeq [1 \quad 0.7047 \quad 0.5916 \quad 0.4169 \quad 0.2938 \quad 0.2070]$$

and

$$\boldsymbol{\eta}_2^{(0)\mathrm{T}} \simeq [1 \quad 0.3679 \quad 0.2231 \quad 0.0821 \quad 0.0302 \quad 0.0111]$$

which with the first of the normal equations gives the o.l.s. estimate for c:

$$\hat{c}^{(0)} = (\boldsymbol{\eta}_1^{(0)} - \boldsymbol{\eta}_2^{(0)})^{\mathrm{T}}\mathbf{y}/(\boldsymbol{\eta}_1^{(0)} - \boldsymbol{\eta}_2^{(0)})^{\mathrm{T}}(\boldsymbol{\eta}_1^{(0)} - \boldsymbol{\eta}_2^{(0)}) \simeq 263.2$$

The model output errors can then be calculated as

$$\mathbf{y} - \mathbf{f}^{(0)} = \mathbf{y} - \hat{a}^{(0)}(\boldsymbol{\eta}_1^{(0)} - \boldsymbol{\eta}_2^{(0)}), \qquad \text{giving } \hat{S}^{(0)} \text{ as } 676.7$$

If c is excluded from the vector of unknown coefficients to be found in the Gauss–Newton step, because $\hat{c}^{(1)}$ will be calculated later from $\boldsymbol{\eta}_1^{(1)}$ and $\boldsymbol{\eta}_2^{(1)}$ in the same way as $\hat{c}^{(0)}$, $J^{(0)}$ is just

$$[\hat{c}^{(0)}\zeta_1^{(0)} \quad -\hat{c}^{(0)}\zeta_2^{(0)}]$$

giving

$$J^{(0)\mathrm{T}}J^{(0)} \simeq \begin{bmatrix} 74390 & -16030 \\ -16030 & 5250 \end{bmatrix}$$

$$[J^{(0)\mathrm{T}}J^{(0)}]^{-1} \simeq \begin{bmatrix} 3.927\times 10^{-5} & 1.199\times 10^{-4} \\ 1.199\times 10^{-4} & 5.564\times 10^{-4} \end{bmatrix}$$

and

$$J^{(0)\mathrm{T}}(\mathbf{y}-\mathbf{f}^{(0)}) \simeq \begin{bmatrix} -1560 \\ -499.2 \end{bmatrix}$$

Rounding six-figure calculations to four figures as above, the new values of $\hat{\lambda}_1$ and $\hat{\lambda}_2$ are found to be

$$\begin{bmatrix} \hat{\lambda}_1^{(1)} \\ \hat{\lambda}_2^{(1)} \end{bmatrix} \equiv \hat{\boldsymbol{\theta}}^{(1)} = \hat{\boldsymbol{\theta}}^{(0)} + [J^{(0)\mathrm{T}}J^{(0)}]^{-1}J^{(0)\mathrm{T}}(\mathbf{y}-\mathbf{f}^{(0)}) \simeq \begin{bmatrix} -0.8211 \\ -2.465 \end{bmatrix}$$

whence

$$\hat{c}^{(1)} \simeq 268.8, \qquad S^{(1)} \simeq 275.8$$

A reasonable improvement has been achieved in this step. It is interesting to find that, if c is retained as an unknown to be adjusted in the Gauss–Newton step, the very poor values

$$\begin{bmatrix} \hat{a}^{(1)} \\ \hat{\lambda}_1^{(1)} \\ \hat{\lambda}_2^{(1)} \end{bmatrix} = \begin{bmatrix} 544.1 \\ -1.238 \\ -1.249 \end{bmatrix}$$

are obtained, at much greater computational cost. △

Ordinary least-squares estimation can be regarded as the single Gauss–Newton step required to reach the optimum $\hat{\boldsymbol{\theta}}$ from a starting guess of zero when the model is linear in $\boldsymbol{\theta}$. For a non-linear model, rapid convergence can be expected only when we are justified in taking J as near-constant over each step and ignoring second derivatives. These assumptions can only be checked at the expense of a great deal of extra computing, and if they are invalid the iterations may not converge at all. There is, therefore, some incentive to look for a more reliable iterative method.

4.3.3 Levenberg–Marquardt Algorithm

One way of ensuring some progress in reducing S would be to take a small step in the local downhill gradient direction from $\hat{\boldsymbol{\theta}}^{(k-1)}$, to

$$\hat{\boldsymbol{\theta}}^{(k)} = \hat{\boldsymbol{\theta}}^{(k-1)} - \alpha(\partial S/\partial \boldsymbol{\theta})|_{\boldsymbol{\theta}=\hat{\boldsymbol{\theta}}^{(k-1)}} \tag{4.3.9}$$

with α a scalar small enough to make the effects of second and higher derivatives of $S(\boldsymbol{\theta})$ over the step negligible. The weakness of this idea is that to find as large as possible a value for α one would have to examine the local shape of $S(\boldsymbol{\theta})$, and having gone to that trouble would have no excuse for not using the shape information in a more ambitious algorithm. If, instead, a conservative value of α were used, progress would be slow, particularly as the gradient became small near the optimum.

The basic idea of the *Levenberg–Marquardt algorithm* (Wolfe, 1978) is to compromise between the downhill gradient direction and the direction given by the Gauss–Newton algorithm, by finding the step that satisfies

$$[J^{(k-1)\mathrm{T}}J^{(k-1)}+\mu^{(k-1)}I](\hat{\boldsymbol{\theta}}^{(k)}-\hat{\boldsymbol{\theta}}^{(k-1)}) = -\partial S/\partial\boldsymbol{\theta}|_{\boldsymbol{\theta}=\hat{\boldsymbol{\theta}}^{(k-1)}} \qquad (4.3.10)$$

with $\mu^{(k-1)}$ a positive scalar. If $\mu^{(k-1)}$ is chosen to be small, the step is almost a Gauss–Newton step, and if $\mu^{(k-1)}$ is large it is almost a downhill gradient step with $1/\mu^{(k-1)}$ for α. Notice that for any sensible model, $J^{\mathrm{T}}J$ is positive-definite, since otherwise $J\,\delta\boldsymbol{\theta}$ would have to be zero for some non-zero $\delta\boldsymbol{\theta}$, implying that two different values of $\boldsymbol{\theta}$ give precisely the same output values. Positive-definiteness guarantees that $J^{\mathrm{T}}J$ can be inverted in the Gauss–Newton algorithm and, with μ positive and I positive-definite, also guarantees that the Levenberg–Marquardt step is feasible. The factorisation methods discussed in Section 4.2 are also applicable here.

The scalar μ in (4.3.10) is adjusted at each step according to how progress in the previous step compared with what was expected. Implementations of the Levenberg–Marquardt algorithm differ in their rules for adjusting μ, aiming to avoid an excessive number of trial evaluations of S and steps in $\hat{\boldsymbol{\theta}}$, and in their safeguards when μ becomes very small. The algorithm is very widely used, but as with most search algorithms cases can be found in which it "hangs up" before reaching an acceptable $\hat{\boldsymbol{\theta}}$. Although this sounds serious, a disposition to hang up is often a good indication that something is wrong with the form of the model or the starting guesses $\hat{\boldsymbol{\theta}}^{(0)}$. Moreover, it is far easier to be fooled into accepting an ill-chosen model structure by an acceptable fit on one set of records than to be deceived by a hung-up search into thinking an optimum $\hat{\boldsymbol{\theta}}$ has been found for the given model when it has not. In the latter case more runs with different starting guesses will often resolve the uncertainty; in the former, more records are required, and even then it may not be easy to recognise an uneconomical model as such.

4.4 WHY BOTHER WITH STATISTICS?

Appealing though the idea of minimising the sum of the squared model-output errors is, other possibilities exist. To reach a proper assessment of least-squares estimation, we must ask how its performance compares with that of other estimators. Conclusions about estimator performance based on one set of records, or even several, are not likely to be entirely reliable. Questions about performance are therefore essentially probabilistic, asking what will happen *on average* over a set of estimation experiments specified in statistical terms.

It may seem a shame to start talking in probabilistic terms when, after all, the purely deterministically motivated least-squares methods often give perfectly acceptable results. There are several reasons why the effort is worthwhile. We shall find that sometimes the performance of ordinary least-squares can be improved by straightforward modifications, for instance when something is known or can be estimated about the correlation structure of the noise present in the observations. We shall see that least-squares estimation thus modified has attractive statistical properties in addition to its algebraic simplicity and relative computational convenience. We shall encounter, and learn to avoid, problems arising when some of the regressors contain noise correlated with the observation noise. Finally, we may find that in the broader methodology of identification, the ability of a statistically motivated estimation method to provide not just $\hat{\theta}$ but also an estimate of its reliability, in the form of its covariance (Section 5.3), is valuable.

The next chapter describes how the probabilistic behaviour of an estimator is characterised, and how least-squares estimation looks from a probabilistic viewpoint.

FURTHER READING

Econometrics texts such as Goldberger (1964) and Johnston (1972) give some of the clearest accounts of least squares. Draper and Smith (1981) provide a detailed basic account of least squares, including non-linear least squares, with many examples and exercises. Another text with plenty of examples is Chatterjee and Price (1977). Computational methods based on singular-value decomposition and Householder transformation are covered in detail by Lawson and Hanson (1974), with FORTRAN listings. They also consider equality and inequality constraints such as prior knowledge that model coefficients are non-negative, and selective deletion of regressors from a tentative model. Matrix factorisation techniques for least squares are described by Bierman (1977) with recursive (sequential) processing of the observations in mind, as in Chapter 7. Sorenson (1970) reviews the history of least squares.

Least-squares routines are available in many software libraries, but it is unwise to use them without an appreciation of the techniques they use and the potential problems of ill-conditioning. Van den Bos (1983) discusses ill-conditioning in non-linear least squares.

Hamming (1973) sums up the point of exercises such as least squares fitting pithily: "The purpose of computing is insight, not numbers".

REFERENCES

Bierman, G. J. (1977). "Factorization Methods for Discrete Sequential Estimation". Academic Press, New York and London.

Brown, F. R., Godfrey, K. R., and Knell, A. (1979). Compartmental modelling based on methionine tolerance test data: a case study. *Med. Biol. Eng. Comput.* **17**, 223–229.

Chatterjee, S., and Price, B. (1977). "Regression Analysis by Example". Wiley, New York.

Draper, N. R., and Smith, H. (1981). "Applied Regression Analysis", 2nd ed. Wiley, New York.

Forsythe, G. E., Malcolm, M. A., and Moler, C. B. (1977). "Computer Methods for Mathematical Computations". Prentice-Hall, Englewood Cliffs, New Jersey.

Gauss, K. F. (1809, transl. 1963). "Theory of the Motion of the Heavenly Bodies about the Sun in Conic Sections". Dover, New York.

Goldberger, A. S. (1964). "Econometric Theory". Wiley, New York.

Golub, G. (1965). Numerical methods for solving least squares problems. *Numer. Math.* **7**, 206–216.

Hamming, R. W. (1973). "Numerical Methods for Scientists and Engineers", 2nd ed. McGraw-Hill, New York.

Johnston, J. (1972). "Econometric Methods", 2nd ed. McGraw-Hill, New York.

Lawson, C. L., and Hanson, R. J. (1974). "Solving Least Squares Problems". Prentice-Hall, Englewood Cliffs, New Jersey.

Sorenson, H. W. (1970). Least squares estimation: from Gauss to Kalman. *IEEE Spectrum* **7**, 63–68.

van den Bos, A. (1983). Limits to resolution in nonlinear least squares model fitting. *IEEE Trans. Autom. Control* **AC-28**, 1118–1120.

Wolfe, M. A. (1978). "Numerical Methods for Unconstrained Optimization". Van Nostrand Reinhold, Wokingham.

PROBLEMS

4.1 A data-logging run on an industrial process gives

Time t	0	1	2	3	4	5	6	7	8	9	10
Input u_t	−0.64	0.36	0.52	0.49	−0.58	−0.36	−0.32	0.72	—	—	—
Output y_t	0.43	−0.41	−1.32	−1.05	−0.21	0.27	0.40	0.09	−0.10	−0.12	0.44

Find the o.l.s. estimates of the unit-pulse-response ordinates h_1 and h_2 in the model $y_t = h_1 u_{t-1} + h_2 u_{t-2} + e_t$, using as many of the observations as possible.

4.2 Repeat Problem 4.1, but including a constant term in the model. Use the Choleski method to invert the normal matrix.

4.3 An impulse-response test gives

Time t	0+	0.2	0.4	0.6	0.8
Response $h(t)$	3.4	2.3	1.7	1.2	0.9

By o.l.s., find the estimates of K and τ in the model $h(t) = K\exp(-t/\tau) + e(t)$ which give a least-squares fit to the decibel value $20\log_{10} h(t)$, i.e. which minimise the sum of the squared proportional or percentage, rather than absolute, errors in $h(t)$.

4.4 Verify algebraically that for any column vector $\mathbf{u}_i^c$ from U, $P(U)\mathbf{u}_i^c$ is $\mathbf{u}_i^c$, where $P(U)$ is $U[U^T U]^{-1} U^T$. What is the geometrical reason?

4.5 Show that $P(U)$ is idempotent. Why is it, geometrically? Defining $P'(U)$ as $I - P(U)$, show that $P'(U)$ is also idempotent. What are the geometrical interpretations of $P(U)\xi$ and $P'(U)\xi$, where ξ is any real vector conformable with $P(U)$?

4.6 By rewriting the transfer-function model

$$Y(z^{-1}) = \frac{b_1 z^{-1} + b_2 z^{-2} + \cdots + b_n z^{-n}}{1 + a_1 z^{-1} + \cdots + a_n z^{-n}} U(z^{-1})$$

as a difference equation relating output sample y_t to earlier samples of the input and output, produce a regression equation which would, in principle, allow o.l.s. estimation of the transfer-function coefficients. [This idea will be pursued in Section 7.2.]

4.7 What happens to the expression for the o.l.s. estimate if the columns of the regressor matrix U form an orthogonal set? What happens if they are orthonormal? [*Note the connection with singular-value decomposition.]

4.8 Find the least-squares estimates of acceleration, initial velocity and initial position as in Example 4.1.1, but weighting the squared errors linearly from weight 1 at time zero to weight 6 at time 1.0, i.e. penalising recent errors more heavily. Use the estimates to predict target position at time 1.2, and compare the prediction error (the actual position being at 312.2 m) with that in Example 4.1.1.

4.9 Express the o.l.s. sum of squared errors $S(\hat{\theta})$ in terms of $\mathbf{y}$ and $P(U)$.

4.10 Compute the second derivatives $\partial^2 f_l / \partial\theta_i \partial\theta_j$ which appear in (4.3.6), for the model and observations of Example 4.3.2, and check the effects of omitting them from the calculation of $[J_{\theta} g]$ by (4.3.7).

Chapter 5

Statistical Properties of Estimators

5.1 INTRODUCTION

Whenever noise is present in the observations from which a model is estimated, the parameter estimates are affected by it and are therefore random variables; taking another set of observations would not give precisely the same results. Any output prediction making use of the parameter estimates is also a random variable. We may choose to regard the actual parameters as deterministic or as themselves random variables. The latter view implies that a range of possible systems as well as signals should be considered.

When dealing with random variables, it is natural to ask how the estimator will perform *on average* over all possible noise realisations, and perhaps all possible actual parameter values. Our measure of how the estimator performs should be consistent with the intended use of the model. For instance, we may be interested in how well it predicts future output values. The accuracy of individual parameters may not then be of direct interest, particularly for parameters with no clear physical significance. Conversely, the whole aim may be to find good values for certain parameters. We need statistical measures of model accuracy which are flexible enough to suit either situation.

While thinking about the accuracy of parameter estimates, we should bear in mind that the model structure will rarely coincide exactly with the mechanism generating the observations. The structure is usually a compromise between simplicity and power to account for observed behaviour. Moreover, the ultimate test of the model is adequacy for a specified purpose, not optimality and still less truth. In these circumstances it is not strictly to the point to speak of 'true' or 'optimal' parameter values, although we shall often do so for convenience. Adequacy is a difficult attribute to analyse or to generalise to a variety of applications.

Ideally, we should enquire into the statistical behaviour of an estimator by examining the joint probability density function (p.d.f.) of the estimates. This is the basis of Bayes estimation, discussed in Section 6.2. Practically, we virtually always settle for knowing about the mean and scatter, because further analysis is either too hard or requires unrealistic quantities of prior information to be supplied, such as the entire noise probability density function. Confining our attention to the estimates' mean and scatter (covariance, defined shortly) may not be as big a limitation as it seems. In the special case of a Gaussian p.d.f., which we can sometimes accept as realistic, the mean and covariance are enough to define the p.d.f. shape completely. The Gaussian assumption simplifies a great deal of estimation theory, as we shall see, but cannot be made uncritically.

Initially we shall regard the true parameters $\boldsymbol{\theta}$ as unknown constants. Later we shall treat them as random variables. Throughout the chapter, statistical properties of estimators will be discussed by reference to least-squares estimation, as it is simple, familiar and important in practice. However, the ideas apply to any estimator, and will recur in connection with other estimators in later chapters. We shall also assume that every sampled noisy waveform, i.e. every *random process* or family of random variables (the samples, indexed by time) is *wide-sense stationary*. That is, at least its mean and variance (or, for a vector process, covariance: see Section 5.3.1) are finite and constant, and the correlation between its values at any two times depends only on the difference between the two times. Hence we would, for instance, treat a noisy sinusoid as a deterministic sinusoid plus a constant-mean random waveform, not as a varying-mean random waveform. For a Gaussian random process, wide-sense stationarity implies *strict stationarity*, i.e. totally time-invariant statistics, since the p.d.f. is completely defined by the mean and variance (or covariance for a vector).

5.1.1 Bias of Estimators

Our first statistical question is whether estimates obtained from similar experiments will cluster about the true value.

For a single fixed but unknown parameter θ, the *bias* is the difference between the expected value of its estimate $\hat{\theta}$ and θ:

$$b \triangleq \underset{\hat{\theta}}{E}[\hat{\theta}] - \theta \triangleq \int \hat{\theta} p(\hat{\theta})\, d\hat{\theta} - \theta \tag{5.1.1}$$

Here the expectation operator E indicates taking the mean of its argument, and $p(\hat{\theta})$ denotes the p.d.f. of $\hat{\theta}$. The integration is over the range of all possible $\hat{\theta}$ values. The definition of bias extends readily to a vector $\boldsymbol{\theta}$ of parameters:

$$\mathbf{b} \triangleq \underset{\hat{\boldsymbol{\theta}}}{E}[\hat{\boldsymbol{\theta}}] - \boldsymbol{\theta} = \int \cdots \int \hat{\boldsymbol{\theta}} p(\hat{\theta}_1, \hat{\theta}_2, \ldots, \hat{\theta}_p)\, d\hat{\theta}_1 \cdots d\hat{\theta}_p - \boldsymbol{\theta}$$

$$\equiv \int \hat{\boldsymbol{\theta}} p(\hat{\boldsymbol{\theta}})\, d\hat{\boldsymbol{\theta}} - \boldsymbol{\theta} \tag{5.1.2}$$

with $p(\hat{\theta}_1, \hat{\theta}_2, \ldots, \hat{\theta}_p)$ the joint p.d.f. of the parameters. When $\boldsymbol{\theta}$ is treated as random, the *conditional bias*

$$\mathbf{b}(\boldsymbol{\theta}) \triangleq \underset{\hat{\boldsymbol{\theta}}|\boldsymbol{\theta}}{E}[\hat{\boldsymbol{\theta}} \mid \boldsymbol{\theta}] - \boldsymbol{\theta} \triangleq \int \hat{\boldsymbol{\theta}} p(\hat{\boldsymbol{\theta}} \mid \boldsymbol{\theta})\, d\hat{\boldsymbol{\theta}} - \boldsymbol{\theta} \tag{5.1.3}$$

is defined, based on the conditional p.d.f. $p(\hat{\boldsymbol{\theta}} \mid \boldsymbol{\theta})$. We can then consider a larger collection of experiments, covering all possible values of $\boldsymbol{\theta}$ as well as all realisations of the signals and noise. The overall *unconditional bias* is then

$$\bar{\mathbf{b}} \triangleq \underset{\boldsymbol{\theta}}{E}[\mathbf{b}(\boldsymbol{\theta})] \equiv \underset{\boldsymbol{\theta}}{E}\left[\underset{\hat{\boldsymbol{\theta}}|\boldsymbol{\theta}}{E}[\hat{\boldsymbol{\theta}} \mid \boldsymbol{\theta}] - \boldsymbol{\theta}\right] \tag{5.1.4}$$

Example 5.1.1 The gain α of a device is estimated from N measurements of its input u and output y with additive noise e, the model being

$$y_t = \alpha u_t + e_t, \qquad t = 1, 2, \ldots, N$$

The estimator, which does not use any very small values of u, is

$$\hat{\alpha} = \left(\sum_{t=1}^{N} \frac{y_t}{u_t}\right) \Big/ N$$

Each u_t will be regarded as deterministic and known exactly, any uncertainty being included in e_t. Each y_t is a random variable, as it contains noise. The bias in $\hat{\alpha}$ is

$$b = E[\hat{\alpha}] - \alpha = \sum_{t=1}^{N} \left\{\frac{E[\alpha u_t + e_t]}{u_t}\right\} \Big/ N - \alpha$$

$$= \sum_{t=1}^{N} \left\{\alpha + \frac{E[e_t]}{u_t}\right\} \Big/ N - \alpha = \sum_{t=1}^{N} \left\{\frac{E[e_t]}{u_t}\right\} \Big/ N$$

Swapping E and $\sum$ like this is allowed by the linearity of the expectation operation; from its definition, the expected value of a sum is the sum of the expected values of its parts, whether or not they are independent.

The bias here depends on the noise mean and input values but not on α. If the noise has a constant mean $\bar{e}$, the bias is $\bar{e}$ times the mean of $1/u$ through the experiment, so unless $\bar{e}$ is zero, the larger the u values the better, as intuition suggests. If $\bar{e}$ is zero, $\hat{\alpha}$ is unbiased for any selection of non-zero input samples and any number of measurements N. △

As records are of finite length in real life, we wish to know the *finite-sample bias* of an estimate based on N samples of each variable, as in Example 5.1.1. We may make do with the *asymptotic bias* in the limit as N tends to infinity, as second best, since it may be possible to determine whether an estimator is asymptotically biased even though its finite-sample bias is difficult or impossible to evaluate.

Example 5.1.2 The variance $s \triangleq E[(v - \bar{v})^2]$ of a wide-sense stationary signal $v(t)$ with mean $\bar{v}$ is estimated from N independent samples v_1 to y_N by

$$\hat{s} = \sum_{t=1}^{N} \frac{\{(v_t - \hat{\bar{v}})^2\}}{N}$$

where $\hat{\bar{v}}$ is the sample mean $\sum_{t=1}^{N} v_t/N$. The bias in $\hat{s}$ is

$$b = E\left[\sum_{t=1}^{N} \{(v_t - \hat{\bar{v}})^2\}\right] \Big/ N - E[(v - \bar{v})^2]$$

$$= \sum_{t=1}^{N} \{E[v_t^2 - 2v_t\hat{\bar{v}} + \hat{\bar{v}}^2]\}/N - E[v^2 - 2v\bar{v} + \bar{v}^2]$$

$$= E[v^2] - 2\sum_{t=1}^{N}\sum_{s=1}^{N} E[v_t v_s]/N^2 + E\left[\sum_{t=1}^{N}\sum_{s=1}^{N} v_t v_s/N^2\right] - E[v^2] + 2\bar{v}^2 - \bar{v}^2$$

$$= -E\left[\sum_{t=1}^{N}\sum_{s=1}^{N} v_t v_s\right] \Big/ N^2 + \bar{v}^2 = -(NE[v^2] + N(N-1)\bar{v}^2)/N^2 + \bar{v}^2$$

$$= -(E[v^2] - \bar{v}^2)/N = -s/N$$

The bias is asymptotically zero since s is finite, but is non-zero for a finite N, in spite of the plausible look of $\hat{s}$. Although the bias depends on the unknown s, a finite-sample-unbiased estimator is obtainable just by rescaling $\hat{s}$ to $N\hat{s}/(N-1)$. △

So long as the finite-sample bias can be found in terms of quantities we can evaluate or estimate well, we can choose the sample size N to make the bias acceptably small. Even so, the estimation error in any one experiment may be large; to accept an estimate is an act of faith. We normally wish to bolster our faith with some assurance that the scatter of the estimates produced by the estimator is, on average, small, and even then there is some risk that we shall be disappointed.

It seems reasonable to select, whenever possible, an unbiased estimator. However, we shall see in Section 5.3.6 that a biased estimator may produce estimates with so much less scatter than an alternative unbiased estimator that it has smaller mean-square error, and is preferable.

5.1.2 Unbiased Linear Estimator for Linear-in-Parameters Model

Estimators which are linear in the output observations forming $\mathbf{y}$ are attractively simple, computationally and algebraically. When the model relating y to explanatory variables through parameters $\boldsymbol{\theta}$ is also linear, as in o.l.s. and w.l.s., it it is easy to find the conditions for the estimator to be unbiased. The model and estimator are

$$\mathbf{y} = U\boldsymbol{\theta} + \mathbf{e}, \qquad \hat{\boldsymbol{\theta}} = A\mathbf{y} \tag{5.1.5}$$

For a fixed $\boldsymbol{\theta}$, the bias is

$$\mathbf{b} = E[A\mathbf{y}] - \boldsymbol{\theta} = E[AU - I]\boldsymbol{\theta} + E[A\mathbf{e}] \tag{5.1.6}$$

Here A will depend in some perhaps complicated way on samples of the explanatory variables. If all those variables are uncorrelated with $\mathbf{e}$, which is so if the model is good at explaining the systematic behaviour of the output, and if also $\mathbf{e}$ is zero-mean, the last term in (5.1.6) is zero, so the bias is zero if

$$E[AU] = I \tag{5.1.7}$$

This restriction on A will be invoked in Chapter 7 when deriving recursive estimators, but more immediately we check if it is obeyed by least-squares estimators in the next section.

An important property of linear-in-the-parameters models with zero-mean $\mathbf{e}$ uncorrelated with U is that unbiased parameter estimates imply unbiased model predictions $\hat{\mathbf{y}}$ of the output due to any specified U, since then

$$E[\hat{\mathbf{y}}] - \mathbf{y} = E[U(\hat{\boldsymbol{\theta}} - \boldsymbol{\theta}) - \mathbf{e}] = U\mathbf{b} - E[\mathbf{e}] \tag{5.1.8}$$

5.2 BIAS OF LEAST-SQUARES ESTIMATE

To analyse the bias of least-squares estimates, we must distinguish between deterministic and random regressors, and between regressors correlated and uncorrelated with the regression-equation error.

5.2.1 Bias with Regressors Deterministic

The o.l.s. estimate (4.1.12) is linear in $\mathbf{y}$ and based on the linear-in-parameters model (4.1.4), so (5.1.5) applies, with $[U^{\mathrm{T}}U]^{-1}U^{\mathrm{T}}$ for A. Treating both $\boldsymbol{\theta}$ and the regressors forming U as deterministic,

$$E[AU] = [U^{\mathrm{T}}U]^{-1}U^{\mathrm{T}}U = I \tag{5.2.1}$$

so from (5.1.6) the bias in $\boldsymbol{\theta}$ is

$$\mathbf{b} = [U^{\mathrm{T}}U]^{-1}U^{\mathrm{T}}E[\mathbf{e}] \tag{5.2.2}$$

Provided $\mathbf{e}$ is zero-mean, the o.l.s. estimate is therefore unbiased, for any number N of samples making up $\mathbf{y}$. As in Example 4.1.1, the mean of $\mathbf{e}$ can be made zero even when y and/or the regressors contain constant components, by including a constant term in the regression model.

For a w.l.s. estimate, (4.1.34) has $[U^{\mathrm{T}}WU]^{-1}U^{\mathrm{T}}W$ as A, so again AU is I, and the bias is

$$\mathbf{b} = [U^{\mathrm{T}}WU]^{-1}U^{\mathrm{T}}WE[\mathbf{e}] \tag{5.2.3}$$

which is zero so long as $\mathbf{e}$ is zero-mean.

If $\boldsymbol{\theta}$ is taken to be random, the conditional bias $\mathbf{b}(\boldsymbol{\theta})$ defined by (5.1.3) must be considered, but in fact the o.l.s. and w.l.s. biases we have just found are independent of $\boldsymbol{\theta}$, so $\mathbf{b}(\boldsymbol{\theta})$ and its mean $\bar{\mathbf{b}}$ coincide with $\mathbf{b}$.

Example 5.2.1 Observations of $y(t)$ are affected by linear drift $d(t) = \alpha t + \beta$ as well as zero-mean noise $n(t)$. A regression with a constant term β but no αt term is tried. The corresponding regression-equation error is

$$\mathbf{e} = \alpha\mathbf{t} + \mathbf{n}$$

where $\mathbf{t}$ comprises the sampling instants. Since the mean of $\mathbf{e}$ is not zero, the o.l.s. and w.l.s. estimates will be biased. The o.l.s. residuals $\hat{\mathbf{e}} \triangleq \mathbf{y} - U\hat{\boldsymbol{\theta}}$ have the property that

$$U^{\mathrm{T}}\hat{\mathbf{e}} = U^{\mathrm{T}}(\mathbf{y} - U[U^{\mathrm{T}}U]^{-1}U^{\mathrm{T}}\mathbf{y}) = U^{\mathrm{T}}\mathbf{y} - U^{\mathrm{T}}\mathbf{y} = \mathbf{0}$$

so the bias could be regarded as caused by o.l.s. forcing $U^{\mathrm{T}}\hat{\mathbf{e}}$ to be zero when $U^{\mathrm{T}}\mathbf{e}$ is not. Weighted-least-squares similarly forces $U^{\mathrm{T}}W\hat{\mathbf{e}}$ to be zero. Since the row of U^T due to the constant term β consists wholly of 1's, we see from $U^T\hat{\mathbf{e}}$ being zero that the residuals add up to zero even though the error samples do not. The presence of the constant term in the model ensures that the constant component of the residuals, but not e in this case, is zero. △

5.2.2 Bias with Regressors Random

When U is partly or wholly random, the bias has to be averaged over U as well as $\mathbf{e}$ and for o.l.s. becomes

$$\mathbf{b} = \underset{\mathbf{e},U}{E}[[U^TU]^{-1}U^T\mathbf{e}] \tag{5.2.4}$$

Little extra complication ensues if $\mathbf{e}$ is independent of U, since then (5.1.7) is still true,

$$\mathbf{b} = \underset{U}{E}[[U^TU]^{-1}U^T]\underset{\mathbf{e}}{E}[\mathbf{e}] \tag{5.2.5}$$

and once more $\hat{\theta}$ is unbiased if $\mathbf{e}$ is zero-mean. On the other hand, if U and $\mathbf{e}$ are not independent, the bias is not generally zero. The bias in these circumstances is investigated in Section 5.2.5, using probability limits.

Two common causes of dependence between regressors and regression-equation error are noise in observing the regressors, and inclusion of earlier samples of the output (regressand) among the regressors in a dynamical model. They are the subjects of the next two sections.

5.2.3 Bias due to Noisily Observed Regressors: The "Errors in Variables" Problem†

Let us first examine o.l.s. estimation of a scalar parameter θ in a model

$$y'_t = u'_t\theta + n_t \tag{5.2.6}$$

from measurements

$$u_t = u'_t + w_t, \qquad y_t = y'_t + v_t, \qquad t = 1, 2, 3, \ldots, N \tag{5.2.7}$$

affected by mutually uncorrelated, zero-mean noises w_t and v_t. The modelling error n_t is uncorrelated with w_t and v_t and has mean zero. The regression equation is

$$y_t = u_t\theta + e_t \tag{5.2.8}$$

in which, from (5.2.6–5.2.8),

$$e_t = n_t + v_t - w_t\theta \tag{5.2.9}$$

† Kendall and Stuart, 1979; Johnston, 1972.

so there is dependence between e_t and regressor u_t through w_t. The o.l.s. estimate of θ is

$$\hat{\theta} = \sum_{t=1}^{N} \{u_t y_t\} \Big/ \sum_{t=1}^{N} \{u_t^2\} \tag{5.2.10}$$

so the bias is

$$\begin{aligned} b &= E\left[\sum_{t=1}^{N} \{u_t(u_t\theta + e_t)\} \Big/ \sum_{t=1}^{N} \{u_t^2\}\right] - \theta \\ &= E\left[\sum_{t=1}^{N} \{u_t e_t\} \Big/ \sum_{t=1}^{N} \{u_t^2\}\right] \\ &= E\left[\sum_{t=1}^{N} \{(u_t' + w_t)(n_t + v_t - w_t\theta)\} \Big/ \sum_{t=1}^{N} \{u_t^2\}\right] \\ &= -E\left[\sum_{t=1}^{N} \{(u_t' + w_t)w_t\} \Big/ \sum_{t=1}^{N} \{(u_t' + w_t)^2\}\right]\theta \end{aligned} \tag{5.2.11}$$

where the last line follows from the assumed uncorrelatedness and zero means. The estimate is clearly biased in general, for deterministic or random u_t' and for any N. The o.l.s. estimate of a vector $\boldsymbol{\theta}$ is similarly biased in the same circumstances:

$$\mathbf{b} = E[[U^{\mathrm{T}}U]^{-1}U^{\mathrm{T}}\mathbf{e}] = -E[[(U' + W)^{\mathrm{T}}(U' + W)]^{-1}[U' + W]^{\mathrm{T}}W]\boldsymbol{\theta} \tag{5.2.12}$$

Note that the bias is due to dependence between U and $\mathbf{e}$, so noise in the inputs to a system being identified by o.l.s. causes bias only when it appears both in the input measurements and in $\mathbf{e}$. Bias would not arise, for instance, from actuator noise affecting the input u' actually applied when a known test signal u was intended, since then u would be uncorrelated with w and hence e, even though u' would be correlated with w.

5.2.4 Bias due to Presence of Output among Regressors

Cross-correlation between regressors and the regression-equation error may also arise when a z-transform transfer-function model

$$Y(z^{-1}) = (z^{-k}B(z^{-1})/(1 + A(z^{-1})))\,U(z^{-1}) + V(z^{-1}) \tag{5.2.13}$$

is rewritten in regression-equation form via

$$(1 + A(z^{-1}))Y(z^{-1}) = z^{-k}B(z^{-1})U(z^{-1}) + (1 + A(z^{-1}))V(z^{-1}) \qquad (5.2.14)$$

Here k is the dead time in sample intervals, $V(z^{-1})$ represents the output noise, and

$$A(z^{-1}) = a_1 z^{-1} + a_2 z^{-2} + \cdots + a_n z^{-n}, \quad B(z^{-1}) = b_1 z^{-1} + \cdots + b_m z^{-m} \qquad (5.2.15)$$

Interpreting z^{-1} as the one-sample-delay operator, we obtain the difference equation

$$\begin{aligned} y_t &= -a_1 y_{t-1} - a_2 y_{t-2} - \cdots - a_n y_{t-n} + b_1 u_{t-k-1} + \cdots + b_m u_{t-k-m} + e_t \\ &\equiv \mathbf{u}_t^{\mathrm{T}} \boldsymbol{\theta} + e_t, \qquad t = 1, 2, 3, \ldots \end{aligned} \qquad (5.2.16)$$

The parameter vector $\boldsymbol{\theta}$ to be estimated is

$$[-a_1 \quad -a_2 \quad \cdots \quad -a_n \quad b_1 \quad \cdots \quad b_m]^{\mathrm{T}}$$

and the regressor vector $\mathbf{u}_t^{\mathrm{T}}$ is

$$[y_{t-1} \quad \cdots \quad y_{t-n} \quad u_{t-k-1} \quad \cdots \quad u_{t-k-m}]$$

From (5.2.14), e_t is seen to be a moving average of $n + 1$ successive samples of the original noise sequence $\{v\}$ in (5.2.13), so $\{e\}$ is autocorrelated even if $\{v\}$ is not.

We shall see in Section 5.3.5 that autocorrelation of $\{e\}$ affects the scatter of least-squares estimates of $\boldsymbol{\theta}$. Our present concern is bias, though. Each y_{t-i} among the regressors in (5.2.16) is directly affected by e_{t-i}. We also see, by writing (5.2.16) with $t - i$ in place of t, that y_{t-i} depends indirectly on e_{t-i-1} to e_{t-i-n} through y_{t-i-1} to y_{t-i-n} and yet more indirectly on all earlier samples from $\{e\}$. Thus correlation of e_t with any earlier e_{t-i} leads to correlation between e_t and one or more regressor y_{t-1} to y_{t-n}. Bias will result in $\hat{\boldsymbol{\theta}}$, as shown by (5.2.4).

One aim of the instrumental variable method described in Section 5.3.6 is to avoid such bias.

Example 5.2.2 A system is described by the first-order discrete-time model

$$Y(z^{-1}) = (b_1 z^{-1}/(1 + a_1 z^{-1}))U(z^{-1}) + V(z^{-1})$$

The model is rewritten as a regression equation

$$y_t = -a_1 y_{t-1} + b_1 u_{t-1} + e_t, \qquad t = 1, 2, 3, \ldots$$

where

$$e_t = v_t + a_1 v_{t-1}$$

Suppose that $\{v\}$ is zero-mean, uncorrelated and of constant variance σ^2. Even then, $\{e\}$ is autocorrelated:

$$r_{ee}(0) = E[(v_t + a_1 v_{t-1})^2] = E[v_t^2] + a_1^2 E[v_{t-1}^2] = (1 + a_1^2)\sigma^2$$

$$r_{ee}(1) = E[(v_t + a_1 v_{t-1})(v_{t+1} + a_1 v_t)] = a_1 E[v_t^2] = a_1\sigma^2 = r_{ee}(-1)$$

$$r_{ee}(i) = E[(v_t + a_1 v_{t-1})(v_{t+i} + a_1 v_{t+i-1})] = 0 = r_{ee}(-i) \qquad \text{for all} \quad i \geq 2$$

Regressor y_{t-1} is correlated with e_t since both depend on v_{t-1}. Specifically,

$$E[y_{t-1}e_t] = E[((\text{function of } u\text{'s up to } u_{t-2}) + v_{t-1})(v_t + a_1 v_{t-1})] = a_1\sigma^2$$

when, as usual, the input sequence $\{u\}$ is independent of $\{v\}$. △

5.2.5 Convergence, Probability Limits and Consistency

Analysis of bias soon runs up against the problem of finding expectations of relatively complicated functions of random variables, as in (5.2.4) if U contains random variables. The problem can be avoided by considering asymptotic bias and employing *probability limits*. Probability limits refer to one particular way in which estimates may settle down as the number N of observations they are based on is increased. A sequence of random variables $\xi(N)$ with N increasing (for instance, parameter estimates computed from longer and longer records) is said to *converge in probability* to x if for any positive real numbers ε and η we can find a value N_0 of N such that

$$\text{prob}(|\xi(N) - x| < \varepsilon) > 1 - \eta \qquad \text{for all} \quad N > N_0 \tag{5.2.17}$$

which implies that

$$\lim_{N \to \infty} \text{prob}(|\xi(N) - x| > \varepsilon) = 0 \tag{5.2.18}$$

Put less formally, the chance that $\xi(N)$ is further than ε from x becomes, then remains, as small as we like as N increases past N_0. Be careful not to interpret this as implying that almost every realisation of $\xi(N)$ converges within ε of x and remains there; the tiny proportion of realisations not within ε may consist of different realisations at different values of N. Nor does convergence in probability to x mean that every realisation has x as the limit of $\xi(N)$.

Permanent convergence of almost every realisation to within ε of x is called *convergence with probability* 1 (*w.p.*1) or *almost sure* (*a.s.*) *convergence*. An alternative form of convergence is *mean-square* (*m.s.*) *convergence* or *convergence in quadratic mean*, defined as convergence of the m.s. deviation of

$\xi(N)$ from x to zero as N tends to infinity. Almost sure and mean square convergence each imply convergence in probability; neither is implied by it. Convergence in probability is a weaker property than a.s. or m.s. convergence, but is usually easier (and certainly no harder!) to prove.

If $\xi(N)$ converges in probability to x, x is said to be the *probability limit* $\text{plim}(\xi)$ of $\xi(N)$. The big attraction of probability limits is the property (Wilks, 1962) that for any continuous function $f(\xi)$,

$$\text{plim}\, f(\xi) = f(\text{plim}\,\xi) \tag{5.2.19}$$

so, for example

$$\text{plim}\,\xi^2 = (\text{plim}\,\xi)^2 \tag{5.2.20}$$

and for two matrices A and B, both functions of the same random variables, we can find the probability limit of each element of AB from

$$\text{plim}(AB) = \text{plim}\,A\;\text{plim}\,B \tag{5.2.21}$$

We can now enquire into the probability limits of o.l.s. parameter estimates. Assuming that the probability limits

$$R \triangleq \text{plim}\left(\frac{1}{N}U^{\text{T}}U\right), \qquad \mathbf{c} \triangleq \text{plim}\left(\frac{1}{N}U^{\text{T}}\mathbf{e}\right) \tag{5.2.22}$$

exist, the o.l.s. estimate has a probability limit

$$\begin{aligned}\text{plim}\,\hat{\boldsymbol{\theta}} &= \text{plim}\{[U^{\text{T}}U]^{-1}U^{\text{T}}(U\boldsymbol{\theta} + \mathbf{e})\}\\ &= \boldsymbol{\theta} + \text{plim}\left\{\left(\frac{1}{N}U^{\text{T}}U\right)^{-1}\right\}\text{plim}\left\{\frac{1}{N}U^{\text{T}}\mathbf{e}\right\}\\ &= \boldsymbol{\theta} + R^{-1}\mathbf{c}\end{aligned} \tag{5.2.23}$$

By its definition, R is positive-definite, and therefore invertible, except in the degenerate case where exact linear dependence between the regressors makes $U\boldsymbol{\alpha}$ zero for some non-zero $\boldsymbol{\alpha}$. As R^{-1} is non-singular, the bias $R^{-1}\mathbf{c}$ is zero if and only if $\mathbf{c}$ is zero. With the regressors and $\{e\}$ stationary, $\mathbf{c}$ turns out to be the vector of cross-correlations $E[u_i e]$ between regressors and $\{e\}$, so the necessary and sufficient condition for $\hat{\boldsymbol{\theta}}$ to be asymptotically unbiased, in the sense that it converges in probability to $\boldsymbol{\theta}$, is that every regressor is uncorrelated with the regression-equation error.

An estimator $\hat{\boldsymbol{\theta}}(N)$ which converges in probability to $\boldsymbol{\theta}$ is said to be (weakly) *consistent*. We have just seen, then, that the consistency of $U^{\text{T}}\mathbf{e}/N$ as an estimator of the cross-correlations between the regressors and $\{\mathbf{e}\}$ guarantees consistency of $\hat{\boldsymbol{\theta}}$, provided the cross-correlations are all zero.

5.3 COVARIANCE OF ESTIMATES

5.3.1 Definition of Covariance

The average scatter of a scalar random variable x about its mean is described by its variance

$$\operatorname{var} x \equiv \sigma^2(x) \triangleq E[(x - Ex)^2] \tag{5.3.1}$$

or standard deviation $\sigma(x)$. For a vector variable $\mathbf{x}$, the counterpart is the *covariance matrix*

$$\operatorname{cov} \mathbf{x} \equiv R(\mathbf{x}) \triangleq [(\mathbf{x} - E\mathbf{x})(\mathbf{x} - E\mathbf{x})^{\mathrm{T}}] \tag{5.3.2}$$

Element r_{ij} of $R(\mathbf{x})$ is the covariance $E[(x_i - Ex_i)(x_j - Ex_j)]$ between elements i and j of $\mathbf{x}$, so the variances of individual elements of $\mathbf{x}$ make up the principal diagonal of $R(\mathbf{x})$. A *random process* $\mathbf{x}_t$, i.e. a random variable with a time argument t, has its covariance defined as

$$\operatorname{cov}(\mathbf{x}, s, t) \triangleq E[(\mathbf{x}_s - E\mathbf{x}_s)(\mathbf{x}_t - E\mathbf{x}_t)^{\mathrm{T}}] \tag{5.3.3}$$

A function of two time arguments like this is cumbersome, and we more often deal with the covariance $\operatorname{cov}(\mathbf{x}, t - s)$ of a wide-sense stationary process. In that case element (i, j) is the cross-correlation at lag $t - s$ between x_i and x_j. For the moment we are concerned mainly with the simplest definition (5.3.2), applied to $\hat{\boldsymbol{\theta}}$.

The covariance matrix is easily expressed in terms of the "mean-square value" matrix and the mean:

$$\operatorname{cov} \mathbf{x} = E[\mathbf{x}\mathbf{x}^{\mathrm{T}} - E\mathbf{x} \cdot \mathbf{x}^{\mathrm{T}} - \mathbf{x}E\mathbf{x}^{\mathrm{T}} + E\mathbf{x} \cdot E\mathbf{x}^{\mathrm{T}}] = E[\mathbf{x}\mathbf{x}^{\mathrm{T}}] - E\mathbf{x}E\mathbf{x}^{\mathrm{T}} \tag{5.3.4}$$

so for a zero-mean random variable, $E[\mathbf{x}\mathbf{x}^{\mathrm{T}}]$ and $\operatorname{cov} \mathbf{x}$ can be used interchangeably.

In addition to the covariance of an estimate, we shall often be interested in the covariance of input, noise or error samples which have been written as a vector, like $\mathbf{e}$ in the regression model, usually comprising successive samples uniform in time. Element (i, j) of $\operatorname{cov} \mathbf{x}$ is then autocorrelation $r_{xx}(|i - j|)$ at lag $|i - j|$ sampling intervals, so

$$R_{xx} = \operatorname{cov} \mathbf{x} = \begin{bmatrix} r_{xx}(0) & r_{xx}(1) & r_{xx}(2) & \cdots & r_{xx}(N) \\ r_{xx}(1) & r_{xx}(0) & \cdots & & r_{xx}(N-1) \\ \vdots & & & & \vdots \\ r_{xx}(N) & \cdots & & & r_{xx}(0) \end{bmatrix} \tag{5.3.5}$$

A special case is when $\{x\}$ is an uncorrelated (white) sequence, with $r_{xx}(0)$ equal to σ^2 and the autocorrelation zero at all other lags, so that R_{xx} is $\sigma^2 I$.

Another common situation occurs when $\mathbf{x}$ consists of simultaneous samples of a collection of separate variables, all white, and

$$\text{cov}(\mathbf{x}, t-s) = R\,\delta(t-s) \tag{5.3.6}$$

i.e. the covariance is zero except at lag $t-s$ zero, when it is R. This occurs in the process- or observation-noise part of state-space models, which we shall be employing in Section 8.1.

Example 5.3.1 If the noise sequence $\{v\}$ in the transfer-function model of Example 5.2.2 is white and has constant variance σ^2 and zero-mean, the regression-equation error vector $\mathbf{e}$ has

$$R_{ee} = \sigma^2 \begin{bmatrix} 1+a_1^2 & a_1 & & & & \\ a_1 & 1+a_1^2 & a_1 & & 0 & \\ & a_1 & 1+a_1^2 & a_1 & & \\ & & \ddots & \ddots & \ddots & \\ & 0 & & & & a_1 \\ & & & & a_1 & 1+a_1^2 \end{bmatrix}$$

If we decided instead to use a state-space model

$$x_t = -a_1 x_{t-1} + b_1 u_{t-1} \qquad y_t = x_t + v_t$$

the (co)variance of the scalar observation noise sequence $\{v\}$ would be

$$\text{cov}(v, t-s) = \sigma^2\,\delta(t-s)$$

△

5.3.2 Covariance of Linear Functions of Estimate; Significance of Minimum-Covariance Estimate

In comparing two estimators or assessing the quality of a model, it is often necessary to examine the covariance of a fixed linear function of parameter estimates making up a vector $\hat{\boldsymbol{\theta}}$. Keeping it as general as possible, the function is $C\hat{\boldsymbol{\theta}}$ with C a matrix, although most often C will only be a row vector. An example is a row vector $\mathbf{u}^{\mathrm{T}}$ of specified regressor values for which we want to know the model-output mean-square error $E[(\mathbf{u}^{\mathrm{T}}(\hat{\boldsymbol{\theta}} - \boldsymbol{\theta}))^2]$ attributable to parameter error. The covariance of $C\hat{\boldsymbol{\theta}}$ is

$$\text{cov}(C\hat{\boldsymbol{\theta}}) = E[(C\hat{\boldsymbol{\theta}} - E[C\hat{\boldsymbol{\theta}}])(C\hat{\boldsymbol{\theta}} - E[C\hat{\boldsymbol{\theta}}])^{\mathrm{T}}] = C(\text{cov}\,\hat{\boldsymbol{\theta}})C^{\mathrm{T}} \tag{5.3.7}$$

Example 5.3.2 An unbiased estimate $\hat{\theta}^{(i)}$ with variance p is obtained from batch i of some observations. We enquire how much the variance could be reduced by taking M batches of observations and computing as $\hat{\theta}$ the mean of $\hat{\theta}^{(1)}$ to $\hat{\theta}^{(M)}$, assuming the batches are independent and statistically identical.

The variance of the mean $\bar{\hat{\theta}}$ is

$$\operatorname{var}\bar{\hat{\theta}} = E\left[\left(\frac{\sum_{i=1}^{M}\hat{\theta}^{(i)}}{M} - \theta\right)^2\right] = \sum_{i=1}^{M}\sum_{j=1}^{M} E\left[\left(\frac{\hat{\theta}^{(i)}-\theta}{M}\right)\left(\frac{\hat{\theta}^{(j)}-\theta}{M}\right)\right]$$

$$= Mp/M^2 = p/M$$

The next-to-last step above recognises that $\hat{\theta}^{(i)} - \theta$ is independent of $\hat{\theta}^{(j)} - \theta$ and has zero mean, since $\hat{\theta}^{(i)}$ is unbiased, so all the cross-product terms in the sum are zero.

In the notation of (5.3.7), C is a row vector with every entry $1/M$, $\hat{\boldsymbol{\theta}}$ is the column vector containing $\hat{\theta}^{(1)}$ to $\hat{\theta}^{(M)}$ and cov $\hat{\boldsymbol{\theta}}$ is pI. Notice that the variance of $\hat{\theta}$ tends to zero as M is increased indefinitely. △

Armed with (5.3.7) we can investigate how to get the minimum-variance estimator of any scalar linear function $\mathbf{c}^T\boldsymbol{\theta}$. We denote a candidate unbiased estimate of $\boldsymbol{\theta}$ by $\hat{\boldsymbol{\theta}}^*$ and its covariance by P^*. We shall compare the variance of $\mathbf{c}^T\hat{\boldsymbol{\theta}}^*$ with that of $\mathbf{c}^T\hat{\boldsymbol{\theta}}$, where $\hat{\boldsymbol{\theta}}$ is any unbiased estimate, with covariance P. Putting $\mathbf{c}^T$ for C in (5.3.7),

$$\operatorname{var}(\mathbf{c}^T\hat{\boldsymbol{\theta}}) - \operatorname{var}(\mathbf{c}^T\hat{\boldsymbol{\theta}}^*) = \mathbf{c}^T P\mathbf{c} - \mathbf{c}^T P^*\mathbf{c} = \mathbf{c}^T(P - P^*)\mathbf{c} \tag{5.3.8}$$

so $\mathbf{c}^T\hat{\boldsymbol{\theta}}^*$ has the lower variance if $P - P^*$ is positive-definite, or, putting it another way, if P^* is smaller than P. What is more, this is true *whatever our choice of* $\mathbf{c}$. As well as implying lower variance of the model output due to any specified regressor values, smaller P^* implies lower variance for each individual element $\hat{\theta}_i^*$ of $\hat{\boldsymbol{\theta}}^*$ than for the corresponding element of $\hat{\boldsymbol{\theta}}$. We see this by choosing as $\mathbf{c}$ a vector which is zero but for one in position i. Among unbiased estimates, smaller variance implies smaller m.s. error, since m.s. error is variance plus (mean error) squared.

The good implications of minimum parameter-estimate covariance are so wide that we are justified in paying covariance close attention as an indicator of estimator quality, particularly in the context of unbiased estimators.

A warning is in order here. It is easy to confuse the variance of an unbiased estimate of a model output $\mathbf{u}^T\boldsymbol{\theta}$ with the mean-square output error actually obtained in fitting the model. The two are not the same. The o.l.s. parameter estimates minimise the actual mean-square model-output error over the record, but do not generally minimise the mean-square model-output error over all regression-equation-error realisations for any specified regressor values, not even for the values occurring in the record. The o.l.s. estimates do, however, minimise it in the special case where the regression-equation error is white. We follow up this point in Sections 5.3.4 and 5.3.5.

5.3.3 Covariance of Ordinary Least-Square and Weighted Least-Square Estimates

When U is independent of $\mathbf{e}$ and $\mathbf{e}$ is zero-mean, the o.l.s. estimate has covariance

$$\begin{aligned}\operatorname{cov}\hat{\boldsymbol{\theta}} &= E[([U^{\mathrm{T}}U]^{-1}U^{\mathrm{T}}\mathbf{y}-\boldsymbol{\theta})([U^{\mathrm{T}}U]^{-1}U^{\mathrm{T}}\mathbf{y}-\boldsymbol{\theta})^{\mathrm{T}}]\\ &= E[([U^{\mathrm{T}}U]^{-1}U^{\mathrm{T}}(U\boldsymbol{\theta}+\mathbf{e})-\boldsymbol{\theta})([U^{\mathrm{T}}U]^{-1}U^{\mathrm{T}}(U\boldsymbol{\theta}+\mathbf{e})-\boldsymbol{\theta})^{\mathrm{T}}]\\ &= E[[U^{\mathrm{T}}U]^{-1}U^{\mathrm{T}}\operatorname{cov}(\mathbf{e})U[U^{\mathrm{T}}U]^{-1}] \qquad (5.3.9)\end{aligned}$$

In the simplest case, $\operatorname{cov}\mathbf{e}$ is $\sigma^2 I$, the elements of $\mathbf{e}$ are uncorrelated and all of variance σ^2. That is so for uniform sampling of a system with stationary $e(t)$, autocorrelated only over lags less than one sampling interval. If so,

$$\operatorname{cov}\hat{\boldsymbol{\theta}} = \sigma^2 E[[U^{\mathrm{T}}U]^{-1}] \qquad (5.3.10)$$

and the expectation operation can be dropped when U is non-random. Since $[U^{\mathrm{T}}U]^{-1}$ can be extracted easily from the o.l.s. computation, $\operatorname{cov}\hat{\boldsymbol{\theta}}$ can be estimated readily providing σ^2 can be estimated. One would guess that the sum of squares of the residuals is proportional to σ^2. In fact

$$\begin{aligned}S = \hat{\mathbf{e}}^{\mathrm{T}}\hat{\mathbf{e}} &= (\mathbf{y}-U\hat{\boldsymbol{\theta}})^{\mathrm{T}}(\mathbf{y}-U\hat{\boldsymbol{\theta}}) = \mathbf{y}^{\mathrm{T}}(I_N - U[U^{\mathrm{T}}U]^{-1}U^{\mathrm{T}})^2\mathbf{y}\\ &= \mathbf{y}^{\mathrm{T}}(I_N - U[U^{\mathrm{T}}U]^{-1}U^{\mathrm{T}})\mathbf{y} = \operatorname{tr}\{(I_N - U[U^{\mathrm{T}}U]^{1}U^{\mathrm{T}})\mathbf{y}\mathbf{y}^{\mathrm{T}}\} \qquad (5.3.11)\end{aligned}$$

where I_N denotes the $N \times N$ identity matrix, the trace tr is the sum of the principal-diagonal elements, and we have used the relation

$$\mathbf{x}^{\mathrm{T}}A\mathbf{y} \equiv \sum_i \sum_k x_i a_{ik} y_k \equiv \operatorname{tr}\{A\mathbf{y}\mathbf{x}^{\mathrm{T}}\} \qquad (5.3.12)$$

Since, when U is non-random and $\operatorname{cov}\mathbf{e}$ is $\sigma^2 I_n$,

$$\begin{aligned}\underset{e}{E}[S] &= \operatorname{tr}\{(I_N - U[U^{\mathrm{T}}U]^{-1}U^{\mathrm{T}})E[\mathbf{y}\mathbf{y}^{\mathrm{T}}]\}\\ &= \operatorname{tr}\{(I_N - U[U^{\mathrm{T}}U]^{-1}U^{\mathrm{T}})(U\boldsymbol{\theta}\boldsymbol{\theta}^{\mathrm{T}}U^{\mathrm{T}}+\sigma^2 I_N)\}\\ &= \sigma^2\operatorname{tr}\{I_N - U[U^{\mathrm{T}}U]^{-1}U^{\mathrm{T}}\} = \sigma^2(\operatorname{tr}\{I_N\} - \operatorname{tr}\{[U^{\mathrm{T}}U]^{-1}U^{\mathrm{T}}U\})\\ &= \sigma^2(N-p) \qquad (5.3.13)\end{aligned}$$

we conclude that $S/(N-p)$ is an unbiased estimator of σ^2.

Example 5.3.3 In Example 4.1.1, the o.l.s. estimates of initial target position x_0, velocity v_0 and acceleration a were $\hat{x}_0 = 4{\cdot}79$, $\hat{v}_0 = 234$, $\hat{a} = 55.4$. The sum of squares of the residuals was $S = 157.9$, so σ^2 is estimated as $S/(N-p) = 157.9/(6-3) = 52.6$, and $\operatorname{cov}\hat{\boldsymbol{\theta}}$ is estimated as

$$52.6[U^{\mathrm{T}}U]^{-1} = \begin{bmatrix} 43.2 & -155 & 235 \\ -155 & 958 & -1763 \\ 235 & -1763 & 3526 \end{bmatrix}$$

The square roots of the principal-diagonal elements give estimated standard deviations 6.57 for $\hat{x}_0$, 31.0 for $\hat{v}_0$ and 59.4 for $\hat{a}$. As N is so small, we should not trust the covariance estimate too much, however. △

For the w.l.s. estimate $\hat{\boldsymbol{\theta}}_{\mathrm{w}}$ given by (4.1.34), the covariance is found by the same steps as in (5.3.9) to be

$$\operatorname{cov}\hat{\boldsymbol{\theta}}_{\mathrm{w}} = E[[U^{\mathrm{T}}WU]^{-1}U^{\mathrm{T}}WRW^{\mathrm{T}}U[U^{\mathrm{T}}WU]^{-1}] \qquad (5.3.14)$$

where R is $\operatorname{cov}\mathbf{e}$. As $[U^{\mathrm{T}}WU]^{-1}U^{\mathrm{T}}W$ is the matrix relating $\hat{\boldsymbol{\theta}}_{\mathrm{w}}$ to $\mathbf{y}$, it might appear that $\operatorname{cov}\hat{\boldsymbol{\theta}}_{\mathrm{w}}$ can be computed cheaply, given R. Direct computation would, on the contrary, be expensive since R is $N \times N$, normally very large. It can be avoided in the most important w.l.s. estimator, the Markov estimator which minimises $\operatorname{cov}\hat{\boldsymbol{\theta}}_{\mathrm{w}}$, as discussed in Section 5.3.5.

5.3.4 Minimum-Covariance Property of Ordinary Least Squares When Error Is Uncorrelated

We shall now discover that when the regression-equation errors forming $\mathbf{e}$ are zero-mean, uncorrelated and all of the same variance, so that R is $\sigma^2 I$, the covariance of the o.l.s. estimate $\hat{\boldsymbol{\theta}}$ is the smallest of any linear, unbiased estimate. If we denote any such estimate $A\mathbf{y}$ by $\hat{\boldsymbol{\theta}}_A$ then, using (5.1.7) to ensure zero bias,

$$\operatorname{cov}\hat{\boldsymbol{\theta}}_A = E[(A\mathbf{y}-\boldsymbol{\theta})(A\mathbf{y}-\boldsymbol{\theta})^{\mathrm{T}}] = E[A\mathbf{e}\mathbf{e}^{\mathrm{T}}A^{\mathrm{T}}] = \sigma^2 E[AA^{\mathrm{T}}] \quad (5.3.15)$$

under the assumption that $\mathbf{e}$ is uncorrelated with the samples making up A. We could now assume that A and U are non-random, allowing us to drop the expectation signs in (5.3.10) and (5.3.15), then show fairly easily (Problem 5.4) that $\sigma^2(AA^{\mathrm{T}} - [U^{\mathrm{T}}U]^{-1})$ is positive-semi-definite, demonstrating that no $\hat{\boldsymbol{\theta}}_A$ gives $\operatorname{cov}\hat{\boldsymbol{\theta}}_A$ smaller than $\operatorname{cov}\hat{\boldsymbol{\theta}}$. A less restrictive assumption is that A may be stochastic but is of the form $[BU]^{-1}B$, which satisfies the condition (5.1.7) that ensures unbiasedness. All the linear, unbiased estimates we shall encounter are of this form. Keeping A and U stochastic, consider

$$D \triangleq E[(A-[U^{\mathrm{T}}U]^{-1}U^{\mathrm{T}})(A-[U^{\mathrm{T}}U]^{-1}U^{\mathrm{T}})^{\mathrm{T}}]$$
$$=E[AA^{\mathrm{T}}]-E[AU[U^{\mathrm{T}}U]^{-1}]-E[[U^{\mathrm{T}}U]^{-1}U^{\mathrm{T}}A^{\mathrm{T}}]+E[[U^{\mathrm{T}}U]^{-1} \tag{5.3.16}$$

Replacing A by $[BU]^{-1}B$, we see that

$$E[AU[U^{\mathrm{T}}U]^{-1}]=E[[U^{\mathrm{T}}U]^{-1}]=E[[U^{\mathrm{T}}U]^{-1}U^{\mathrm{T}}A^{\mathrm{T}}] \tag{5.3.17}$$

so

$$D=E[AA^{\mathrm{T}}]-E[[U^{\mathrm{T}}U]^{-1}]=\mathrm{cov}\,\hat{\boldsymbol{\theta}}_A-\mathrm{cov}\,\hat{\boldsymbol{\theta}} \tag{5.3.18}$$

and D is positive-semi-definite, being the mean of the positive-semi-definite product of a real matrix and its transpose, so

$$\mathrm{cov}\,\hat{\boldsymbol{\theta}}_A \geq \mathrm{cov}\,\hat{\boldsymbol{\theta}} \tag{5.3.19}$$

This minimum-covariance property is a powerful incentive to employ o.l.s. when the regression-equation error *is* uncorrelated and of constant variance. It would be nice to have a comparably simple minimum-covariance estimate when the error-vector elements are correlated and/or of differing variances. Such an estimate is found in the next section.

5.3.5 Minimum-Covariance Estimate When Error Is Autocorrelated/Non-Stationary: The Markov Estimate

When the covariance of the regression-equation error $\mathbf{e}$ is not of the form $\sigma^2 I$, we no longer have any reason to suppose that the o.l.s. estimate of $\boldsymbol{\theta}$ has the smallest covariance of all linear, unbiased estimates. We can, however, still obtain the minimum-covariance estimate if we first operate on the regression equation so as to turn it into an equation with an error vector which does have covariance $\sigma^2 I$. The required operation is linear filtering. That is, $\mathbf{y}$ and U are pre-multiplied by some $N \times N$ matrix Q to give

$$\mathbf{y}'=Q\mathbf{y}, \qquad U'=QU \tag{5.3.20}$$

From the original regression equation we then have

$$\mathbf{y}'=Q(U\boldsymbol{\theta}+\mathbf{e})=U'\boldsymbol{\theta}+\mathbf{e}' \tag{5.3.21}$$

where $\mathbf{e}'$ is $Q\mathbf{e}$. This new error is still zero-mean, but its covariance is

$$\mathrm{cov}\,\mathbf{e}'=E[Q\mathbf{e}\mathbf{e}^{\mathrm{T}}Q^{\mathrm{T}}]=QRQ^{\mathrm{T}} \tag{5.3.22}$$

where R is $\mathrm{cov}\,\mathbf{e}$. By choosing Q such that

$$Q^{-1}Q^{-\mathrm{T}}=R \tag{5.3.23}$$

we make the covariance

$$\operatorname{cov} \mathbf{e}' = QQ^{-1}Q^{-\mathrm{T}}Q^{\mathrm{T}} = I \tag{5.3.24}$$

It is always possible to factorize R as in (5.3.23), because R is positive-definite (unless the elements of $\mathbf{e}$ are always linearly dependent, not a practical possibility); the factorisation amounts to rewriting the quadratic form $\boldsymbol{\xi}^{\mathrm{T}}R\boldsymbol{\xi}$ as $(Q^{-\mathrm{T}}\boldsymbol{\xi})^{\mathrm{T}}(Q^{-\mathrm{T}}\boldsymbol{\xi})$, and positive-definiteness ensures that at least one element of $Q^{-1}\boldsymbol{\xi}$ is non-zero, so Q exists. The Choleski factorisation in Chapter 4 is one such factorisation.

The o.l.s. estimate based on the filtered equation (5.3.21) is

$$\begin{aligned}\hat{\boldsymbol{\theta}}' &= [U'^{\mathrm{T}}U']^{-1}U'^{\mathrm{T}}\mathbf{y}' = [U^{\mathrm{T}}Q^{\mathrm{T}}QU]^{-1}U^{\mathrm{T}}Q^{\mathrm{T}}Q\mathbf{y}\\ &= [U^{\mathrm{T}}R^{-1}U]^{-1}U^{\mathrm{T}}R^{-1}\mathbf{y}\end{aligned} \tag{5.3.25}$$

so it turns out that *the minimum-covariance, linear, unbiased estimate is w.l.s., with the inverse of the covariance of the regression-equation error as the weighting matrix*. The estimate is called the *Markov*, *Aitken* or *generalized-least squares* (*g.l.s.*) estimate. Since the filtering of the regression equation not only uncorrelates the error $\{e\}$ but also makes its variance unity, the covariance of $\hat{\boldsymbol{\theta}}'$ is, from (5.3.10), $E[[U'^{\mathrm{T}}U']^{-1}]$ or $E[[U^{\mathrm{T}}R^{-1}U]^{-1}]$.

Direct implementation of (5.3.25) is unattractive, as R is $N \times N$, normally large. A better solution is to find a low-order linear filter with the required "noise whitening" effect on $\{e\}$. This may be done iteratively, processing all the regression-equation errors at once as in Section 7.2.2, or recursively, running through the observations and regression-equation errors one at a time, as in Section 7.4. In either method we find the required filter by identifying the structure of the regression-equation error. The problem is a special case of the usual input–output identification problem, with "output" e_t modelled as a linear function of earlier samples of itself and white-noise forcing. The sequence $\{e\}$ is, of course, not directly available and must be approximated by residuals $\{y_t - \mathbf{u}_t^{\mathrm{T}}\hat{\boldsymbol{\theta}}\}$ with $\hat{\boldsymbol{\theta}}$ the best estimate of $\boldsymbol{\theta}$ at hand.

Example 5.3.4 We return to the plant and model of Example 5.2.2. The transfer-function model was rewritten as a regression equation

$$y_t = -ay_{t-1} + bu_{t-1} + e_t \qquad t = 1, 2, \ldots, N$$

with

$$e_t = v_t + av_{t-1}$$

Suppose that in reality a is -0.8 and the noise sequence $\{v\}$ of the transfer-function model is uncorrelated. If the autocorrelation function of $\{e\}$ could be calculated exactly, it would be

$$r_{ee}(0) = (1 + a^2)\sigma^2 = 1.64\sigma^2, \quad r_{ee}(\pm 1) = a\sigma^2 = -0.8\sigma^2, \quad r_{ee}(\pm i) = 0, \quad i \geq 2$$

as found in Example 5.2.2. In practice, $\{e\}$ is not accessible from $\{y\}$ and $\{u\}$ without exact knowledge of a and b, which would do away with the need for identification. Instead, we have to use the o.l.s. residuals $\{\hat{e}\}$, for instance, which differ systematically from $\{e\}$ because the o.l.s. estimates $\hat{a}$ and $\hat{b}$ are inexact. We find, we hope,

$$\operatorname{cov}\hat{\mathbf{e}} \simeq R = \sigma^2 \begin{bmatrix} 1.64 & -0.8 & 0 & 0 & \dots & 0 \\ -0.8 & 1.64 & -0.8 & 0 & \dots & 0 \\ 0 & & & & & \\ \vdots & & & & & \\ 0 & & & & \dots & 1.64 \end{bmatrix}$$

Gaussian elimination inverts this tridiagonal matrix quite quickly for use in (5.3.25) (Fenner, 1974), but a neater and more informative solution is to notice that R can be factorized as SS^{T} where S is the $N \times (N+1)$ matrix

$$\begin{bmatrix} \sigma & -0.8\sigma & 0 & \cdots & 0 & 0 \\ 0 & \sigma & -0.8\sigma & \cdots & 0 & 0 \\ \vdots & & & & & \\ 0 & \cdots & & & \sigma & -0.8\sigma \end{bmatrix}$$

Generally we can see that when e_t is a linear combination of $n+1$ successive samples v_t to v_{t-n}, R has $2n+1$ non-zero diagonals containing $r_{ee}(0)$ to $r_{ee}(n)$, and each row of S has $n+1$ non-zero elements, proportional to the coefficients of v_t to v_{t-n} in e_t. Equating elements of SS^{T} and R gives those coefficients uniquely. We then want to invert the linear relation between $\{e\}$ and $\{v\}$ so as to find the filter that turns $\{u\}$ into $\{u'\}$ and $\{y\}$ into $\{y'\}$, just as Q^{-1} was inverted to give Q to premultiply U and $\mathbf{y}$. As S is not square, it is not obvious how to do so (S^{-1} does not exist). We can, however, invert $U = SU'$ and $\mathbf{y} = S\mathbf{y}'$ one sample at a time. In this example

$$u_t = u'_t - 0.8u'_{t-1}, \qquad y_t = y'_t - 0.8y'_{t-1}, \qquad t = 1, 2, \dots, N$$

gives

$$u'_t = 0.8u'_{t-1} + u_t, \qquad y'_t = 0.8y'_{t-1} + y_t, \qquad t = 1, 2, \dots, N$$

The only difficulty is that we do not know the initial conditions, u'_0 and y'_0 here, and must choose them arbitrarily. So long as the filtering from $\{u\}$ and $\{y\}$ to $\{u'\}$ and $\{y'\}$ is stable, the effects of incorrect initial conditions eventually die out, so we need only discard the first few (we hope) values in $\{u'\}$ and $\{y'\}$. △

We saw in Section 5.2.4 that the combination of autocorrelated $\{e\}$ and inclusion of earlier outputs among the regressors (like y_{t-1} in Example 5.3.4) causes bias in $\hat{\theta}$, due to correlation between e_t and those earlier-output regressors. The bias does not arise if $\{e\}$ has its autocorrelation removed by linear filtering, since then the correlation between regressors and regression-equation error vanishes. An alternative method of avoiding such bias is the subject of the next section.

5.3.6 Instrumental Variables

The bias due to correlation between regressors and regression-equation error, described in Section 5.2.4, can be avoided by modifying the o.l.s. estimate into the *instrumental variable estimate*

$$\hat{\theta}_Z = [Z^T U]^{-1} Z^T \mathbf{y} \tag{5.3.26}$$

where Z is a matrix in which the error-correlated regressors of U are replaced by other variables (the *instrumental variables*, or just *instruments*) not correlated with the error. A suitable choice of Z will make $\hat{\theta}_Z$ a consistent estimator of θ, for

$$\begin{aligned} \operatorname{plim} \hat{\theta}_Z &= \operatorname{plim}([Z^T U]^{-1} Z^T (U\theta + \mathbf{e})) = \theta + \operatorname{plim}([Z^T U]^{-1} Z^T \mathbf{e}) \\ &= \theta + \operatorname{plim}(([Z^T U]/N)^{-1}) \operatorname{plim}(Z^T \mathbf{e}/N) = \theta + \bar{R}_{ZU}^{-1} \bar{r}_{Ze} \end{aligned} \tag{5.3.27}$$

We are assuming here that putting in the $1/N$ is enough to make the probability limits $\bar{R}_{ZU}$ and $\bar{r}_{Ze}$ exist. The effect is that, for instance, element (i,j) of $\bar{R}_{ZU}$ becomes the probability limit of the mean, over the N samples in column i of Z and column j of U, of the product of the variables forming those columns. As one would guess, in many cases each such N-sample mean tends with increasing N to the expected value of the product. Hence, if the variables in Z, U and $\mathbf{e}$ are zero-mean, the elements of $\bar{R}_{ZU}$ and $\bar{r}_{Ze}$ are the covariances between the corresponding variables. (Caution is needed here, for situations can be devised in which an N-sample mean is asymptotically biased but still consistent. A proportion of realisations might be biased, the proportion decreasing asymptotically to zero as N increases and so not preventing consistency, but the bias in that proportion might rise more rapidly with N and produce non-zero nett bias. We cannot therefore assume blindly that for any variable ξ_N with a probability limit, $\operatorname{plim} \xi_N$ and $\lim_{N\to\infty} E\xi_N$ coincide.)

Returning to (5.3.27) we see that plim $\hat{\theta}_Z$ is θ if $\bar{r}_{Ze}$ is zero. When $\bar{r}_{Ze}$ is the vector of covariances between the instrumental variables and $\{\mathbf{e}\}$ as discussed above, we conclude that $\hat{\theta}_Z$ is consistent if the instrumental variables are uncorrelated with $\{e\}$.

Our choice of Z is further guided by wanting $\hat{\theta}_Z$ to have a small asymptotic covariance. Taking cov **e** to be $\sigma^2 I$, the simplest possibility, and using the fact that in general

$$\underset{\xi,\eta}{E}[f(\xi,\eta)]$$

is

$$\underset{\xi}{E}\left[\underset{\eta|\xi}{E}[f(\xi,\eta)\,|\,\xi]\right]$$

(Melsa and Sage, 1973, p. 162), we find that

$$\begin{aligned}
\operatorname{cov}\hat{\boldsymbol{\theta}}_Z &\triangleq \underset{Z,U,e}{E}[(\hat{\boldsymbol{\theta}}_Z-\boldsymbol{\theta})(\hat{\boldsymbol{\theta}}_Z-\boldsymbol{\theta})^{\mathrm{T}}] \\
&= \underset{Z,U,e}{E}[[Z^{\mathrm{T}}U]^{-1}Z^{\mathrm{T}}\mathbf{e}\mathbf{e}^{\mathrm{T}}Z[U^{\mathrm{T}}Z]^{-1}] \\
&= \underset{Z,U}{E}[[Z^{\mathrm{T}}U]^{-1}Z^{\mathrm{T}}\underset{e|Z,U}{E}[\mathbf{e}\mathbf{e}^{\mathrm{T}}]Z[U^{\mathrm{T}}Z]^{-1}] \\
&= \sigma^2\underset{Z,U}{E}[[Z^{\mathrm{T}}U]^{-1}Z^{\mathrm{T}}Z[U^{\mathrm{T}}Z]^{-1}] \\
&= \sigma^2\underset{Z,U}{E}[[Z^{\mathrm{T}}U/N]^{-1}[Z^{\mathrm{T}}Z/N][U^{\mathrm{T}}Z/N]^{-1}]/N \qquad (5.3.28)
\end{aligned}$$

Like $Z^{\mathrm{T}}U$, $Z^{\mathrm{T}}Z$ has to be divided by N to get a finite probability limit $\bar{R}_{ZZ}$. Plim $U^{\mathrm{T}}Z/N$ is just $\bar{R}_{ZU}^{\mathrm{T}}$, and the inversion of $Z^{\mathrm{T}}U/N$ and $U^{\mathrm{T}}Z/N$ inverts $\bar{R}_{ZU}$ and $\bar{R}_{ZU}^{\mathrm{T}}$, so altogether

$$\begin{aligned}
\operatorname{plim}\operatorname{cov}\hat{\boldsymbol{\theta}}_Z &= \sigma^2\operatorname{plim}[Z^{\mathrm{T}}U/N]^{-1}\operatorname{plim}[Z^{\mathrm{T}}Z/N]\operatorname{plim}[U^{\mathrm{T}}Z/N]^{-1}/N \\
&= \sigma^2\bar{R}_{ZU}^{-1}\bar{R}_{ZZ}\bar{R}_{ZU}^{-\mathrm{T}}/N \qquad (5.3.29)
\end{aligned}$$

Now (5.3.29) reveals a danger. If, in making Z uncorrelated with $\{e\}$, we should render Z almost uncorrelated with U, $\bar{R}_{ZU}$ and $\bar{R}_{UZ}$ would be small and their inverses large. With $\bar{R}_{ZZ}$ not particularly small, we could then have an undesirably large asymptotic covariance for $\hat{\boldsymbol{\theta}}_Z$, even though $\hat{\boldsymbol{\theta}}_Z$ might be consistent. It seems Z must consist of instrumental variables correlated as little as possible with $\{e\}$ but as much as possible with U. There need be no conflict, in principle; the error-correlated variables in U are functions of both the noise-free variables driving the system ($\{u\}$ here) and the regression-equation error, and we are merely asking for the former to be emphasised and the latter suppressed in Z. However, not knowing **e** exactly, we cannot check accurately how closely Z approaches the desired correlation behaviour.

The required correlation properties of Z do not prescribe explicitly what we should choose as instrumental variables. Several possible choices will now be reviewed in an example.

Example 5.3.5 Bias was caused in Examples 5.2.2 and 5.3.4 by the presence of y_{t-1}, correlated with e_t, as one of the regressors. To avoid bias, we replace y_{t-1} by an instrumental variable z_{t-1} uncorrelated with e_t. We must also take care that

$$\bar{R}_{ZU} = \begin{bmatrix} \operatorname{plim} \sum_{t=1}^{N} \frac{z_{t-1}y_{t-1}}{N} & \operatorname{plim} \sum_{t=1}^{N} \frac{z_{t-1}u_{t-1}}{N} \\ \operatorname{plim} \sum_{t=1}^{N} \frac{u_{t-1}y_{t-1}}{N} & \operatorname{plim} \sum_{t=1}^{N} \frac{u_{t-1}^2}{N} \end{bmatrix}$$

$$= \begin{bmatrix} r_{zy}(0) & r_{zu}(0) \\ r_{uy}(0) & r_{uu}(0) \end{bmatrix}$$

is not near-singular, inflating $\operatorname{plim} \operatorname{cov} \hat{\theta}_Z$. Let us examine $\bar{R}_{ZU}$ for three choices of z_{t-1}, each uncorrelated with e_t:

(i) $z_{t-1} = u_{t-1}$, so that $r_{zy}(0) = r_{uy}(0)$ and $r_{zu}(0) = r_{uu}(0)$. Clearly, $\bar{R}_{ZU}$ becomes singular. We should have foreseen this disaster, as we have introduced exact linear dependence between the columns of Z, namely column 1 − column 2 = 0. Linear dependence or near-dependence among columns of Z might be much harder to foresee in an example with more regressors.

(ii) $z_{t-1} = y_{t-2}$, so that $r_{zy}(0) = r_{yy}(1)$ and $r_{zu}(0) = r_{yu}(1)$. With

$$y_t = -ay_{t-1} + bu_{t-1} + e_t, \qquad e_t = v_t + av_{t-1}, \qquad t = 1, 2, \ldots, N$$

y_{t-2} depends on u's up to u_{t-3} and e's up to e_{t-2}, and thus on v's up to v_{t-2}. As $\{u\}$ is uncorrelated with $\{v\}$, $r_{yu}(1)$ is then zero providing $r_{uu}(i)$ is zero for all $i > 1$. Similarly, $r_{uy}(0)$ depends on r_{uu} and is zero if $\{u\}$ is white. Hence $|\bar{R}_{ZU}| = r_{yy}(1)r_{uu}(0) -$ (a contribution which depends on r_{uu} and is zero if $r_{uu}(i) = 0$ for all $i > 1$). It is not difficult to see that $r_{yy}(1)$ is not generally zero, so $|\bar{R}_{ZU}|$ is non-zero and $\bar{R}_{ZU}^{-1}$ exists. Problem 5.5 examines the resulting probability limit of $N \operatorname{cov} \hat{\theta}_Z$ when $\{u\}$ is not strongly autocorrelated.

(iii) $z_{t-1} = y_{t-1} - e_{t-1}$. Now z_{t-1} is not obtainable exactly, since e_{t-1} is not known exactly, but it could be approximated by $-\hat{a}y_{t-2} + \hat{b}u_{t-2}$ using tentative estimates of a and b, for instance from o.l.s. This choice of z_{t-1} is appealing because it modifies the troublesome regressor y_{t-1} as little as necessary to uncorrelate it from e_t. If $\{u\}$ is white, it is fairly easy to show (Problem 5.6) that $\bar{R}_{ZU}$, $\bar{R}_{ZZ}$ and $\bar{R}_{UZ}$ are all diagonal, with principal-diagonal elements $(a^2g^2 + b^2)r_{uu}(0)$ and $r_{uu}(0)$, where g^2 is the "power gain" from $r_{uu}(0)$ to $r_{yy}(0)$, obtainable by calculating from a and b the unit-pulse response then squaring and summing its ordinates, as in Problem 3.2. Hence,

$$\text{plim}\, N \operatorname{cov} \hat{\boldsymbol{\theta}}_Z = \frac{(1+a^2)r_{vv}(0)}{r_{uu}(0)} \begin{bmatrix} 1/(a^2g^2+b^2) & 0 \\ 0 & 1 \end{bmatrix}$$

and high input power compared with noise power, i.e. low $r_{vv}(0)/r_{uu}(0)$, is seen to be beneficial, as one would expect. △

5.3.7 Minimum-Mean-Square-Error Estimation: Ridge Regression

Up to now we have asked first that an estimator should be unbiased, then that it should have minimum covariance among unbiased estimators. Reasonable as this seems, it is not always the best thing to do, as it does not guarantee minimum mean-square error (m.s.e.) in the estimates. The m.s.e. matrix for estimate $\hat{\boldsymbol{\theta}}$ of $\boldsymbol{\theta}$, with mean $E\hat{\boldsymbol{\theta}}$ equal to $\bar{\boldsymbol{\theta}}$, is (treating $\boldsymbol{\theta}$ as non-random)

$$\begin{aligned} M &\triangleq E[(\hat{\boldsymbol{\theta}}-\boldsymbol{\theta})(\hat{\boldsymbol{\theta}}-\boldsymbol{\theta})^T] = E[(\hat{\boldsymbol{\theta}}-\bar{\boldsymbol{\theta}}+\bar{\boldsymbol{\theta}}-\boldsymbol{\theta})(\hat{\boldsymbol{\theta}}-\bar{\boldsymbol{\theta}}+\bar{\boldsymbol{\theta}}-\boldsymbol{\theta})^T] \\ &= E[(\hat{\boldsymbol{\theta}}-\bar{\boldsymbol{\theta}})(\hat{\boldsymbol{\theta}}-\bar{\boldsymbol{\theta}})^T] + (\bar{\boldsymbol{\theta}}-\bar{\boldsymbol{\theta}})(\bar{\boldsymbol{\theta}}-\boldsymbol{\theta})^T + (\bar{\boldsymbol{\theta}}-\boldsymbol{\theta})(\bar{\boldsymbol{\theta}}-\bar{\boldsymbol{\theta}})^T + (\bar{\boldsymbol{\theta}}-\boldsymbol{\theta})(\bar{\boldsymbol{\theta}}-\boldsymbol{\theta})^T \\ &= \operatorname{cov}\hat{\boldsymbol{\theta}} + \mathbf{b}\mathbf{b}^T \end{aligned} \tag{5.3.30}$$

where $\mathbf{b}$ is the bias in $\hat{\boldsymbol{\theta}}$. This matrix counterpart of the familiar "mean-square value equals variance plus mean squared" indicates that a finite bias may be worth exchanging for a reduced covariance.

Reduction of m.s.e. is the aim of *ridge regression*, which modifies the o.l.s. estimate to

$$\hat{\boldsymbol{\theta}}_R = [U^T U + K]^{-1} U^T \mathbf{y} \tag{5.3.31}$$

with K some symmetric matrix. Several forms have been suggested for K (Hoerl and Kennard, 1970; Goldstein and Smith, 1974), the simplest being kI with k a positive scalar. To see how a reduction in m.s.e. comes about, consider a scalar θ, for which (5.3.31) becomes

$$\begin{aligned} \hat{\theta}_R &= \sum_{t=1}^{N} u_t y_t \Big/ \left(\sum_{t=1}^{N} u_t^2 + k\right) \\ &= \left(\sum_{t=1}^{N} u_t(u_t\theta + e_t) + k\theta - k\theta\right) \Big/ \left(\sum_{t=1}^{N} u_t^2 + k\right) \\ &= \theta + \left(\sum_{t=1}^{N} u_t e_t - k\theta\right) \Big/ \left(\sum_{t=1}^{N} u_t^2 + k\right) \end{aligned} \tag{5.3.32}$$

Assuming $\{e\}$ is not autocorrelated or correlated with $\{u\}$, and has mean zero and variance σ^2, and writing $\sum_{t=1}^{N} u_t^2$ as $\sum$, the m.s.e. of $\hat{\theta}_R$ is

$$M = \underset{u,e}{E}\left[\left(\frac{\sum_{t=1}^{N} u_t e_t - k\theta}{\sum + k}\right)^2\right] = \underset{u}{E}\left[\frac{\sigma^2 \sum + k^2\theta^2}{(\sum + k)^2}\right] \tag{5.3.33}$$

A stationary value of M is achieved when

$$\frac{\partial M}{\partial k} = \underset{u}{E}\left[\frac{2k\theta^2}{(\sum + k)^2} - \frac{2(\sigma^2 \sum + k^2\theta^2)}{(\sum + k)^3}\right] = 0 \tag{5.3.34}$$

which requires k to be σ^2/θ^2. The stationary value is, in fact, a minimum since $\partial^2 M/\partial k^2$ is entirely composed of positive terms. We cannot choose the best k in advance even in the scalar case, not knowing σ or θ, but we could find an acceptable k by trial and error (Hoerl and Kennard, 1970), checking that the estimates are credible and the sum of squares of residuals is not unduly inflated by using k.

For a vector $\hat{\boldsymbol{\theta}}$, analysis of how k affects M is not easy, and Hoerl and Kennard avoid it by considering the mean sum of squares of estimate errors $E[(\hat{\boldsymbol{\theta}} - \boldsymbol{\theta})^{\mathrm{T}}(\hat{\boldsymbol{\theta}} - \boldsymbol{\theta})]$. This is less satisfactory, particularly if the elements of $\boldsymbol{\theta}$ are of differing orders of magnitude, as all the squared errors are weighted equally. Goldstein and Smith justify the choice kI through its effect in reducing ill-conditioning. The singular-value decomposition described in Section 4.2.4 is applied to U, so

$$U = PRQ^{\mathrm{T}} \tag{5.3.35}$$

where P and Q are orthogonal matrices and R is $N \times p$ and zero but for non-negative singular values r_{ii}, $1 \le i \le p$. As in Section 4.2.4, P, Q and R are used to transform the normal equations to

$$R^{\mathrm{T}}R\hat{\boldsymbol{\theta}}^* = R^{\mathrm{T}}\mathbf{y}^*, \qquad \text{i.e.} \quad r_{ii}^2\hat{\theta}_i^* = r_{ii}y_i^*, \quad i = 1, 2, \ldots, p \tag{5.3.36}$$

where

$$\hat{\boldsymbol{\theta}}^* \triangleq Q^{\mathrm{T}}\hat{\boldsymbol{\theta}}, \qquad \mathbf{y}^* \triangleq P^{\mathrm{T}}\mathbf{y} \tag{5.3.37}$$

Ill-conditioning appears as at least one r_{ii} being very small, making the sum of squares of residuals (model-output errors)

$$S = \sum_{i=1}^{p} \{(y_i^* - r_{ii}\hat{\theta}_i^*)^2\} + \sum_{i=p+1}^{N} \{y_i^{*2}\} \tag{5.3.38}$$

from the transformed regression equation

$$\mathbf{y}^* = P^{\mathrm{T}}(U\boldsymbol{\theta} + \mathbf{e}) = P^{\mathrm{T}}PRQ^{\mathrm{T}}\boldsymbol{\theta} + P^{\mathrm{T}}\mathbf{e} = R\boldsymbol{\theta}^* + \mathbf{e}^* \tag{5.3.39}$$

very insensitive to the related $\hat{\theta}_i^*$. Since S is the same as in the original problem, the $\hat{\theta}_i^*$ is a linear combination of elements of $\hat{\boldsymbol{\theta}}$ which has little influence on the model fit, and is consequently poorly estimated. The ill-conditioning can be alleviated by replacing every $1/r_{ii}$ by $r_{ii}/(r_{ii}^2+k)$ with k positive, effectively preventing any r_{ii} from being too small. The result is to modify each $\hat{\theta}_i^*$ from y_i^*/r_{ii} (which minimises S in (5.3.38)) to $r_{ii}y_i^*/(r_{ii}^2+k)$, say $\hat{\theta}_{Ri}^*$. The diagonal matrix R^TR in (5.3.36) is thereby modified to R^TR+kI, so solving (5.3.36),

$$\begin{aligned}\hat{\boldsymbol{\theta}}_R^* &= [R^TR+kI]^{-1}R^T\mathbf{y}^* \\ &= [Q^TU^TUQ+kQ^TQ]^{-1}R^TP^T\mathbf{y}\end{aligned} \tag{5.3.40}$$

where we have used the fact that P and Q are both orthogonal. Hence

$$\begin{aligned}\hat{\boldsymbol{\theta}}_R &= Q^{-T}\hat{\boldsymbol{\theta}}_R^* = Q\hat{\boldsymbol{\theta}}_R^* = [Q^TU^TU+kQ^T]^{-1}R^TP^T\mathbf{y} \\ &= [U^TU+kI]^{-1}QR^TP^T\mathbf{y} = [U^TU+kI]^{-1}U^T\mathbf{y}\end{aligned} \tag{5.3.41}$$

and ridge regression emerges as the result of preventing any singular values of U from being very small.

Ridge regression can be shown to be capable of reducing the m.s.e. of each element of $\hat{\boldsymbol{\theta}}$. The proof consists of writing the derivative of the m.s.e. at $k=0$, i.e. at the point where ridge regression departs from o.l.s., as a sum of negative terms.

5.3.8 Minimum-Mean-Square-Error Linear Estimator and Orthogonality

Having established that the minimum-mean-square-error (m.m.s.e.) estimator is not generally the minimum-covariance unbiased estimator, let us find out what it is. Initially we shall consider the scalar weighted m.s.e.

$$Q = E[(\hat{\boldsymbol{\theta}}-\boldsymbol{\theta})^TW(\hat{\boldsymbol{\theta}}-\boldsymbol{\theta})] \tag{5.3.42}$$

rather than the m.s.e. matrix M in (5.3.30). The difference is less significant than it might seem, as we are usually interested in m.s. weighted errors of the form

$$\begin{aligned}E[(\mathbf{c}^T(\hat{\boldsymbol{\theta}}-\boldsymbol{\theta}))^2] &= E[(\hat{\boldsymbol{\theta}}-\boldsymbol{\theta})^T\mathbf{c}\mathbf{c}^T(\hat{\boldsymbol{\theta}}-\boldsymbol{\theta})] \\ &= E[\mathbf{c}^T(\hat{\boldsymbol{\theta}}-\boldsymbol{\theta})(\hat{\boldsymbol{\theta}}-\boldsymbol{\theta})^T\mathbf{c}] = \mathbf{c}^TM\mathbf{c}\end{aligned} \tag{5.3.43}$$

We also restrict the estimator to the form

$$\hat{\boldsymbol{\theta}} = A\mathbf{y} \tag{5.3.44}$$

linear in the observations, like all the least-squares estimators, for computational and analytical simplicity.

Each element a_{ij} of A must give

$$\frac{\partial Q}{\partial a_{ij}} = E\left[\left(\frac{\partial}{\partial \hat{\boldsymbol{\theta}}}((\hat{\boldsymbol{\theta}} - \boldsymbol{\theta})^{\mathrm{T}} W(\hat{\boldsymbol{\theta}} - \boldsymbol{\theta}))^{\mathrm{T}} \frac{\partial \hat{\boldsymbol{\theta}}}{\partial a_{ij}}\right]$$

$$= E[(\text{element } i \text{ of } 2W(\hat{\boldsymbol{\theta}} - \boldsymbol{\theta}))y_j] = 0 \qquad (5.3.45)$$

so, writing out $\partial Q/\partial a_{ij}$ for all rows i and columns j of A,

$$\partial Q/\partial A = 2WE[(\hat{\boldsymbol{\theta}} - \boldsymbol{\theta})\mathbf{y}^{\mathrm{T}}] = 0 \qquad (5.3.46)$$

Whatever the value of W, $\partial Q/\partial A$ will be zero if A gives

$$E[(\hat{\boldsymbol{\theta}} - \boldsymbol{\theta})\mathbf{y}^{\mathrm{T}}] = 0 \qquad (5.3.47)$$

These *orthogonality conditions* say that, on average over all possible values of the observations, the error in each parameter should be unrelated to each observation. The orthogonality conditions (4.1.20) for the output estimate based on o.l.s. are rather similar; they imply that there is no relation between the output error and each regressor, on average over the samples in one record. The output estimate is linear in the regressors, just as $\hat{\boldsymbol{\theta}}$ is linear in the observations here.

We can find A for the m.m.s.e. estimator explicitly where the observations are generated by

$$\mathbf{y} = U\boldsymbol{\theta} + \mathbf{e}, \qquad E\mathbf{e} = \mathbf{0}, \qquad \operatorname{cov} \mathbf{e} = R \qquad (5.3.48)$$

with U and $\boldsymbol{\theta}$ deterministic. The weighted m.s.e. in $\hat{\boldsymbol{\theta}}$ is calculated over all realisations of $\mathbf{e}$ for a particular value of U and $\boldsymbol{\theta}$. Substituting (5.3.48) into (5.3.47),

$$E[((AU - I)\boldsymbol{\theta} + A\mathbf{e})(U\boldsymbol{\theta} + \mathbf{e})^{\mathrm{T}}] = (AU - I)\boldsymbol{\theta}\boldsymbol{\theta}^{\mathrm{T}} U^{\mathrm{T}} + AR = 0 \quad (5.3.49)$$

so

$$A = \boldsymbol{\theta}\boldsymbol{\theta}^{\mathrm{T}} U^{\mathrm{T}}[U\boldsymbol{\theta}\boldsymbol{\theta}^{\mathrm{T}} U^{\mathrm{T}} + R]^{-1} \qquad (5.3.50)$$

The inverse in this expression exists, since R is a positive-definite covariance and $U\boldsymbol{\theta}\boldsymbol{\theta}^{\mathrm{T}} U^{\mathrm{T}}$ is non-negative-definite. Although such an expression for A in terms of the unknown $\boldsymbol{\theta}$ is not directly usable, it does allow us to work out the bias and m.s. error of the theoretically optimal estimator. We can then compare this estimator with the g.l.s. minimum-covariance estimator.

First, we must simplify A. We can easily verify that

$$A = \frac{\boldsymbol{\theta}\boldsymbol{\theta}^{\mathrm{T}} U^{\mathrm{T}} R^{-1}}{1 + \boldsymbol{\theta}^{\mathrm{T}} U^{\mathrm{T}} R^{-1} U\boldsymbol{\theta}} = \frac{\boldsymbol{\theta}\boldsymbol{\theta}^{\mathrm{T}} U^{\mathrm{T}} R^{-1}}{1 + \alpha} \quad \text{say} \qquad (5.3.51)$$

by postmultiplying this expression and the original one (5.3.50) by $U\boldsymbol{\theta}\boldsymbol{\theta}^{\mathrm{T}} U^{\mathrm{T}} + R$. Hence the bias is

$$E[A\mathbf{y} - \boldsymbol{\theta}] = E[AU\boldsymbol{\theta} + A\mathbf{e} - \boldsymbol{\theta}] = AU\boldsymbol{\theta} - \boldsymbol{\theta} = \frac{\boldsymbol{\theta}\alpha}{1 + \alpha} - \boldsymbol{\theta} = -\frac{\boldsymbol{\theta}}{1 + \alpha} \qquad (5.3.52)$$

so the m.m.s.e. estimator is indeed biased. Its m.s.e. matrix is

$$\begin{aligned} M &= E[(\hat{\boldsymbol{\theta}} - \boldsymbol{\theta})(\hat{\boldsymbol{\theta}} - \boldsymbol{\theta})^{\mathrm{T}}] \\ &= E[((AU - I)\boldsymbol{\theta} + A\mathbf{e})(\mathbf{e}^{\mathrm{T}}A^{\mathrm{T}} + \boldsymbol{\theta}^{\mathrm{T}}(U^{\mathrm{T}}A^{\mathrm{T}} - I))] \\ &= (AU - I)\boldsymbol{\theta}\boldsymbol{\theta}^{\mathrm{T}}(U^{\mathrm{T}}A^{\mathrm{T}} - I) + ARA^{\mathrm{T}} \end{aligned} \tag{5.3.53}$$

and from (5.3.51),

$$ARA^{\mathrm{T}} = \boldsymbol{\theta}\alpha\boldsymbol{\theta}^{\mathrm{T}}/(1+\alpha)^2$$

so

$$M = \frac{\boldsymbol{\theta}\boldsymbol{\theta}^{\mathrm{T}}}{(1+\alpha)^2} + \frac{\alpha\boldsymbol{\theta}\boldsymbol{\theta}^{\mathrm{T}}}{(1+\alpha)^2} = \frac{\boldsymbol{\theta}\boldsymbol{\theta}^{\mathrm{T}}}{1+\alpha} \tag{5.3.54}$$

The g.l.s. estimate $\hat{\boldsymbol{\theta}}'$ given by (5.3.25) is unbiased, since

$$\begin{aligned} E[\boldsymbol{\theta}' - \boldsymbol{\theta}] &= E[[U^{\mathrm{T}}R^{-1}U]^{-1}U^{\mathrm{T}}R^{-1}(U\boldsymbol{\theta} + \mathbf{e}) - \boldsymbol{\theta}] \\ &= E[[U^{\mathrm{T}}R^{-1}U]^{-1}U^{\mathrm{T}}R^{-1}\mathbf{e}] = \mathbf{0} \end{aligned} \tag{5.3.55}$$

and its m.s.e. matrix is

$$M' = E[[U^{\mathrm{T}}R^{-1}U]^{-1}U^{\mathrm{T}}R^{-1}\mathbf{e}\mathbf{e}^{\mathrm{T}}R^{-1}U[U^{\mathrm{T}}R^{-1}U]^{-1}] = [U^{\mathrm{T}}R^{-1}U]^{-1} \tag{5.3.56}$$

We show that M' is larger than M, in the sense that $M' - M$ is positive-definite, by first noting that

$$\begin{aligned} U^{\mathrm{T}}R^{-1}U(M' - M)U^{\mathrm{T}}R^{-1}U &= U^{\mathrm{T}}\left(R^{-1} - \frac{R^{-1}U\boldsymbol{\theta}\boldsymbol{\theta}^{\mathrm{T}}U^{\mathrm{T}}R^{-1}}{1+\alpha}\right)U \\ &= U^{\mathrm{T}}[R + U\boldsymbol{\theta}\boldsymbol{\theta}^{\mathrm{T}}U^{\mathrm{T}}]^{-1} \quad U > 0 \end{aligned} \tag{5.3.57}$$

The last step can be verified by multiplying the inner matrices by $R + U\boldsymbol{\theta}\boldsymbol{\theta}^{\mathrm{T}}U^{\mathrm{T}}$, and the inequality follows from $[R + U\boldsymbol{\theta}\boldsymbol{\theta}^{\mathrm{T}}U^{\mathrm{T}}]^{-1}$ being the inverse of a positive-definite matrix, as noted earlier, and therefore itself positive-definite. Now since $U^{\mathrm{T}}R^{-1}U$ is invertible, any quadratic form $\boldsymbol{\xi}^{\mathrm{T}}(M' - M)\boldsymbol{\xi}$ can be rewritten $\boldsymbol{\zeta}^{\mathrm{T}}U^{\mathrm{T}}R^{-1}U(M' - M)U^{\mathrm{T}}R^{-1}U\boldsymbol{\zeta}$ where $\boldsymbol{\zeta}$ is $[U^{\mathrm{T}}R^{-1}U]^{-1}\boldsymbol{\xi}$, so (5.3.57) shows that $M' - M$ is positive-definite. The practical conclusion is that the g.l.s. estimate has a larger mean-square weighted error

$$E[(\mathbf{c}^{\mathrm{T}}(\hat{\boldsymbol{\theta}}' - \boldsymbol{\theta}))^2] = \mathbf{c}^{\mathrm{T}}M'\mathbf{c} \tag{5.3.58}$$

than the theoretical m.m.s.e. estimator, as the penalty for being unbiased.

5.4 EFFICIENCY

5.4.1 Cramér–Rao Bound

Besides establishing that an estimator converges, we may wish to measure its performance against some standard. A standard for estimation covariance is provided by the *Cramér–Rao bound.* The bound applies to any unbiased estimator $\hat{\boldsymbol{\theta}}(y)$ of a parameter vector $\boldsymbol{\theta}$ using measurements y. For instance, y might comprise all the elements of y and U in the usual regression model, but the bound is not restricted to any particular model or any particular estimator form. Some at least of the measurements are random variables (noise is present) so they are described by their joint probability density function $p(y|\boldsymbol{\theta})$, which is influenced by $\boldsymbol{\theta}$. Subject to some conditions on $p(y|\boldsymbol{\theta})$, discussed below, the covariance of $\hat{\boldsymbol{\theta}}(y)$ cannot be less than the Cramér–Rao bound F^{-1} where

$$F \triangleq \underset{y|\boldsymbol{\theta}}{E}\left\{\frac{\partial}{\partial\boldsymbol{\theta}}\ln p(y\,|\,\boldsymbol{\theta})\left(\frac{\partial}{\partial\boldsymbol{\theta}}\ln p(y\,|\,\boldsymbol{\theta})\right)^{\mathrm{T}}\right\} \tag{5.4.1}$$

Matrix F is called the *Fisher information matrix.* Without attempting a detailed interpretation of F, we can accept the name as associating lowest potential covariance F^{-1} of an estimate with most information about it. It also seems reasonable that the information about $\boldsymbol{\theta}$ is conveyed by its influence on the measurements through $p(y\,|\,\boldsymbol{\theta})$.

The *Cramér–Rao inequality*

$$\operatorname{cov}\hat{\boldsymbol{\theta}}(y) \geq F^{-1} \tag{5.4.2}$$

is proved by considering the covariance of the augmented vector

$$\boldsymbol{\phi} \triangleq \begin{bmatrix} \hat{\boldsymbol{\theta}}(y) \\ \dfrac{\partial}{\partial\boldsymbol{\theta}}\ln p(y\,|\,\boldsymbol{\theta}) \end{bmatrix} \tag{5.4.3}$$

First we find $E\{(\partial/\partial\boldsymbol{\theta})\ln p(y\,|\,\boldsymbol{\theta})\}$:

$$E\left\{\frac{\partial}{\partial\boldsymbol{\theta}}\ln p(y\,|\,\boldsymbol{\theta})\right\} = \int \frac{\partial}{\partial\boldsymbol{\theta}}\ln p(y\,|\,\boldsymbol{\theta})\cdot p(y\,|\,\boldsymbol{\theta})\,dy$$

$$= \int \frac{\partial p(y\,|\,\boldsymbol{\theta})}{\partial\boldsymbol{\theta}}\,dy = \frac{\partial}{\partial\boldsymbol{\theta}}\int p(y\,|\,\boldsymbol{\theta})\,dy = \frac{\partial}{\partial\boldsymbol{\theta}}(1) = \mathbf{0} \tag{5.4.4}$$

assuming that $p(y\,|\,\boldsymbol{\theta})$ is well enough behaved to allow reversal of the order of differentiation and integration. To be more specific, the regularity conditions on $p(y\,|\,\boldsymbol{\theta})$ are that the range of integration (over which $p(y\,|\,\boldsymbol{\theta})$ is non-zero) must not depend on $\boldsymbol{\theta}$, and the integral must converge in spite of the differentiation in its integrand. Next let us examine $E\{\hat{\boldsymbol{\theta}}(y)(\partial/\partial\boldsymbol{\theta}) \ln p(y \mid \boldsymbol{\theta})\}$ in the same fashion:

$$E\left\{\hat{\boldsymbol{\theta}}(y)\frac{\partial}{\partial\boldsymbol{\theta}} \ln p(y \mid \boldsymbol{\theta})\right\} = \int \hat{\boldsymbol{\theta}}(y)\frac{\partial p(y \mid \boldsymbol{\theta})}{\partial\boldsymbol{\theta}}\, dy = \frac{\partial}{\partial\boldsymbol{\theta}}\int \hat{\boldsymbol{\theta}}(y) p(y \mid \boldsymbol{\theta})\, dy$$

$$= \frac{\partial}{\partial\boldsymbol{\theta}} \hat{E}\{\hat{\boldsymbol{\theta}}(y)\} = \frac{\partial}{\partial\boldsymbol{\theta}}(\boldsymbol{\theta}) = I \qquad (5.4.5)$$

since $\hat{\boldsymbol{\theta}}(y)$ is by assumption unbiased. Hence

$$\operatorname{cov}\boldsymbol{\phi} = \begin{bmatrix} \operatorname{cov}\hat{\boldsymbol{\theta}}(y) & I \\ I & E\left\{\frac{\partial}{\partial\boldsymbol{\theta}} \ln p(y \mid \boldsymbol{\theta})\left(\frac{\partial}{\partial\boldsymbol{\theta}} \ln p(y \mid \boldsymbol{\theta})\right)^{\mathrm{T}}\right\} \end{bmatrix}$$

$$\equiv \begin{bmatrix} \operatorname{cov}\hat{\boldsymbol{\theta}}(y) & I \\ I & F \end{bmatrix} \qquad (5.4.6)$$

Like any other covariance matrix, $\operatorname{cov}\boldsymbol{\phi}$ is positive-semi-definite, so in particular

$$[I \quad -F^{-\mathrm{T}}]\begin{bmatrix} \operatorname{cov}\hat{\boldsymbol{\theta}}(y) & I \\ I & F \end{bmatrix}\begin{bmatrix} I \\ -F^{-1} \end{bmatrix} = \operatorname{cov}\hat{\boldsymbol{\theta}}(y) - F^{-1} \geq 0 \qquad (5.4.7)$$

proving (5.4.2).

Example 5.4.1 We want to estimate parameter α of the probability density $p(x \mid \alpha) = (1/\alpha)\exp(-x/\alpha)$, $x \geq 0$ from N independent samples of x. We shall find the Cramér–Rao bound on the variance of any unbiased estimator of α, and then check whether the unbiased estimator $\hat{\alpha} = \sum_{t=1}^{N} x_t/N$ attains it.

Integrating $xp(x \mid \alpha)$ and $x^2 p(x \mid \alpha)$ by parts, we find the mean $\bar{x}$ to be α and the m.s. value $\overline{x^2}$ to be $2\alpha^2$. The samples of x have a joint probability density

$$p(y \mid \boldsymbol{\theta}) \equiv p(x_1, x_2, \ldots, x_N \mid \alpha) = \prod_{t=1}^{N} p(x_t \mid \alpha) = \alpha^{-N} \exp\left(\frac{-\sum_{t=1}^{N} x_t}{\alpha}\right)$$

for x_1 to x_n non-negative, so

$$\ln p(y \mid \boldsymbol{\theta}) = -N \ln \alpha - \frac{\sum_{t=1}^{N} x_t}{\alpha}$$

and therefore

$$\frac{\partial}{\partial\boldsymbol{\theta}} \ln p(y \mid \boldsymbol{\theta}) = -\frac{N}{\alpha} + \frac{\sum_{t=1}^{N} x_t}{\alpha^2}$$

From this we find, given the independence of the samples, that

$$F = E\left\{\left(\frac{\partial}{\partial\alpha}\ln p(y\,|\,\alpha)\right)^2\right\}$$

$$= \frac{N^2}{\alpha^2} - \frac{2N}{\alpha^3}E\left\{\sum_{t=1}^{N} x_t\right\} + \frac{1}{\alpha^4}E\left\{\left(\sum_{t=1}^{N} x_t\right)^2\right\}$$

$$= \frac{N^2}{\alpha^2} - \frac{2N}{\alpha^3}N\alpha + \frac{1}{\alpha^4}(NE\{x^2\} + N(N-1)(Ex)^2) = \frac{N}{\alpha^2}$$

so the Cramér–Rao bound on the variance of $\hat{\alpha}$ is α^2/N.

The mean of $\hat{\alpha}$ is α and the variance of $\hat{\alpha}$ is

$$\operatorname{var}\hat{\alpha} = E\{(\hat{\alpha} - \alpha)^2\} = E\left\{\frac{(\sum_{t=1}^{N} x_t)^2}{N^2} - \frac{2\alpha\sum_{t=1}^{N} x_t}{N} + \alpha^2\right\} = \frac{\alpha^2}{N}$$

so $\hat{\alpha}$ does attain the Cramér–Rao bound; we couldn't do better with any other unbiased estimator, however ingenious. △

We next examine an equally simple example in which things are not so straightforward.

Example 5.4.2 The gain g of a plant modelled by

$$y_t = gu_t + e_t$$

is to be estimated from N independent pairs of measurements (u_t, y_t). The model error e_t is believed to be uniformly distributed over $[-r, r]$. What is the Cramér–Rao bound for $\hat{g}$? Here

$$p(y\,|\,\boldsymbol{\theta}) = \prod_{t=1}^{N}\left\{\frac{1}{2r} \text{ for } |y_t - gu_t| \le r \text{ only}\right\}$$

$$= \begin{cases} \dfrac{1}{(2r)^N} & \text{if } \{y\} \text{ is such that } gu_t - r \le y_t \le gu_t + r \\ & \text{for } t = 1, 2, \ldots, N \\ 0 & \text{otherwise} \end{cases}$$

The range over which $p(y\,|\,\boldsymbol{\theta})$ is non-zero clearly depends on $\boldsymbol{\theta}$ (g here), so the regularity conditions are not all satisfied, and the Cramér–Rao bound is inapplicable. A moment's thought reveals that in principle the variance of $\hat{g}$ can be made as small as you please by using large enough absolute values for the input samples $\{u\}$. △

The Fisher information matrix relates easily to the second-derivative matrix of $\ln p(y\,|\,\boldsymbol{\theta})$ with respect to $\boldsymbol{\theta}$, for

$$\frac{\partial^2}{\partial\theta_i\,\partial\theta_j}\ln p(y\,|\,\boldsymbol{\theta}) = \frac{\partial}{\partial\theta_i}\left(\frac{1}{p(y\,|\,\boldsymbol{\theta})}\frac{\partial p(y\,|\,\boldsymbol{\theta})}{\partial\theta_j}\right)$$
$$= \frac{1}{p(y\,|\,\boldsymbol{\theta})}\frac{\partial^2 p(y\,|\,\boldsymbol{\theta})}{\partial\theta_i\,\partial\theta_j} - \frac{1}{p^2(y\,|\,\boldsymbol{\theta})}\frac{\partial p(y\,|\,\boldsymbol{\theta})}{\partial\theta_i}\frac{\partial p(y\,|\,\boldsymbol{\theta})}{\partial\theta_j} \tag{5.4.8}$$

and if once more we can reverse the order of integration and differentiation,

$$E\left\{\frac{1}{p(y\,|\,\boldsymbol{\theta})}\frac{\partial^2 p(y\,|\,\boldsymbol{\theta})}{\partial\theta_i\,\partial\theta_j}\right\} = \int\frac{\partial^2 p(y\,|\,\boldsymbol{\theta})}{\partial\theta_i\,\partial\theta_j}\,dy = \frac{\partial^2}{\partial\theta_i\,\partial\theta_j}\int p(y\,|\,\boldsymbol{\theta})\,dy$$
$$= \frac{\partial^2}{\partial\theta_i\,\partial\theta_j}(1) = 0 \tag{5.4.9}$$

giving

$$E\left\{\frac{\partial^2 \ln p(y\,|\,\boldsymbol{\theta})}{\partial\theta_i\,\partial\theta_j}\right\} = -E\left\{\frac{1}{p^2(y\,|\,\boldsymbol{\theta})}\frac{\partial p(y\,|\,\boldsymbol{\theta})}{\partial\theta_i}\frac{\partial p(y\,|\,\boldsymbol{\theta})}{\partial\theta_j}\right\}$$
$$= -E\left\{\frac{\partial \ln p(y\,|\,\boldsymbol{\theta})}{\partial\theta_i}\frac{\partial \ln p(y\,|\,\boldsymbol{\theta})}{\partial\theta_j}\right\} \tag{5.4.10}$$

which is minus element (i,j) of the Fisher information matrix, so altogether

$$F = -E\left\{\frac{\partial^2}{\partial\boldsymbol{\theta}^2}\ln p(y\,|\,\boldsymbol{\theta})\right\} \tag{5.4.11}$$

We shall re-encounter $\ln p(y|\boldsymbol{\theta})$ when we cover maximum-likelihood estimation in Chapter 6.

5.4.2 Efficiency

An unbiased estimate is said to be efficient if its covariance equals the Cramér–Rao bound. We can define the efficiency of a scalar estimate as the Cramér–Rao bound divided by the estimation variance. The main practical significance of efficiency is in determining whether further efforts to devise a lower-covariance estimate would be futile because the present estimate is efficient or nearly efficient. Even so, efficiency is not always critical, as a far-from-efficient estimate may be the best practicable and, more to the point, may be acceptably accurate.

Investigation of efficiency may require some idealising assumption about the form of $p(y\,|\,\boldsymbol{\theta})$, as in the following example.

Example 5.4.3 We shall test the efficiency of the Markov estimate introduced in Section 5.3.5. With the usual regression model, and considering U not to be random,

$$p(y\,|\,\boldsymbol{\theta}) = \text{prob}(\mathbf{y} = \text{observed value}\,|\,\boldsymbol{\theta})$$
$$= \text{prob}(\mathbf{e} = \mathbf{y} - U\boldsymbol{\theta}\,|\,\boldsymbol{\theta}) \qquad \text{with observed } \mathbf{y} \text{ and } U \text{ inserted}$$

To get any further we have to assume some form for the p.d.f. of $\mathbf{e}$. A popular assumption which makes the algebra easy and may bear some resemblance to the truth is that $\mathbf{e}$ is a zero-mean, Gaussian random variable with p.d.f.

$$p(\mathbf{e}\,|\,\boldsymbol{\theta}) = \exp(-\tfrac{1}{2}\mathbf{e}^{\mathrm{T}}R^{-1}\mathbf{e})/((2\pi)^{N/2}|R|)$$

where R is cov $\mathbf{e}$ as usual. Putting $\mathbf{y} - U\boldsymbol{\theta}$ for $\mathbf{e}$ and then using (5.4.11), we obtain the Fisher information matrix

$$\ln p(y\,|\,\boldsymbol{\theta}) = -\tfrac{1}{2}(\mathbf{y} - U\boldsymbol{\theta})^{\mathrm{T}}R^{-1}(\mathbf{y} - U\boldsymbol{\theta}) + \text{const independent of } \boldsymbol{\theta}$$

$$\frac{\partial}{\partial\boldsymbol{\theta}}\ln p(y\,|\,\boldsymbol{\theta}) = U^{\mathrm{T}}R^{-1}(\mathbf{y} - U\boldsymbol{\theta}), \qquad \frac{\partial^2}{\partial\boldsymbol{\theta}^2}\ln p(y\,|\,\boldsymbol{\theta}) = -U^{\mathrm{T}}R^{-1}U$$

so, with U non-random,

$$F = U^{\mathrm{T}}R^{-1}U, \qquad F^{-1} = [U^{\mathrm{T}}R^{-1}U]^{-1}$$

The Markov estimate $\hat{\boldsymbol{\theta}}'$ given by (5.3.25):

$$\hat{\boldsymbol{\theta}}' = [U^{\mathrm{T}}R^{-1}U]^{-1}U^{\mathrm{T}}R^{-1}\mathbf{y}$$

has covariance $[U^{\mathrm{T}}R^{-1}U]^{-1}$ as we saw in Section (5.3.5), so $\hat{\boldsymbol{\theta}}'$ has covariance equal to the Cramér–Rao bound F^{-1} and is efficient. △

FURTHER READING

Wadsworth and Bryan (1974) and Helstrom (1984) among many other books give the basic material on probability in detail and are generous with examples and problems. Whittle (1970) is more advanced but very concise and readable, and discusses the convergence of random sequences. This topic is also introduced by Papoulis (1965), who provides a wide background in stochastic processes and looks at least-squares estimation in a stochastic setting.

Silvey (1975) covers minimum-covariance unbiased estimation and the main topics of Chapter 6, and is beautifully concise. A good selection of the estimation theory we need is summarised by Goodwin and Payne (1977), and the appendices of that book contain several standard results we shall find useful later.

REFERENCES

Fenner, R. T. (1974). "Computing for Engineers". Macmillan, London.

Goldstein, M., and Smith, A. F. M. (1974). Ridge-type estimators for regression analysis. *J. Roy. Stat. Soc. B* **36**, 284–291.

Goodwin, G. C., and Payne, R. A. (1977). "Dynamic System Identification Experiment Design and Data Analysis". Academic Press, New York and London.

Helstrom, C. W. (1984). "Probability and Stochastic Processes for Engineers". Macmillan, New York.

Hoerl, A. E., and Kennard, R. W. (1970). Ridge regression: biased estimation for nonorthogonal problems. *Technometrics* **12**, 55–68.

Johnston, J. (1972). "Econometric Methods", 2nd ed. McGraw-Hill, New York.

Kendall, M. G., and Stuart, A. (1979). "The Advanced Theory of Statistics", 4th ed., Vol. 2. Griffin, London.

Melsa, J. L., and Sage, A. P. (1973). "An Introduction to Probability and Stochastic Processes". Prentice-Hall, Englewood Cliffs, New Jersey.

Papoulis, A. (1965). "Probability, Random Variables, and Stochastic Processes". McGraw-Hill Kogakusha, New York and Tokyo.

Silvey, S. D. (1975). "Statistical Inference". Chapman & Hall, London.

Wadsworth, G. P., and Bryan, J. G. (1974). "Applications of Probability and Random Variables", 2nd ed. McGraw-Hill, New York.

Whittle, P. (1970). "Probability". Penguin, Harmondsworth, England.

Wilks, S. (1962). "Mathematical Statistics". Wiley, New York.

PROBLEMS

5.1 Show that, in the notation of Example 5.1.2, $\hat{\bar{v}}^2$ is a biased estimator of $\bar{v}^2$, with bias s/N. If two independent batches of N samples of v have sample means $\hat{\bar{v}}^{(1)}$ and $\hat{\bar{v}}^{(2)}$, is $\hat{\bar{v}}^{(1)}\hat{\bar{v}}^{(2)}$ an unbiased estimator of $\bar{v}^2$? Show that $\sum_{t=1}^{N} \{v_t(v_t - \hat{\bar{v}}^{(2)})\}/N$ is an unbiased estimator of s, where $\hat{\bar{v}}^{(2)}$ is based on N samples independent of v_1 to v_N. Does it matter whether v_1 to v_N are mutually independent in this estimator?

5.2 Find the variance of $\hat{\bar{v}}^{(1)}\hat{\bar{v}}^{(2)}$ in Problem 5.1.

5.3 If, in the "errors in variables" situation described in Section 5.2.3, $\boldsymbol{\theta}$ is a vector but only one regressor is affected by noise, does that noise cause bias in all of $\boldsymbol{\theta}$, only the element multiplying the affected regressor, or none of $\boldsymbol{\theta}$? Would your answer change if $\boldsymbol{\theta}$ were estimated by w.l.s., with any positive-definite weighting matrix?

5.4 In Problem 4.5, $I - U[U^{\mathrm{T}}U]^{-1}U^{\mathrm{T}}$ was found to be idempotent. Verify that, because it is idempotent and symmetric, it is positive-semi-definite. With A a matrix such that $AU = I$, write $AA^{\mathrm{T}} - [U^{\mathrm{T}}U]^{-1}$ as a symmetric expression in A and $I - U[U^{\mathrm{T}}U]^{-1}U^{\mathrm{T}}$ and hence show that it is positive-semi-definite. [Section 5.3.4 brings out the relevance of this problem to the minimum-covariance property of o.l.s.]

5.5 Assuming that the input sequence $\{u\}$ in Example 5.3.4 is uncorrelated, show (by expressing $r_{yy}(1)$ in terms of $r_{yy}(0)$ and $r_{vv}(0)$ and then $r_{yy}(0)$ in terms of $r_{uu}(0)$ and $r_{vv}(0)$) that $r_{yy}(1)$ is not generally zero. Hence find the probability limit of $N \operatorname{cov} \hat{\theta}_Z$ in part (ii) of Example 5.3.4 in terms of $r_{uu}(0)$, $r_{vv}(0)$, a and b.

5.6 Verify that in part (iii) of Example 5.3.5, $\bar{R}_{ZU}$ and $\bar{R}_{ZZ}$ are as claimed there.

5.7 Is u_{t-2} a suitable instrumental variable to replace y_{t-1} in the model in Example 5.3.4? Specifically, is it uncorrelated with e_t but strongly correlated with U?

5.8 Is $y_{t-1} - u_{t-1}$ an acceptable instrumental variable to replace y_{t-1} in Example 5.3.4?

Chapter 6

Optimal Estimation, Bayes and Maximum-Likelihood Estimators

6.1 INTRODUCTION

A large number of identification methods will be described in Chapter 7, yet they represent only a fraction of the methods available. We must somehow classify and compare the throng of competing methods, whether we want a technique for one application or a broad perspective on the whole field. A framework which will accommodate many identification methods is set up in this chapter; Chapters 7–10 go on to see how the methods are implemented.

It is convenient to categorise methods initially according to what measure of estimation goodness they try to optimise. Model structure and computational tactics can be considered later. Our basis for categorising will be Bayes estimation, which is easy to appreciate, almost all-embracing and appealing to common sense. Other frameworks are possible, and one in particular, the prediction-error formulation (discussed briefly in Chapter 7) is a very useful basis for analysis of asymptotic properties of parameter estimators. Least-squares methods for models linear in their parameters will fit into either framework, and will be the main object of our attention in later chapters.

We shall be asking of each method "Is it simple and computationally cheap?" and "Are its assumptions realistic?". The answers will often be "no", and will lead us to simplify the methods and be wary of their results.

6.2 BAYESIAN APPROACH TO OPTIMAL ESTIMATION

The least-squares estimators we have concentrated on were motivated by the simple idea of fitting model output to observed output as closely as possible. They proved to have attractive statistical properties under suitable assumptions, including optimality in the sense of minimum covariance among all linear, unbiased estimators. The following questions about them remain unanswered, however:

(1) Can we improve on them by using some non-linear estimator?

(2) What estimator is best if the criterion is something other than minimum output m.s. error or parameter-estimate covariance?

(3) How can we bring in prior information on the most likely values for the parameters?

(4) Regarding the parameters as random variables, can we estimate their joint p.d.f. rather than just their means?

The aim of Sections 6.2 and 6.3 is to answer these questions. We start by setting in a broader context the definition of the best estimator.

6.2.1 Optimality: Loss Functions and Risk

Think for the moment of a scalar parameter θ. We could express how seriously we take estimation errors of different sizes by nominating a scalar *loss function* $L(\hat{\theta}, \theta)$, larger for a worse error. Some possible loss functions are $(\hat{\theta} - \theta)^2$, which we have already met; $|\hat{\theta} - \theta|$, which gives less weight than $(\hat{\theta} - \theta)^2$ to large errors; $((\hat{\theta} - \theta)/\theta)^2$, which implies that proportional rather than absolute error is important; $\max_\theta|\hat{\theta} - \theta|$, a pessimist's choice, which weighs only the worst error; 0 for $|\hat{\theta} - \theta| \leq \alpha$ and 1 for $|\hat{\theta} - \theta| > \alpha$, which indicates indifference to errors up to α and equal dislike of all larger errors, i.e. classifies each error as "serious" or "not serious"; or $(\hat{\theta} - \theta)^2$ for $|\hat{\theta} - \theta| \leq \alpha$ and $2\alpha|\hat{\theta} - \theta| - \alpha^2$ for $|\hat{\theta} - \theta| > \alpha$, a compromise between $(\hat{\theta} - \theta)^2$ and $|\hat{\theta} - \theta|$. Evidently choosing a loss function is a subjective matter, and depends on how bad the consequences of an error of any given size are perceived to be. Sometimes the ultimate application of the model makes the choice easy, but more often not.

Example 6.2.1 A river-flow predictor based on an estimated model of the catchment dynamics and rainfall measurements is required to give warning of any flow f likely to overtop a coffer dam protecting civil engineering works. If the overtopping flow is f_0 and the predicted flow $\hat{f}$, a prediction error is fairly important when $\hat{f} > f_0$ and $f < f_0$, as it will precipitate an unnecessary and expensive halt and evacuation, extremely important when $f > f_0$ and $\hat{f} < f_0$ as everything and everybody will get wet, and unimportant in all other cases. The simplest loss function reflecting this situation is 0 when $(f - f_0)(\hat{f} - f_0) > 0$, α when $f < f_0$ and $\hat{f} > f_0$, and β when $f > f_0$ and $\hat{f} < f_0$, with $\beta > \alpha > 0$. Refinements are possible, and the practicability of designing an estimator with this loss function is an open question. △

Once a loss function is chosen we can begin to design an estimator to minimise the scalar *risk*

$$r(\boldsymbol{\theta}) = \underset{Y}{E} L(\hat{\boldsymbol{\theta}}, \boldsymbol{\theta}) \tag{6.2.1}$$

defined as the average loss over all possible realisations Y of the measured output and explanatory-variable values ($\mathbf{y}$ and U together, in regression). The minimum-covariance estimator, for instance, minimises the risk with $(\mathbf{c}^{\mathrm{T}}(\hat{\boldsymbol{\theta}} - \boldsymbol{\theta}))^2$ as $L(\hat{\boldsymbol{\theta}}, \boldsymbol{\theta})$, whatever the real, non-zero value of $\mathbf{c}$. At this point, we can open up an entirely new possibility by extending the aim of the estimator to minimising the *average risk* $\bar{r}$ over all possible values of $\boldsymbol{\theta}$:

$$\bar{r} = \underset{\boldsymbol{\theta}}{E} r(\boldsymbol{\theta}) \tag{6.2.2}$$

The crucial importance of this extension is that it makes use of a *prior p.d.f.* $p(\boldsymbol{\theta})$ embodying all the available background knowledge about the likely parameter values. The need to provide a prior p.d.f. for the parameters characterises *Bayes estimation* and, we shall see presently, is responsible for both its power and its practical weakness.

Example 6.2.2 In compartmental models of drug metabolism in the body, each rate constant for transfer of a drug between compartments is non-negative by definition. An experienced investigator may be able to quote maximum credible values for each. In the absence of any further information, each rate constant k_{ij} can be assigned a uniform prior probability density $1/a_{ij}$ over the range zero to its maximum credible value a_{ij}. The rate constants do in fact vary from subject to subject, so it is reasonable to treat them as random variables. △

6.2.2 Posterior Probability Density of Parameters; Bayes' Rule

The easiest way to find the *minimum-risk estimator* which minimises $\bar{r}$ is to think of Y as fixed and determine the estimator $\hat{\boldsymbol{\theta}}(Y)$ which minimises the average loss over all possible $\boldsymbol{\theta}$ for that Y. If the estimator minimises $\underset{\boldsymbol{\theta}}{E} L(\hat{\boldsymbol{\theta}}(Y), \boldsymbol{\theta})$ for every realisation Y, it minimises the average risk over all $\boldsymbol{\theta}$ and Y. With Y fixed, averaging over $\boldsymbol{\theta}$ requires use of the *posterior p.d.f.* $p(\boldsymbol{\theta} \mid Y)$ of $\boldsymbol{\theta}$ given Y:

$$\underset{\boldsymbol{\theta}|Y}{E} L(\hat{\boldsymbol{\theta}}(Y), \boldsymbol{\theta}) = \int \cdots \int_{-\infty}^{\infty} L(\hat{\boldsymbol{\theta}}(Y), \boldsymbol{\theta}) p(\boldsymbol{\theta} \mid Y)\, d\boldsymbol{\theta} \tag{6.2.3}$$

Subsequent averaging over Y would produce $\bar{r}$:

$$\begin{aligned}
\underset{Y}{E}\left\{\underset{\theta|Y}{E} L(\hat{\theta}(Y),\theta)\right\} &= \int_{-\infty}^{\infty}\cdots\int_{-\infty}^{\infty}\left\{\int_{-\infty}^{\infty}\cdots\int_{-\infty}^{\infty} L(\hat{\theta}(Y),\theta)p(\theta\,|\,Y)\,d\theta\right\}p(Y)\,dY \\
&= \int_{-\infty}^{\infty}\cdots\int_{-\infty}^{\infty}\int_{-\infty}^{\infty}\cdots\int_{-\infty}^{\infty} L(\hat{\theta}(Y),\theta)p(\theta, Y)\,d\theta\,dY \\
&= \int_{-\infty}^{\infty}\cdots\int_{-\infty}^{\infty}\int_{-\infty}^{\infty}\cdots\int_{-\infty}^{\infty} L(\hat{\theta}(Y),\theta)p(Y\,|\,\theta)p(\theta)\,dY\,d\theta \\
&= \int_{-\infty}^{\infty}\cdots\int_{-\infty}^{\infty} \underset{Y|\theta}{E} L(\hat{\theta}(Y),\theta)p(\theta)\,d\theta \\
&= \underset{\theta}{E}\left\{\underset{Y|\theta}{E} L(\hat{\theta}(Y),\theta)\right\} = \bar{r} \qquad (6.2.4)
\end{aligned}$$

Buried in (6.2.4) is the relation between the posterior p.d.f. $p(\theta\,|\,Y)$ and the prior p.d.f. $p(\theta)$ given by

$$\boxed{\textbf{Bayes' Rule} \quad p(\theta\,|\,Y) = p(Y\,|\,\theta)p(\theta)/p(Y)} \qquad (6.2.5)$$

Before seeing in detail how Bayes' rule is employed in minimum-risk estimation, let us pause to weigh up the idea of finding $p(\theta\,|\,Y)$.

The most we could ask of any estimation method is that it should find the entire p.d.f. of the parameters, given the measurements. The p.d.f. says much more about the parameters than a point estimate $\hat{\theta}$ and its covariance could. Figure 6.2.1 exemplifies $p(\theta\,|\,Y)$ for a scalar parameter. It indicates that in this instance too little is yet known to locate $\hat{\theta}$ confidently, values over a considerable range being estimated as about equally likely. We should want to refine $p(\theta\,|\,Y)$ by adding more measurements to Y. Nonetheless, it is already clear that θ is unlikely to be negative. Notice the danger of relying on a point

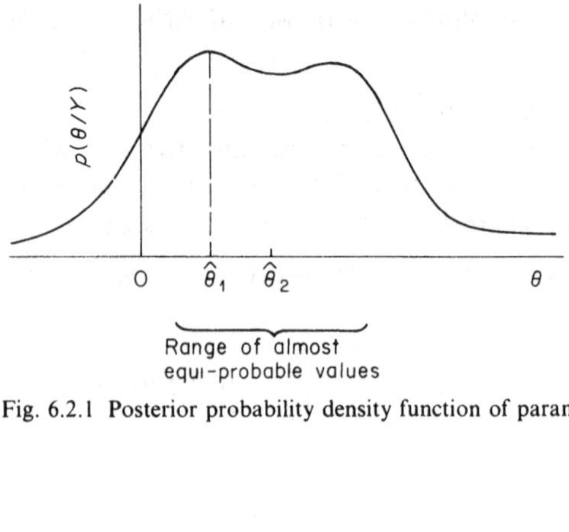

Fig. 6.2.1 Posterior probability density function of parameter.

estimate. The "most likely" value of θ might reasonably be taken as $\hat{\theta}_1$, at the global maximum of $p(\theta \mid Y)$, but a good case could be argued for the notably different centroid $\hat{\theta}_2$ and for other measures of the middle of $p(\theta \mid Y)$. An associated variance estimate, even if accurate, would fail to warn adequately of the uncertainty in $\hat{\theta}$, as we might be quite happy with that variance if the p.d.f. were unimodal.

Granted that $p(\boldsymbol{\theta} \mid Y)$ is highly desirable in itself and will also take us along the road to a minimum-risk estimate, how is it computed?

6.2.3 Bayes Estimation: Details

Once the measurements Y have been taken, $p(Y)$ in Bayes' rule (6.2.5) is just a number, and serves only to scale $p(\boldsymbol{\theta} \mid Y)$ so that its integral over $\boldsymbol{\theta}$ is unity. According to what estimator is used, $p(Y)$ may or may not have to be computed. When it must, it is found by integrating $p(Y|\boldsymbol{\theta})p(\boldsymbol{\theta})$, i.e. $p(\boldsymbol{\theta}, Y)$, over all $\boldsymbol{\theta}$. The prior density $p(\boldsymbol{\theta})$ is provided by the user from background knowledge or guesswork, and $p(Y|\boldsymbol{\theta})$ comes from the model relating $\boldsymbol{\theta}$ to Y, together with the p.d.f. of each random variable influencing Y.

Example 6.2.3 A plant is modelled by

$$y_t = g u_{t-1} + e_t$$

and its gain g is to be estimated from N independent measurements y_1 to y_N. Each noise sample e_t is believed to be a Gaussian random variable with mean zero and variance σ^2. The input sequence u_0 to u_{N-1} is known exactly, and so can be treated as deterministic, leaving only y_1 to y_N as Y. Thus with g as $\boldsymbol{\theta}$, $p(Y \mid \boldsymbol{\theta})$ is

$$p_y(y_1, \ldots, y_N \mid g) = \prod_{t=1}^{N} p_y(y_t \mid g) = \prod_{t=1}^{N} p_e(y_t - g u_{t-1})$$

$$= (2\pi\sigma^2)^{-N/2} \exp\left(- \prod_{t=1}^{N} (y_t - g u_{t-1})^2/(2\sigma^2) \right)$$

where each p.d.f. refers to the random variable indicated by its subscript.

Before the measurements are made, g is known on physical grounds to lie between 1 and 5, but that is all, so we assign to g the uniform p.d.f.

$$p(g) = \tfrac{1}{4}, \qquad 1 \leq g \leq 5$$

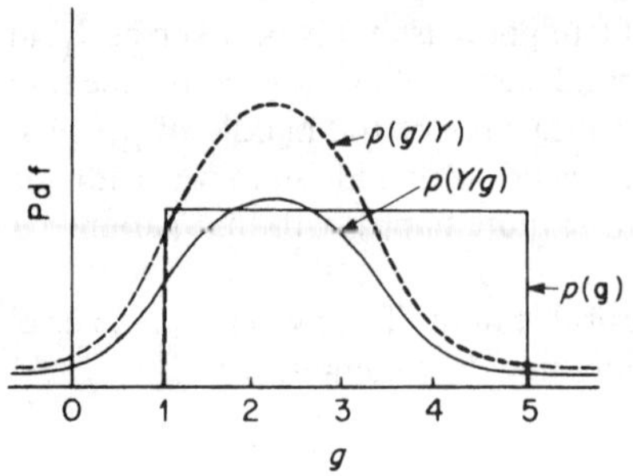

Fig. 6.2.2 Probability density functions for Example 6.2.3.

Inserting the known values of u and measurements of y, we can compute

$$p(g \mid Y) = p(Y \mid g)p(g)/p(Y)$$

$$= \frac{1}{4}(2\pi\sigma^2)^{-N/2} \exp\left(-\prod_{t=1}^{N}(y_t - gu_{t-1})^2/(2\sigma^2)\right)/p(Y), \qquad 1 \le g \le 5$$

We could calculate $p(Y)$ by integrating $p(Y \mid g)p(g)$ over all g, but it is unnecessary as $p(Y)$ does not affect the shape of $p(g|Y)$. The shape is easily seen (Fig. 6.2.2) to be a Gaussian p.d.f., with peak at $g = \sum_{t=1}^{N} y_t u_{t-1} / \sum_{t=1}^{N} u_{t-1}^2$ and variance $\sigma^2 / \sum_{t=1}^{N} u_{t-1}^2$, chopped off at $g = 1$ and $g = 5$. △

One of the most valuable features of Bayes estimation is its aptness for estimation in steps, bringing in new measurements at each. Chapter 7 covers stepwise estimation, but we should note here that it is necessary whenever measurements are received and must be processed in real time, and that it is a convenient way to estimate a time-varying model, even off-line. Bayes estimation in steps involves using the posterior p.d.f. from each step as the prior p.d.f. for the next, after allowing, usually straightforwardly, for any dynamics of the parameters themselves if the parameters are time-varying.

Example 6.2.4 The measurements in Example 6.2.3 could have been processed one at a time. As g is assumed constant, it is not necessary to update $\hat{g}$ at each time to account for its evolution; we need only bring in the information conveyed by the new measurement. On receiving y_k, we update $p(g \mid y_1, \ldots, y_{k-1})$ to

$$p(g \mid y_1, \ldots, y_k) = p(y_k \mid g)p(g \mid y_1, \ldots, y_{k-1})/p(y_k)$$

The effect is to adjust the location of the peak of the posterior p.d.f. and sharpen it, reducing its variance (discounting the truncation at $g=1$ and $g=5$) from

$$\sigma^2 \Big/ \sum_{t=1}^{k-1} u_{t-1}^2 \quad \text{to} \quad \sigma^2 \Big/ \sum_{t=1}^{k} u_{t-1}^2 \qquad \triangle$$

Having recognised the attractions of Bayes estimation, let us pass on to its drawbacks. They result from its membership of the luxury class, providing everything you could want but at a high price in information and computation. The need to provide a prior p.d.f. is hardest to meet; indeed, some statisticians find themselves unable to do so with a clear conscience, because it is subjective. Before you dismiss these scruples, consider an example. In Example 5.4.1 we estimated the parameter α of the p.d.f. $p(x)=\exp(-x/\alpha)/\alpha$. If we had opted for Bayes estimation, but knew in advance only that α lay between 1 and 2, we might well have taken the prior $p(\alpha)$ as uniform at 1 from 1 to 2. If alternatively we had written the model as $p(x)=\beta\exp(-\beta x)$ with β for $1/\alpha$, then knowing only that β was between $\frac{1}{2}$ and 1 we should have taken a uniform p.d.f. with $p(\beta)$ equal to 2 over that range. However, this is equivalent to a prior p.d.f. $p(\alpha)=p(\beta)|d\beta/d\alpha|=2/\alpha^2$. The paradox would not arise if we had enough previous experience of parameter values in similar cases to guess their relative frequencies of occurrence, in other words, if we had an empirical prior p.d.f. Rarely is this so. Instead we have to interpret the prior p.d.f. as stating degrees of belief, however shakily founded, in each possible parameter value.

A further factor is that often the information conveyed by the measurements far outweighs that contained in the prior p.d.f., so the final parameter estimate is not very sensitive to the prior p.d.f. A fair question is then "Why use a prior p.d.f. at all, if it has an insignificant final effect?". The answer is that Bayes estimation is still an appealing conceptual framework even if its use of a prior p.d.f. is not, in the upshot, numerically significant.

The other fundamental drawback of Bayes estimation is the amount of work entailed in forming the posterior p.d.f. and then extracting the minimum-risk estimate. A short-cut procedure which forms or updates the estimate without computing the entire posterior p.d.f. is more likely to be acceptable, especially when many parameters must be estimated at once.

For these reasons full-blown Bayes estimators are seldom implemented (Moore and Jones, 1978), but the Bayes framework is often helpful in interpreting other algorithms. An important special case is that of Gaussian prior and posterior p.d.f.'s, which are completely defined by their means and variances (or covariance matrices, for vector r.v.'s). We shall see in Chapter 7 that several recursive estimators operate by updating a parameter estimate and its estimated covariance each time a new measurement is made. For a

Gaussian-distributed estimate, the updating can be viewed as computing the posterior p.d.f., via its mean and covariance, from a prior p.d.f. also described by its mean (the old estimate) and covariance.

The next section examines some minimum-risk estimators.

6.3 MINIMUM-RISK ESTIMATORS

6.3.1 Minimum Quadratic Cost

The estimate $\hat{\theta}$ minimising the expected value of $(\hat{\theta}-\theta)^2$ over all possible θ, given measurements Y, is found from

$$\frac{\partial}{\partial\hat{\theta}}\left\{\int_{-\infty}^{\infty}(\hat{\theta}-\theta)^2 p(\theta\,|\,Y)\,d\theta\right\} = 2\int_{-\infty}^{\infty}(\hat{\theta}-\theta)p(\theta\,|\,Y)\,d\theta = 0 \qquad (6.3.1)$$

giving

$$\hat{\theta}\int_{-\infty}^{\infty} p(\theta\,|\,Y)\,d\theta = \hat{\theta} = \int_{-\infty}^{\infty}\theta p(\theta\,|\,Y)\,d\theta = E[\theta\,|\,Y] \qquad (6.3.2)$$

A minimum is found by (6.3.2) since the second derivative equals 2, which is positive. Equation (6.3.2) says that *the minimum-quadratic-cost estimator is the conditional (posterior) mean.*

Example 6.3.1 The posterior p.d.f. of the gain g in Example 6.2.3 was Gaussian, but truncated at $g = 1$ and 5. In one experiment, the input values, measurements and noise variance put the peak at $g = 2.2$ and make the variance before truncation 4. A table of the cumulative Gaussian distribution allows us to calculate that for the area under the truncated p.d.f. to be 1, the p.d.f. between $g = 1$ and 5 must be 3.101 times the Gaussian p.d.f. with mean 2.2 and variance 4. Numerical integration of $gp(g\,|\,Y)$, using a table of the Gaussian p.d.f., then gives the conditional mean as $E[g|Y] \simeq 2.73$. △

For a vector $\boldsymbol{\theta}$, the minimum-quadratic-cost estimator is found in much the same way. For any weighting matrix W,

$$\begin{aligned}
&\frac{\partial}{\partial\hat{\boldsymbol{\theta}}}\left\{\int_{-\infty}^{\infty}\cdots\int_{-\infty}^{\infty}(\hat{\boldsymbol{\theta}}-\boldsymbol{\theta})^{\mathrm{T}}W(\hat{\boldsymbol{\theta}}-\boldsymbol{\theta})p(\boldsymbol{\theta}\,|\,Y)\,d\boldsymbol{\theta}\right\}\\
&= 2W\int_{-\infty}^{\infty}\cdots\int_{-\infty}^{\infty}(\hat{\boldsymbol{\theta}}-\boldsymbol{\theta})p(\boldsymbol{\theta}\,|\,Y)\,d\boldsymbol{\theta}\\
&= 2W(\hat{\boldsymbol{\theta}}-E[\boldsymbol{\theta}\,|\,Y]) = \mathbf{0} \qquad (6.3.3)
\end{aligned}$$

The second-derivative matrix is W, so if the minimum is to be unique, W must be positive-definite, and to satisfy (6.3.3), $\hat{\boldsymbol{\theta}}$ must equal $E[\boldsymbol{\theta} \mid Y]$: again the minimum-quadratic-cost estimator is the conditional mean.

We should pause here to note the distinction between the minimum-quadratic-cost estimate and the minimum-covariance estimate in Section 5.3.2, which also minimised the expected value of a quadratic cost. There the m.s. error of any linear function $\mathbf{c}^T\boldsymbol{\theta}$ was minimised, i.e.

$$\begin{aligned} \mathbf{c}^T \operatorname{cov} \hat{\boldsymbol{\theta}}\, \mathbf{c} &= \mathbf{c}^T E[(\hat{\boldsymbol{\theta}} - \boldsymbol{\theta})(\hat{\boldsymbol{\theta}} - \boldsymbol{\theta})^T]\mathbf{c} = E[(\mathbf{c}^T(\hat{\boldsymbol{\theta}} - \boldsymbol{\theta}))^2] \\ &= E[(\hat{\boldsymbol{\theta}} - \boldsymbol{\theta})^T \mathbf{c}\mathbf{c}^T (\hat{\boldsymbol{\theta}} - \boldsymbol{\theta})] \end{aligned} \tag{6.3.4}$$

which is the expected quadratic cost with $\mathbf{c}\mathbf{c}^T$ for W. The strong similarity of the two estimators is only superficial. The averaging in (6.3.4) is over measurement realisations with $\boldsymbol{\theta}$ fixed, but in (6.3.3) it is over realisations of $\boldsymbol{\theta}$ with the measurements fixed. In (6.3.4) the random variable is $\hat{\boldsymbol{\theta}}$; in (6.3.3) it is $\boldsymbol{\theta}$. Although the two estimates might coincide in particular cases, they are not the same in general.

The conditional mean has been shown (Sherman, 1958; Deutsch, 1965) to be optimal for a broader class of loss functions than quadratic cost. The conditional mean is the minimum-risk estimator provided that

(i) The loss function (of $\hat{\theta} - \theta$ only) is symmetrical about zero error and monotonically non-decreasing each side of zero error;
(ii) the posterior p.d.f. is symmetrical about the mean; and
(iii) either the loss function is also convex or the posterior cumulative distribution function is also convex below the mean.

(A convex function $f(x)$ satisfies $f(\lambda x_1 + (1-\lambda)x_2) \le \lambda f(x_1) + (1-\lambda) f(x_2)$ for all x_1 and x_2 and any $0 \le \lambda \le 1$. That is, the section of $f(x)$ between any two points on $f(x)$ lies entirely under or on the straight line joining them.) Of the list of loss functions in Section 6.2.1, the first two are convex and symmetrical about zero error, but not the fifth, which is not convex. Asymmetrical p.d.f.'s abound, but many of them are not skewed enough to make the conditional mean a bad estimator. Convexity of the cumulative distribution function is destroyed if the p.d.f. increases at any point as you move away from the mean; such behaviour may well result from the presence of two or more sources of estimation error with different means, or from a conflict between the prior p.d.f. and the evidence in the measurements. However, the cumulative distribution function need not be convex nor the p.d.f. symmetrical for the conditional mean to be the minimum mean-square-error estimate (Deutsch, 1965).

6.3.2 Minimum Expected Absolute Error

For a vector $\boldsymbol{\theta}$, absolute error does not lend itself to forming a single scalar risk function. Instead, we can minimise the risks $E[|\hat{\theta}_i - \theta_i| \mid Y]$ for all the elements of $\boldsymbol{\theta}$ at once, by requiring their derivatives with respect to $\hat{\boldsymbol{\theta}}$ to be zero:

$$
\begin{aligned}
&\frac{\partial}{\partial \hat{\boldsymbol{\theta}}} \int_{-\infty}^{\infty} \cdots \int_{-\infty}^{\infty} |\hat{\theta}_i - \theta_i| p(\boldsymbol{\theta} \mid Y)\, d\boldsymbol{\theta} \\
&= \frac{\partial}{\partial \hat{\boldsymbol{\theta}}} \int_{-\infty}^{\infty} \cdots \int_{-\infty}^{\hat{\theta}_i} \cdots \int_{-\infty}^{\infty} (\hat{\theta}_i - \theta_i) p(\boldsymbol{\theta} \mid Y)\, d\boldsymbol{\theta} \\
&\quad + \frac{\partial}{\partial \hat{\boldsymbol{\theta}}} \int_{-\infty}^{\infty} \cdots \int_{\hat{\theta}_i}^{\infty} \cdots \int_{-\infty}^{\infty} (\theta_i - \hat{\theta}_i) p(\theta \mid Y)\, d\boldsymbol{\theta} \\
&= \boldsymbol{\delta}_i \int_{-\infty}^{\infty} \cdots \int_{-\infty}^{\hat{\theta}_i} \cdots \int_{-\infty}^{\infty} p(\boldsymbol{\theta} \mid Y)\, d\boldsymbol{\theta} \\
&\quad - \boldsymbol{\delta}_i \int_{-\infty}^{\infty} \cdots \int_{\hat{\theta}_i}^{\infty} \cdots \int_{-\infty}^{\infty} p(\boldsymbol{\theta} \mid Y)\, d\boldsymbol{\theta} \\
&= \mathbf{0}, \qquad i = 1, 2, \ldots, p
\end{aligned}
\tag{6.3.5}
$$

where $\boldsymbol{\delta}_i$ is zero but for one as element i. In (6.3.5), the contribution to the derivative due to variation of $\hat{\theta}_i$ in the integration limits is zero since $\hat{\theta}_i - \theta_i$ is zero at that point. The integrals in the last expression of (6.3.5) are the cumulative probabilities of θ_i being respectively below and above $\hat{\theta}_i$. To satisfy (6.3.5) they must be equal, so we conclude that $\hat{\theta}_i$ must be the median of the marginal posterior p.d.f. of θ; *the minimum-expected-absolute-error estimator is the conditional (posterior) median.*

Example 6.3.2 Refer again to Example 6.2.3, with numbers as in Example 6.3.1. The minimum-expected-absolute-error estimate of g cuts the area under the posterior p.d.f. in half. From a table of the Gaussian distribution, we find that 0.2743 of the area under the untruncated p.d.f. is chopped off at $g = 1$ (0.6σ below the peak) and 0.0808 at $g = 5$ (1.4σ above the peak), leaving 0.6449. The proportion of the area under the untruncated p.d.f. below $\hat{g}$ is therefore $0.2743 + \frac{1}{2}(0.6449) = 0.5968$. The table then gives $\hat{g}$ as 0.245σ above the peak, so $\hat{g} = 2.2 + 0.245(2.0) = 2.69$. This conditional-median estimate differs from the conditional mean found in Example 6.3.1 since $p(g \mid Y)$ is asymmetrical, but the difference is quite small. △

Here the normalising constant $p(Y)$ in Bayes' rule need not be calculated, whereas it must be to get the conditional mean.

6.3.3 Minimax Estimator

The most pessimistic choice is the minimax estimator, which minimises the expected maximum possible error. The idea makes sense only with regard to a scalar parameter, unless we are prepared to measure the error in a vector parameter by some additional cost function. Leaving aside such complications for the moment, an instant's thought shows that the minimax $\hat{\theta}$ is the midpoint of the range of possible values of θ as indicated by the extremities of $p(\theta \mid Y)$.

Minimax estimation fits into a Bayes context by virtue of the fact that it minimises the loss function $\lim_{q \to \infty}(\hat{\theta} - \theta)^{2q}$, weighting extreme values infinitely more heavily than all others. Bayes estimation does not, however, seem a natural context, since we need not compute the whole of $p(\theta \mid Y)$ if we are interested only in its endpoints. Furthermore, the most obvious reason for restricting attention to the ends of the range of θ would be that the range was the only convincing information, and too little was known to determine $p(Y \mid \theta)$ and $p(\theta)$.

Example 6.3.3 If in the problem of Example 6.2.3, we did not know the noise p.d.f. but knew only that the noise in each observation was between -5 and 5, we could still establish the range of possible values of θ, i.e. g, as follows. Since

$$p(Y \mid \theta) \equiv p_y(y_1, y_2, \ldots, y_n \mid g) = \prod_{t=1}^{N} p_e(y_t - gu_{t-1})$$

we know that $p(Y \mid \theta)$ is zero outside the range

$$gu_{t-1} - 5 \le y_t \le gu_{t-1} + 5, \qquad 1 \le t \le N$$

which implies that

$$\max_{1 \le t \le N}\left\{\frac{y_t - 5}{u_{t-1}}\right\} \le g \le \min_{1 \le t \le N}\left\{\frac{y_t + 5}{u_{t-1}}\right\}, \qquad \text{say} \quad g_1 \le g \le g_2$$

As in Example 6.2.3, our prior information on g is that $1 \le g \le 5$, so $p(\theta)$ is zero for $\theta < 1$ and $\theta > 5$. We conclude from Bayes' rule that $p(\theta|Y)$ is zero below $\max(g_1, 1)$ and above $\min(g_2, 5)$. These values define the possible range for g, and the minimax estimate of g is their mean. △

The problem presented by Example 6.3.3 is to identify the range of parameter values consistent with given prior bounds and with measurements containing noise described only by bounds. Only relatively recently has this problem, estimation based on a bare minimum of statistical information, received much attention in the engineering literature (Schweppe, 1973; Fogel and Huang, 1982). Unlike the more general minimax problem, it generalises readily to cover a vector parameter, becoming the problem of identifying the region in parameter space consistent with a prior region and with the bounded-noise measurements. We consider this further in Section 8.6.

6.3.4 "Most Likely" Estimate

The simplest idea of all is to take as $\hat{\boldsymbol{\theta}}$ the value giving the largest value of $p(\boldsymbol{\theta} \mid Y)$: *the conditional (posterior) mode or maximum a posteriori estimator.* Figure 6.2.1 shows that this is not always a good idea. The location of the mode may give an incomplete or misleading impression when the posterior p.d.f. is strongly skewed or has two or more peaks of not very different heights. To some extent the same criticism can be made of any other point estimate, as the information in the p.d.f. cannot always be adequately summarised by a single $\hat{\boldsymbol{\theta}}$. Even so, unless the peak is very high it seems wise to choose an estimator which is at the middle, in some defined sense, of the p.d.f.

The conditional-mode estimator is the limit, as α tends to zero, of a minimum-risk estimator with loss function zero for $\hat{\boldsymbol{\theta}}$ within a distance α of $\boldsymbol{\theta}$ and unity everywhere else.

Example 6.3.4 The conditional-mode estimate based on the p.d.f. of Fig. 6.2.2 is $\hat{g} = 2.2$, quite a way from the conditional-mean and conditional-median estimates. Had the peak been sharper or the truncation more nearly symmetrical, the three estimates would have differed less. △

6.3.5 Bayes Estimation with Gaussian Probability Density Function

A Gaussian posterior p.d.f. simplifies analysis greatly, and often approximates the truth well enough. For parameter $\boldsymbol{\theta}$ with p elements, a Gaussian posterior p.d.f. is

$$p(\boldsymbol{\theta} \mid Y) = ((2\pi)^p |R|)^{-1/2} \exp(-\tfrac{1}{2}(\boldsymbol{\theta} - \bar{\boldsymbol{\theta}})^T R^{-1} (\boldsymbol{\theta} - \bar{\boldsymbol{\theta}})) \qquad (6.3.6)$$

where R^{-1} is a positive-definite matrix, and $\bar{\boldsymbol{\theta}}$ and R depend on the prior p.d.f. and Y.

The conditional mean, median and mode can be found for this p.d.f. by a transformation of variables. Since R^{-1} is positive-definite, it can be factorised into QQ^T where Q is a square non-singular matrix. If we define $\boldsymbol{\phi}$ as $Q^T(\boldsymbol{\theta} - \bar{\boldsymbol{\theta}})$, we obtain

$$p(\boldsymbol{\theta} \mid Y) = ((2\pi)^p |R|)^{-1/2} \exp(-\tfrac{1}{2}\boldsymbol{\phi}^T \boldsymbol{\phi}) \qquad (6.3.7)$$

The exponential is at its maximum, maximising $p(\boldsymbol{\theta} \mid Y)$, where $\boldsymbol{\phi}$ is zero and

$$\boldsymbol{\theta} = \bar{\boldsymbol{\theta}} + Q^{-\mathrm{T}}\boldsymbol{\phi} = \bar{\boldsymbol{\theta}} \tag{6.3.8}$$

The conditional mode is thus at $\bar{\boldsymbol{\theta}}$, and because $p(\boldsymbol{\theta} \mid Y)$ is totally symmetrical about $\boldsymbol{\phi} = \mathbf{0}$, the conditional mean and median also occur at $\bar{\boldsymbol{\theta}}$. We conclude that $\bar{\boldsymbol{\theta}}$ is simultaneously the "most likely", minimum-quadratic-cost and minimum-expected-absolute-error estimate of $\boldsymbol{\theta}$.

The coincidence of these three optimal estimates is an incentive to assume, or even pretend, that the p.d.f. is Gaussian. For the estimates to coincide, the posterior p.d.f. need only be symmetrical about its peak. The Gaussian p.d.f. is, however, the most popular to assume for other reasons as well, including its significance in maximum-likelihood estimation (Section 6.4.2) and its relative ease of analysis.

6.4 MAXIMUM-LIKELIHOOD ESTIMATION

The reliance of Bayes estimation on a prior p.d.f. is both its principal strength and its most worrying aspect. Maximum-likelihood (m.l.) estimation forgoes the strength but avoids the worry.

6.4.1 Conditional Maximum-Likelihood Estimator

The joint p.d.f. $p(Y \mid \boldsymbol{\theta})$ of the measurements is determined, as before, by the model structure together with the p.d.f.'s of the noise and of the inputs if they are stochastic. Once the measurements have been made and numbers can be substituted for Y, $p(Y \mid \boldsymbol{\theta})$ is a function of the unknown parameters $\boldsymbol{\theta}$ only. The maximum-likelihood (m.l.) estimate of $\boldsymbol{\theta}$ is the value which maximises $p(Y \mid \boldsymbol{\theta})$. That is, once Y is known, $p(Y \mid \boldsymbol{\theta})$ is taken to indicate the likelihood of $\boldsymbol{\theta}$. It may help if we imagine $\boldsymbol{\theta}$ being stepped in very small increments over a wide range, and a fixed large number of sets of measurements being made at each $\boldsymbol{\theta}$. If we then examine only those results where the measurements are very close to a particular set of values Y, we shall find more generated by $\boldsymbol{\theta}$ values such that $p(Y \mid \boldsymbol{\theta})$ is high than by values with $p(Y \mid \boldsymbol{\theta})$ low.

Computation of an m.l. estimate is simple in principle. Given the model form and numerical Y, we write down $p(Y \mid \boldsymbol{\theta})$ and find its global maximum. In practice, $p(Y \mid \boldsymbol{\theta})$ is usually a complicated function of $\boldsymbol{\theta}$, and any trick which simplifies its maximisation is welcome. If the measurement set Y can be arranged to consist of a number of much smaller independent sets Y_1 to Y_N then $p(Y \mid \boldsymbol{\theta})$ is $\prod_{t=1}^{N} p(Y_t \mid \boldsymbol{\theta})$. An effective trick to make this product easier to maximise is to take logs, giving $\sum_{t=1}^{N} \log p(Y_t \mid \boldsymbol{\theta})$ as $\log p(Y \mid \boldsymbol{\theta})$. Since log is a monotonically increasing function of positive values of its argument, the $\boldsymbol{\theta}$ which maximises the *log-likelihood function* $\log p(Y \mid \boldsymbol{\theta})$ also maximises $p(Y \mid \boldsymbol{\theta})$.

Example 6.4.1 A known input sequence $\{u\}$ is applied to a system with known unit-pulse response $\{h\}$ but unknown constant output disturbance μ, and N samples of the output y are observed. The noise $\{e\}$ is zero-mean and has p.d.f. $p_e(e) = \exp(-|e|/\alpha)/2\alpha$ with α positive but unknown. The noise samples affecting successive output observations may be assumed independent. We wish to find m.l. estimates of μ and α.

From the model

$$y_t = h_1 u_{t-1} + h_2 u_{t-2} + \cdots + h_m u_{t-m} + \mu + e_t$$

we can calculate at each observation instant t an effective measurement

$$Y_t = \mu + e_t = y_t - h_1 u_{t-1} - \cdots - h_m u_{t-m}$$

whose p.d.f., given μ and α, is just the p.d.f. of e_t, so

$$p(Y \mid \boldsymbol{\theta}) \equiv \prod_{t=1}^{N} p_e(Y_t \mid \mu, \alpha) = \exp\left\{\frac{-\sum_{t=1}^{N}(|Y_t - \mu|)}{\alpha}\right\} \Big/ (2\alpha)^N$$

The log-likelihood function is then

$$L(\boldsymbol{\theta}) = \log p(Y \mid \boldsymbol{\theta}) = -\sum_{t=1}^{N} (|Y_t - \mu|)/\alpha - N(\log 2 + \log \alpha)$$

The values $\hat{\mu}$ and $\hat{\alpha}$ maximising $L(\boldsymbol{\theta})$ are found by examining

$$\frac{\partial L}{\partial \alpha} = \frac{\sum_{t=1}^{N}(|Y_t - \mu|)}{\alpha^2} - \frac{N}{\alpha}, \qquad \frac{\partial L}{\partial \mu} = \frac{\sum_{t=1}^{N} \operatorname{sgn}(Y_t - \mu)}{\alpha}$$

Setting $\partial L/\partial \alpha$ to zero gives $\hat{\alpha}$ as $\sum_{t=1}^{N}(|Y_t - \hat{\mu}|)/N$. As $\partial L/\partial \mu$ is discontinuous at each Y_t, it requires a little more thought. If N is even, any value of $\hat{\mu}$ between the $(N/2)$th smallest Y_t and the next larger makes $\partial L/\partial \mu$ zero; L is a piecewise linear continuous function of μ with a flat top, for any given α. If N is odd, $\hat{\mu}$ must coincide with the middle-ranking Y_t as, although $\partial L/\partial \mu$ is undefined at that value, a small change in μ in either direction reduces L. For completeness, the whole shape of $L(\mu, \alpha)$ about $\hat{\mu}$ and $\hat{\alpha}$ should be checked to verify that a maximum has been found. △

Even an example as simple as Example 6.4.1 brings out the need for care in maximising the log-likelihood function, particularly where a non-smooth or local maximum may exist.

Before enquiring into the statistical properties of m.l. estimates, we look at two special cases in the next two sections.

6.4.2 Maximum-Likelihood Estimator with Gaussian Measurements and Linear Model

Let us return to the regression model

$$\mathbf{y} = U\boldsymbol{\theta} + \mathbf{e} \tag{6.4.1}$$

with U determinstic and $\mathbf{e}$ zero-mean and of known covariance R. If $\mathbf{e}$ is assumed Gaussian, as is often reasonable, and it has N elements, its p.d.f. is

$$p_e(\mathbf{e}) = ((2\pi)^N|R|)^{-1/2}\exp(-\tfrac{1}{2}\mathbf{e}^{\mathrm{T}}R^{-1}\mathbf{e}) \tag{6.4.2}$$

and the log-likelihood function for $\boldsymbol{\theta}$ is

$$\begin{aligned} L(\boldsymbol{\theta}) &= \ln p_e(\mathbf{y} - U\boldsymbol{\theta} \mid \boldsymbol{\theta}) \\ &= -\tfrac{1}{2}\ln((2\pi)^N|R|) - \tfrac{1}{2}(\mathbf{y} - U\boldsymbol{\theta})^{\mathrm{T}}R^{-1}(\mathbf{y} - U\boldsymbol{\theta}) \end{aligned} \tag{6.4.3}$$

With R independent of $\boldsymbol{\theta}$, we can maximise $L(\boldsymbol{\theta})$ by minimising $(\mathbf{y} - U\boldsymbol{\theta})^{\mathrm{T}}R^{-1}(\mathbf{y} - U\boldsymbol{\theta})$. From Section 5.3.4, the minimising $\boldsymbol{\theta}$ is the Markov estimate

$$\hat{\boldsymbol{\theta}} = [U^{\mathrm{T}}R^{-1}U]^{-1}U^{\mathrm{T}}R^{-1}\mathbf{y} \tag{6.4.4}$$

Hence *the m.l. estimator for a model linear in the parameters and with Gaussian additive noise is identical to the Markov estimator* and shares its properties of zero bias and minimum covariance of all linear, unbiased estimators. For this reason, the Markov estimation algorithms described in Chapter 7 often go under the name of maximum-likelihood, but the name is accurate only if $p(Y \mid \boldsymbol{\theta})$ is Gaussian.

6.4.3 Unconditional Maximum-Likelihood Estimator

The conditional-mode Bayes estimator in Section 6.3.4 bears some similarity to the m.l. estimator in Section 6.4.1, but maximises $p(\boldsymbol{\theta} \mid Y)$ rather than $p(Y \mid \boldsymbol{\theta})$. Now for a given $Y, p(\boldsymbol{\theta} \mid Y)$ differs from $p(Y, \boldsymbol{\theta})$ only by the factor $p(Y)$, a number independent of $\boldsymbol{\theta}$, so the conditional-mode estimator can be viewed as an m.l. estimator based on the unconditional joint p.d.f. $p(Y, \boldsymbol{\theta})$ rather than the conditional p.d.f. $p(Y \mid \boldsymbol{\theta})$.

What is more, if we have no prior information on $\boldsymbol{\theta}$ and so take $p(\boldsymbol{\theta})$ as flat and of unlimited extent, $p(Y, \boldsymbol{\theta})$ is the same shape as $p(Y \mid \boldsymbol{\theta})$, being $p(Y \mid \boldsymbol{\theta})p(\boldsymbol{\theta})$, and the distinction between conditional and unconditional m.l. estimators vanishes.

6.4.4 Properties of Maximum-Likelihood Estimators

Maximum-likelihood estimators have finite-sample bias in some instances. An example is estimating the standard deviation σ of uncorrelated additive noise in the model (6.4.1). Following through (6.4.2) and (6.4.3) with $\sigma^2 I$ for the noise covariance R, and allowing for an unknown noise mean $\bar{e}$, we find

$$L(\boldsymbol{\theta}) = -\frac{N}{2}\ln(2\pi) - N\ln\sigma - \frac{1}{2\sigma^2}\sum_{t=1}^{N}(y_t - \mathbf{u}_t^{\mathrm{T}}\boldsymbol{\theta} - \bar{e})^2 \qquad (6.4.5)$$

where $\mathbf{u}_t^{\mathrm{T}}$ is row t of U. Differentiating,

$$\frac{\partial L}{\partial \bar{e}} = \frac{1}{\sigma^2}\sum_{t=1}^{N}(y_t - \mathbf{u}_t^{\mathrm{T}}\boldsymbol{\theta} - \bar{e}), \quad \frac{\partial L}{\partial \sigma} = -\frac{N}{\sigma} + \frac{1}{\sigma^3}\sum_{t=1}^{N}(y_t - \mathbf{u}_t^{\mathrm{T}}\boldsymbol{\theta} - \bar{e})^2 \qquad (6.4.6)$$

so to make $\partial L/\partial\bar{\theta}$ and $\partial L/\partial\sigma$ zero, the estimates must be

$$\hat{\bar{e}} = \frac{1}{N}\sum_{t=1}^{N}(y_t - \mathbf{u}_t^{\mathrm{T}}\boldsymbol{\theta}), \qquad \hat{\sigma}^2 = \frac{1}{N}\sum_{t=1}^{N}(y_t - \mathbf{u}_t^{\mathrm{T}}\boldsymbol{\theta} - \hat{\bar{e}})^2 \qquad (6.4.7)$$

The sample mean $\hat{\bar{e}}$ is unbiased, but, as we saw in Example 5.1.2, the sample mean-square deviation from the sample mean, $\hat{\sigma}^2$, is biased for finite N. Maximum-likelihood estimates are nevertheless *asymptotically* unbiased in this instance and in general.

Maximum-likelihood estimates from independent, identically distributed measurements are strongly consistent (w.p.1.). The proof is not simple (Wald, 1949). When the measurements are not all identically distributed because the system generating them varies or the distribution of the system's forcing or noise does, the m.l. estimate may well not be consistent (Kendall and Stuart, 1979, Chapter 18). The root of the trouble is that measurements then correspond to realisations of random variables whose p.d.f.'s have different parameter values at each sampling instant. The amount of information about each parameter value no longer increases continually as more measurements are taken, so the small-sample bias persists.

An additional assumption that $L(\boldsymbol{\theta})$ is everywhere differentiable twice enables a consistent m.l. estimate to be proved unique (ibid).

The covariance of m.l. estimates asymptotically reaches the Cramér–Rao bound, so they are asymptotically efficient (Cramér, 1946; Wald, 1943).

Granted that with appropriate assumptions m.l. estimates have good asymptotic statistical properties, a further property becomes significant, namely *invariance*. Invariance is the property that the m.l. estimate of a vector $\mathbf{f}(\boldsymbol{\theta})$ of functions, no more in number than the dimension of $\boldsymbol{\theta}$, is just $\mathbf{f}(\hat{\boldsymbol{\theta}})$ where $\hat{\boldsymbol{\theta}}$ is the m.l. estimate of $\boldsymbol{\theta}$. This applies whether or not the $\boldsymbol{\theta}$ corresponding to any particular value of $\mathbf{f}(\boldsymbol{\theta})$ is unique. The invariance property saves an enormous amount of work enquiring into the behaviour of practically important functions of estimated parameters, as Example 6.4.2 demonstrates.

The explanation for this helpful property is quite simple. The maximised $L(\hat{\boldsymbol{\theta}})$ is no smaller than $L(\boldsymbol{\theta})$ for any other $\boldsymbol{\theta}$, including all those values which give $\mathbf{f}(\boldsymbol{\theta})$ different from $\mathbf{f}(\hat{\boldsymbol{\theta}})$. The m.l. estimate of $\mathbf{f}$ is found by evaluating the same log-likelihood but regarding $\mathbf{f}$ as its argument; at each value of $\mathbf{f}$ we pick the largest log-likelihood given by any $\boldsymbol{\theta}$ which gives the required $\mathbf{f}(\boldsymbol{\theta})$. As we have just remarked, no value of $\mathbf{f}$ different from $\mathbf{f}(\hat{\boldsymbol{\theta}})$ will result in a larger log-likelihood than does $\mathbf{f}(\hat{\boldsymbol{\theta}})$, so $\mathbf{f}(\hat{\boldsymbol{\theta}})$ is the m.l. estimate of $\mathbf{f}$.

Example 6.4.2 Suppose we have found m.l. estimates $\hat{a}_1$ to $\hat{a}_n$ and $\hat{b}_1$ to $\hat{b}_n$ of the coefficients in the model

$$y_t = -a_1 y_{t-1} - \cdots - a_n y_{t-n} + b_1 u_{t-k-1} + \cdots + b_n u_{t-k-n} + e_t$$

and we require estimates of the steady-state (d.c.) gain and poles and zeros of the input–output relation.

In z-transforms the model is

$$Y(z^{-1}) = \frac{z^{-k}(b_1 z^{-1} + \cdots + b_n z^{-n})}{1 + a_1 z^{-1} + \cdots + a_n z^{-n}} U(z^{-1}) + \frac{E(z^{-1})}{1 + \cdots + a_n z^{-n}}$$

Letting z tend to 1, the steady-state gain is

$$(b_1 + \cdots + b_n)/(1 + a_1 + \cdots + a_n)$$

and its m.l. estimate is

$$(\hat{b}_1 + \cdots + \hat{b}_n)/(1 + \hat{a}_1 + \cdots + \hat{a}_n)$$

with negligible further computation and no further analysis. Similarly the m.l. estimates of the poles and zeros are simply the zeros of $1 + \hat{a}_1 z^{-1} + \cdots + \hat{a}_n z^{-n}$ and $z^{-k}(\hat{b}_1 z^{-1} + \cdots + \hat{b}_n z^{-n})$, respectively. Notice that although the poles and zeros correspond to unique values of the original parameters a_1 to a_n and b_1 to b_n, the steady-state gain does not. △

6.4.5 Maximum-Likelihood Estimation with Gaussian Vector Measurements and Unknown Regression-Equation Error Covariance

We conclude our look at maximum-likelihood estimation by finding the m.l. estimate of $\boldsymbol{\theta}$ from r-vector measurements $\mathbf{y}_t$ given by

$$\mathbf{y}_t = U_t\boldsymbol{\theta} + \mathbf{e}_t, \qquad t = 1, 2, \ldots, N \tag{6.4.8}$$

with the covariance R of $\mathbf{e}_t$ independent of t but unknown. Every $\mathbf{e}_t$ is taken as zero-mean, Gaussian and, in contrast to Section 6.4.2, independent of all the others. Note that R is now the covariance between errors in one sample, not different samples as in Section 6.4.2. If also each U_t is deterministic,

$$p(Y \mid \boldsymbol{\theta}, R) = p_e(\mathbf{e}_1, \mathbf{e}_2, \ldots, \mathbf{e}_N) = \prod_{t=1}^{N} \{((2\pi)^r |R|)^{-1/2} \exp(-\tfrac{1}{2}\mathbf{e}_t^{\mathrm{T}} R^{-1} \mathbf{e}_t)\}$$

$$= (2\pi)^{-Nr/2} |R|^{-N/2} \exp\left(-\frac{1}{2}\sum_{t=1}^{N} \mathbf{e}_t^{\mathrm{T}} R^{-1} \mathbf{e}_t\right) \tag{6.4.9}$$

so by taking logs, the log-likelihood function is

$$L(\boldsymbol{\theta}, R) = -\frac{N}{2}(r \ln 2\pi + \ln|R|) - \frac{1}{2}\sum_{t=1}^{N} \mathbf{e}_t^{\mathrm{T}} R^{-1} \mathbf{e}_t \tag{6.4.10}$$

Differentiation with respect to $\boldsymbol{\theta}$, with $\mathbf{e}_t$ related to $\boldsymbol{\theta}$ by (6.4.8), gives one of the conditions for a maximum of L:

$$\frac{\partial L}{\partial \boldsymbol{\theta}} = \sum_{t=1}^{N} U_t^{\mathrm{T}} R^{-1} \mathbf{e}_t = \mathbf{0} \tag{6.4.11}$$

so the m.l. estimates $\hat{R}$ and $\hat{\boldsymbol{\theta}}$ must satisfy

$$\hat{\boldsymbol{\theta}} = \left(\sum_{t=1}^{N} U_t^{\mathrm{T}} \hat{R}^{-1} U_t\right)^{-1} \sum_{t=1}^{N} U_t^{\mathrm{T}} \hat{R}^{-1} \mathbf{y}_t \tag{6.4.12}$$

Also $\partial L/\partial R$ must be zero. To find $\partial L/\partial R$ we need the standard results (Goodwin and Payne, 1977)

$$\frac{\partial}{\partial R}(\ln|R|) = R^{-\mathrm{T}}, \qquad \frac{\partial}{\partial R}(\operatorname{tr} AR^{-1}) = -(R^{-1}AR^{-1})^{\mathrm{T}} \tag{6.4.13}$$

and a little manipulation of $\mathbf{e}_t^T R^{-1} \mathbf{e}_t$:

$$\sum_{t=1}^{N} \mathbf{e}_t^T R^{-1} \mathbf{e}_t = \sum_{t=1}^{N} \text{tr}(\mathbf{e}_t \mathbf{e}_t^T R^{-1}) \tag{6.4.14}$$

Then, remembering that the covariance R is symmetric,

$$\frac{\partial L}{\partial R} = -\frac{NR^{-1}}{2} + \frac{1}{2} \sum_{t=1}^{N} R^{-1} \mathbf{e}_t \mathbf{e}_t^T R^{-1} \tag{6.4.15}$$

The maximum of L is therefore where

$$\text{tr}(NI) = Nr = \text{tr}\left(\sum_{t=1}^{N} \hat{\mathbf{e}}_t \hat{\mathbf{e}}_t^T \hat{R}^{-1}\right) = \sum_{t=1}^{N} \hat{\mathbf{e}}_t^T \hat{R}^{-1} \hat{\mathbf{e}}_t \tag{6.4.16}$$

so back in (6.4.10),

$$L(\hat{\boldsymbol{\theta}}, \hat{R}) = -\frac{N}{2}(r(1 + \ln 2\pi) + \ln|\hat{R}|) \tag{6.4.17}$$

and from (6.4.15) and (6.4.8),

$$\hat{R} = \frac{1}{N} \sum_{t=1}^{N} \hat{\mathbf{e}}_t \hat{\mathbf{e}}_t^T = \frac{1}{N} \sum_{t=1}^{N} (\mathbf{y}_t - U_t \hat{\boldsymbol{\theta}})(\mathbf{y}_t - U_t \hat{\boldsymbol{\theta}})^T \tag{6.4.18}$$

The coupled equations (6.4.12) and (6.4.18) give the m.l. estimates $\hat{\boldsymbol{\theta}}$ and $\hat{R}$. In the scalar case, $\hat{R}^{-1}$ cancels in (6.4.12) leaving the o.l.s. estimate, and the m.l. estimate of the error variance is, from (6.4.18), the sample mean-square error.

6.5 PRACTICAL IMPLICATIONS OF THIS CHAPTER

We have seen that Bayes estimation has a satisfying rationale, and provides a broad framework in which other estimators can usually be seen as simplified versions of a Bayes estimator. The need in Bayes estimation for a prior p.d.f. and a loss function is, depending on your viewpoint, either an advantage, allowing the estimator to be tailored to the problem in hand and to any background information, or a disadvantage, introducing subjective and even arbitrary judgements. Ambitious optimists with time to experiment feel the former, conservative pessimists in a hurry the latter.

Bayes estimators are almost always too demanding in computation to implement fully. A small number of features of the posterior p.d.f. is computed rather than the whole p.d.f., most often the mean because it is optimal in several ways as discussed earlier, and the covariance as an indicator of the spread of possible parameter values. The idea of proceeding from a prior to a posterior p.d.f. is well suited to recursive estimation, considered in Chapter 7.

Maximum-likelihood estimation is popular because of its good asymptotic properties, reasonable computational demands and considerable intuitive appeal. The 'maximum-likelihood' algorithms popular in the identification community are, in fact, Markov least-squares algorithms. This is not to say they are good only for observations with Gaussian random components; the Markov estimator was derived as the minimum-covariance, unbiased, linear, generalised least-squares estimator without reference to any p.d.f.

FURTHER READING

Bayes and maximum-likelihood estimation are introduced in an easy-to-read fashion by Mood *et al.*, (1974), a good general reference for Chapters 5 and 6. The material is also covered by Zacks (1971, 1981) at a rather more advanced level, and an enormous number of other textbooks. Sage and Melsa (1971) give a clear account of estimation theory slanted towards control-engineering applications.

The incorporation of prior background knowledge into estimation and observation-based decision-making has had considerable attention in the engineering literature (Jaynes, 1968; Kashyap, 1971; Potter and Anderson, 1980). Approximate Bayesian computational methods are relatively well developed for state estimation, closely related to identification as we shall see in Chapter 7, and for combined state and parameter estimation, discussed in Section 8.9. The best-known technique (Sorenson and Alspach, 1971) uses a weighted sum of Gaussian p.d.f.'s to approximate the non-Gaussian posterior p.d.f. of the state or parameters. Gaussian sums have been used, for instance, to decide between possible manoeuvres in target tracking. The underlying idea has quite a long history (Magill, 1965).

REFERENCES

Cramér, H. (1946). "Mathematical Methods of Statistics". Princeton Univ. Press, Princeton, New Jersey.

Deutsch, R. (1965). "Estimation Theory". Prentice-Hall, Englewood Cliffs, New Jersey

Fogel, E., and Huang, Y. F. (1982). On the value of information in system identification-bounded noise case. *Automatica* **18**, 229–238.

Goodwin, G. C., and Payne, R. L. (1977). "Dynamic System Identification Experiment Design and Data Analysis". Academic Press, New York and London.
Jaynes, E. T. (1968). Prior probabilities. *IEEE Trans. Syst. Sci. Cybern.* **SSC-4**, 227–241.
Kashyap, R. L. (1971). Probability and uncertainty. *IEEE Trans Inf. Theory* **IT-17**, 641–650.
Kendall, M. G., and Stuart, A. (1979). "The Advanced Theory of Statistics". Vol. 2, 4th ed. Griffin, London.
Magill, D. T. (1965). Optimal adaptive estimation of sampled stochastic processes. *IEEE Trans. Autom. Control* **AC-10**, 434–439.
Mood, A. M., Graybill, F. A., and Boes, D. C. (1974). "Introduction to the Theory of Statistics", 3rd ed. McGraw-Hill, New York.
Moore, R. J., and Jones, D. A. (1978). Coupled Bayesian–Kalman filter estimation of parameters and states of dynamic water quality model. In "Applications of Kalman Filtering and Technique to Hydrology, Hydraulics and Water Resources" (C-L. Chiu, ed.). *Proc. AGU Chapman Conf., Dept. Civ. Eng., Univ. of Pittsburgh, Pittsburgh, Pennsylvania.*
Potter, J. M., and Anderson, B. D. O. (1980). Partial prior information and decisionmaking. *IEEE Trans. Syst. Man Cybern.* **SC-10**, 125–133.
Sage, A. P., and Melsa, J. L. (1971). "Estimation Theory with Applications to Communications and Control". McGraw-Hill, New York.
Schweppe, F. C. (1973). "Uncertain Dynamic Systems". Prentice-Hall, Englewood Cliffs, New Jersey.
Sherman, S. (1958). Non-mean-square error criteria *IRE Trans. IT* **4**, 125–126.
Sorenson, H. W., and Alspach, D. L. (1971). Recursive Bayesian estimation using Gaussian sums. *Automatica* **7**, 465–479.
Wald, A. (1943). Tests of statistical hypotheses concerning several parameters when the number of observations is large. *Trans. Amer. Math. Soc.* **54**, 426–482.
Wald, A. (1949). Note on the consistency of the maximum-likelihood estimate. *Ann. Math. Stat.* **20**, 595–601.
Zacks, S. (1971). "The Theory of Statistical Inference". Wiley, New York.
Zacks, S. (1981). "Parametric Statistical Inference". Pergamon, Oxford.

PROBLEMS

6.1 After observations Y, the information about an unknown θ is in the form of the posterior p.d.f. $p(\theta \mid Y) = e^{x-\theta}$, $\theta \geq x$. Find the minimum-expected-absolute-error, minimum-quadratic-cost and unconditional maximum-likelihood estimates of θ.

6.2 A noisy observation $y = \theta + v = 2.5$ is made of an unknown θ with prior p.d.f. $p(\theta) = 2\theta$, $0 \leq \theta \leq 1$. The noise v is known to have p.d.f. $p(v) = v/2$, $0 \leq v \leq 2$, and is independent of θ. Find the conditional-m.l., posterior-mode, conditional-median and conditional-mean estimates of θ.

6.3 An unknown θ is observed indirectly and noisily by $y = \theta^2 + v$, where noise v has a uniform p.d.f. over 0 to 2 and is independent of θ, and the prior p.d.f. of θ is uniform over 0 to 1. Find the posterior p.d.f. of θ if (i) $y = 2.5$ and (ii) $y = 0.5$.

6.4 A scalar loss function which is sometimes realistic is $L(\hat{\theta},\theta) = 0$, $|\hat{\theta} - \theta| < D$; $L(\hat{\theta}, \theta) = 1$, $|\hat{\theta} - \theta| \geq D$. It implies that errors in $\hat{\theta}$ up to D are acceptable, and larger errors unacceptable. How is the minimum-risk estimate $\hat{\theta}$ based on this loss function found from the posterior p.d.f. of θ? In what circumstances might $\hat{\theta}$ be non-unique? How could the non-uniqueness be removed or avoided in those circumstances? [*Hint:* Think of maximising something.]

6.5 Find the minimum-risk $\hat{\theta}$, according to the loss function of Problem 6.4, given by the observation and p.d.f.'s of Problem 6.2.

6.6 Repeat Problem 6.4 for the loss function $L(\hat{\theta}, \theta) = |\hat{\theta} - \theta|$, $|\hat{\theta} - \theta| < D$; $L(\hat{\theta}, \theta) = 1$, $|\hat{\theta} - \theta| \geq D$.

6.7 Repeat Problem 6.5 for the loss function of Problem 6.6. Note that a straightforward analytical solution is not possible, but numerical search is not necessary.

Chapter 7

Computational Algorithms for Identification

The control-engineering literature of the past two decades describes work on a huge variety of identification methods, problems and applications (Eykhoff, 1974; Bekey and Saridis, 1983; Isermann, 1980; Eykhoff, 1981).

On the theoretical side, one of the greatest successes has been the unification of many algorithms and experimental situations, notably in analyses of asymptotic behaviour. Just the same, we cannot hope to cover more than a small fraction of the available methods, even at an introductory level. Our selection is on the basis of popularity and proven effectiveness, relative simplicity and value as examples. The selection is also influenced, of course, by personal bias. From a user's point of view, a good practical appreciation of a few methods is more valuable than a theoretical acquaintance with a great many.

7.1 ASSUMPTIONS AND MODEL FORM

7.1.1 Assumptions

The algorithms to be described all cater for *s.i.s.o. systems with linear dynamics*. They will accept non-linear functions of the observed variables as explanatory variables, but like least-squares most of them rely on the model being linear in the parameters. The model may have to be split up or rewritten to that end. Methods for non-linear systems are reviewed in Section 8.8.

We pass over m.i.m.o. systems because they raise a new complication. Unlike s.i.s.o. systems. a m.i.m.o. system can be represented by more than one minimum-order "transfer-function" model (matrix-fraction description: see Section 8.7) with exactly the same input–output behaviour, so the first problem is to decide which one to identify. Alternative representations which are input–output equivalent may differ, for instance, in ease of physical interpretation or in how well conditioned the parameter estimation will be. We have too little space to do justice to multivariable representation theory, so we confine ourselves to an example or two in Section 8.7 showing the new features with no s.i.s.o. counterparts. Leaving aside the problem of choosing a model structure, we may extend the s.i.s.o. regression model heuristically by adding terms in further inputs. A m.i.s.o. model with inputs $u^{(1)}$ to $u^{(q)}$ and output y would then be

$$y_t = -a_1 y_{t-1} - a_2 y_{t-2} - \cdots + b_1^{(1)} u_{t-k_1-1}^{(1)} + b_2^{(1)} u_{t-k_1-2}^{(1)} + \cdots + b_1^{(2)} u_{t-k_2-1}^{(2)} + \cdots + b_m^{(q)} u_{t-k_q-m}^{(q)} + e_t \tag{7.1.1}$$

or in transfer-function form

$$Y(z^{-1}) = \frac{(B^{(1)}(z^{-1})U^{(1)}(z^{-1}) + \cdots + B^{(q)}(z^{-1})U^{(q)}(z^{-1}) + E(z^{-1}))}{1 + A(z^{-1})} \tag{7.1.2}$$

Note the restriction that all the input–output relations have the same poles in this model.

Single-input–multi-output models are probably best treated as a collection of s.i.s.o. models. The alternative, employing a vector-output model, might in principle require fewer parameters, as part of the model would be common to more than one output. An example is the state equation of a state-space model. On the other hand, identification of separate s.i.s.o. models, each with no more parameters than necessary to describe its dominant dynamics, could well be computationally cheaper than simultaneous estimation of all the parameters in a s.i.m.o. model. It could also be more convenient because it does not require access to all the outputs at the same time.

Several other assumptions, some dubious, underlie the identification algorithms in this chapter. We assume:

(i) *A quadratic function* of residuals, prediction errors or parameter errors *is the performance criterion*, mainly for mathematical convenience but also because of a lack of well tried rules for choosing other criteria in specific experimental situations. Maximum-likelihood estimation is an exception, as it relates the function to be minimised to the p.d.f. of the observations, and hence to the noise p.d.f. However, the algorithm of that name popular in control engineering and described in Section 7.2.3 assumes a Gaussian p.d.f. and minimises a quadratic cost function.

(ii) *The plant and noise parameters are taken as constant* or at most slowly changing. The design of the algorithms takes no account of changes in dynamics due to common occurrences like variation of feedstock quality, operating point or demand in process plant, and the effects of unmonitored inputs generally. Nor does it consider non-stationary noise dominated at different times by different sources. Some well established techniques for tracking time-varying dynamics are described in Section 8.1.

(iii) *The input and noise are assumed independent.* This natural-looking assumption means we shall have to consider separately the identification of closed-loop systems, where the input depends on the fed-back noisy output. Section 8.3 does so.

(iv) *The model order is assumed fixed,* which may cause difficulty in on-line identification. Uncertainty about the best model order and doubt whether it can be adjusted reliably on line may lead to too high an order being used. The identification then runs a risk of ill-conditioning because of near-redundancy of some parameters. As an example, model terms in too-recent input samples may be included if the dead time is uncertain or variable. Model-order testing is covered in Section 9.4, but existing techniques are aimed mainly at off-line use.

(v) *Asymptotic properties are assumed important.* They are, but the emphasis on them is largely due to a lack of finite-sample theory. An algorithm with better asymptotic properties, such as efficiency, may also perform better on finite records, but it may not. The r.m.l. 2 algorithm of Section 7.4.3 is a case in point, where a modification to improve asymptotic behaviour tends to destabilise the algorithm, ruining its performance on short records (Norton, 1977).

7.1.2 Standard Linear Single-Input–Single-Output Model

All the algorithms described employ one or other specialisation of the model illustrated in Fig. 7.1.1:

$$Y(z^{-1}) = \frac{B(z^{-1})z^{-k}}{1+A(z^{-1})}U(z^{-1}) + \frac{1+C(z^{-1})}{1+D(z^{-1})}V(z^{-1}) \tag{7.1.3}$$

in which

$$\begin{aligned} A(z^{-1}) &\equiv a_1 z^{-1} + \cdots + a_n z^{-n}, \qquad & B(z^{-1}) &\equiv b_1 z^{-1} + \cdots + b_m z^{-m} \\ C(z^{-1}) &\equiv c_1 z^{-1} + \cdots + c_q z^{-q}, \qquad & D(z^{-1}) &\equiv d_1 z^{-1} + \cdots + d_r z^{-r} \end{aligned} \tag{7.1.4}$$

Often m is taken equal to n, but this is not essential. Integer k, the dead time, is specified in advance. The estimation of k is discussed in Chapter 10, but we noted under assumption (iv) that too small a value for k can result in ill-conditioning. If the actual dead time is k_a and the model dead time k, parameters b_1 to b_{k_a+1-k} are redundant. What is more, if the order n of $A(z^{-1})$ is also higher than necessary, near-cancelling pole–zero pairs will be estimated, contributing almost nothing to the input–output behaviour. The spurious poles and zeros have little effect on the model performance so long as the poles remain stable, but the accompanying ill-conditioning may affect the

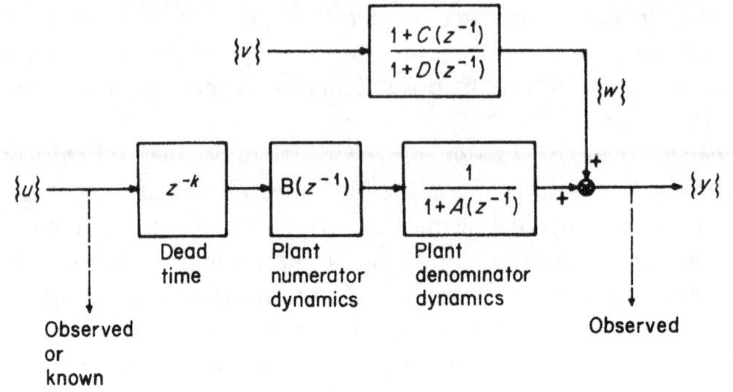

Fig. 7.1.1 Standard z-transform model for linear s.i.s.o. system where $\{v\}$ is the white noise-generating sequence, $\{u\}$ the input, and $\{w\}$ the structured noise.

non-redundant parameters too. Luckily near-cancelling pole–zero pairs are easy to detect, at the cost of factorising $1 + A(z^{-1})$ and $B(z^{-1})$.

The noise in (7.1.3), say $\{w\}$, is represented as the result of filtering a zero-mean stationary white sequence $\{v\}$ through the transfer function $(1 + C(z^{-1}))/(1 + D(z^{-1}))$, thereby shaping the noise autocorrelation function and power spectral density. We assume that the input sequence $\{u\}$ is independent of $\{v\}$ and hence of $\{w\}$, so we do not allow feedback of the noisy output. The connection between the noise-shaping filter coefficients and the a.c.f. of $\{w\}$ can be seen with the help of the unit-pulse response $H(z^{-1})$ of the filter, given by

$$W(z^{-1}) = \frac{1 + C(z^{-1})}{1 + D(z^{-1})} V(z^{-1}) \equiv H(z^{-1})V(z^{-1}) \tag{7.1.5}$$

The a.c.f. of $\{w\}$ at lag i is

$$\begin{aligned} r_{ww}(i) &= E[w_t w_{t+i}] \\ &= E[(h_0 v_t + h_1 v_{t-1} + h_2 v_{t-2} + \cdots \infty)(h_0 v_{t+i} + h_1 v_{t+i-1} + \cdots \infty)] \\ &= \sigma^2 (h_0 h_i + h_1 h_{i+1} + \cdots \infty) \\ &= \sigma^2 \times \text{coefficient of } z^{-i} \text{ in } H(z^{-1})H(z) \\ &= \sigma^2 \times \text{coefficient of } z^{-i} \text{ in } \frac{(1 + C(z^{-1}))(1 + C(z))}{(1 + D(z^{-1}))(1 + D(z))} \end{aligned} \tag{7.1.6}$$

where σ^2 is the m.s. value of $\{v\}$. The final expression in (7.1.6) is a rational polynomial function of z^{-1} or z, so $\{w\}$ is said to be a stochastic process with *rational spectral density*. Numerical a.c.f. values for $\{w\}$ can be found by splitting the rational polynomial function into partial fractions symmetrical in

z^{-1} and z, giving the z-transforms of the a.c.f. for positive and negative lags respectively.

Example 7.1.1 A noise-generating sequence $\{v\}$ of zero mean and m.s. value 1 produces a noise sequence $\{w\}$ through

$$W(z^{-1}) = V(z^{-1})/(1 - 0.5z^{-1})$$

so the a.c.f. of $\{w\}$ can be found from

$$H(z^{-1})H(z) = \frac{1}{(1-0.5z^{-1})(1-0.5z)} = \frac{4}{3}\left(\frac{1}{1-0.5z^{-1}} + \frac{1}{1-0.5z} - 1\right)$$

The infinite-series expansion of the first term gives

$$\tfrac{4}{3}(1 + 0.5z^{-1} + 0.25z^{-2} + \cdots \infty)$$

so

$$r_{ww}(i) = \tfrac{4}{3}(0.5)^i \qquad \text{for} \quad i > 0$$

and the other terms give

$$r_{ww}(i) = \tfrac{4}{3}(0.5)^i \qquad \text{for} \quad i < 0, \qquad r_{ww}(0) = \tfrac{4}{3}(1 + 1 - 1) = \tfrac{4}{3}$$

△

7.1.3 Output-Error and Equation-Error Models

The estimation algorithms in this chapter all identify the coefficients a_1 to a_n and b_1 to b_m in model (7.1.3), but differ in how they treat the noise part. Two basic approaches to the noise can be distinguished, leading to *output-error* and *equation-error algorithms*. Figure 7.1.2 shows the difference. Equation-error methods rewrite (7.1.3) into a form suitable for l.s. estimation. Multiplying (7.1.3) by $1 + A(z^{-1})$ and dropping z^{-1} for brevity, we obtain

$$Y = -AY + Bz^{-k}U + E \tag{7.1.7}$$

However, E is $(1 + A)W$, so even if $\{w\}$ is white, $\{e\}$ is not. Consequently $\{e\}$ is correlated with the noise content of the lagged output samples in AY. As seen in section 5.2.4, such correlation causes bias in l.s. estimates $\hat{A}$ and $\hat{B}$. To avoid bias, equation-error algorithms identify a noise-structure model along with A and B, as we shall see later.

Output-error algorithms (Dugard and Landau, 1980) instead adjust $\hat{A}$ and $\hat{B}$ to minimise the error $\{\hat{w}\}$ in

$$Y = -\hat{A}\hat{Y} + \hat{B}z^{-k}U + \hat{W}; \qquad \hat{Y} \triangleq \frac{\hat{B}z^{-k}}{1+\hat{A}}U \tag{7.1.8}$$

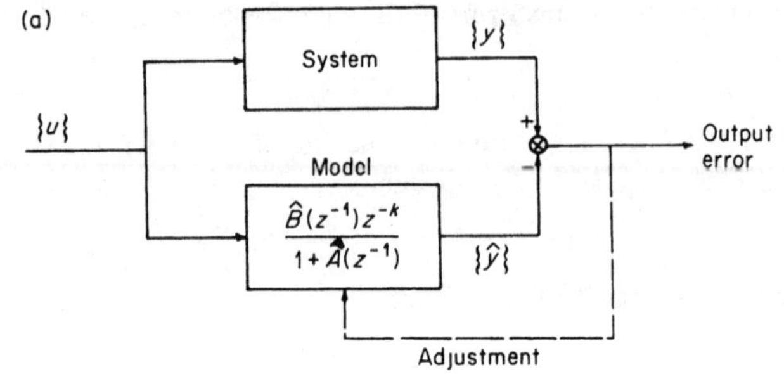

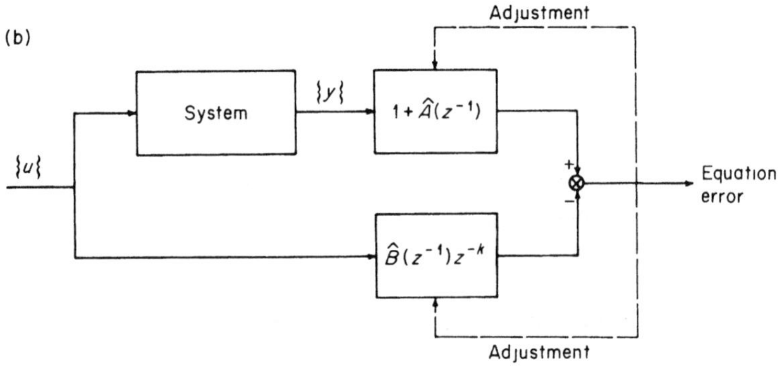

Fig. 7.1.2 (a) Output-error, and (b) equation-error identification, where $\{u\}$ is the input.

The explanatory variables in $A\hat{Y}$ are all lagged versions of $\{u\}$ and thus uncorrelated with $\{w\}$. Bias does not arise since $\{\hat{y}\}$ is free of the noise which affects $\{y\}$. Clearly the lagged versions of $\{\hat{y}\}$ in $A\hat{Y}$ are instrumental variables. Section 7.4.5 discusses instrumental-variable identification further. The output-error approach looks direct, but has its own complication, the need to estimate the unobservable $\{\hat{y}\}$ using some prior $\hat{A}$ and $\hat{B}$. In other words, $\{w\}$ in (7.1.8) is non-linear in $\hat{A}$ and $\hat{B}$, whereas the equation error $\{e\}$ in (7.1.7) is linear in A and B.

7.1.4 A.r.m.a.x. Model

Equation (7.1.7) gives the regression-type equation

$$y_t = -a_1 y_{t-1} - \cdots - a_n y_{t-n} + b_1 u_{t-k-1} + \cdots + b_m u_{t-k-m} + e_t \qquad (7.1.9)$$

The present output y_t is partly an *autoregression* (the $-a_i y_{t-i}$ terms) and partly a *moving average* (the $b_i u_{t-k-i}$ terms) of the *exogenous* (externally generated) input $\{u\}$: altogether an *autoregressive-moving average-exogenous* (*a.r.m.a.x.*) *model*. If no observed exogenous input were present, $\{y\}$ could still be modelled by an *autoregressive-moving average* or *a.r.m.a. model* driven by an unobserved white noise-generating sequence. Several of the identification algorithms we shall discuss are based on least-squares applied to (7.1.9).

7.2 BATCH (OFF-LINE) IDENTIFICATION ALGORITHMS

7.2.1 Role of Batch Algorithms

Batch algorithms process all the observations of y and u simultaneously and produce a single estimate of the parameter vector. By contrast, the recursive methods of Sections 7.3–7.5 process the observations one sampling instant at a time and update the parameter estimates each time. Batch algorithms are suitable only when estimates are required once and for all or at long intervals, or when computing is cheap, since they process the entire record every time. Most real-time applications are better met by recursive algorithms since time and computing power are strictly limited. An important example is microprocessor-based self-tuning control (Åström *et al.*, 1977; Wellstead *et al.*, 1979), in which a new control value is computed by reference to a freshly updated model after each sampling of the output. Batch methods have some advantage in iterative processing, where the estimates are improved by a succession of iterations, each processing the whole record. They allow monitoring of output or equation errors at the end of each iteration, using the model obtained in that iteration. Progress can be checked and anomalies like large isolated errors due to untrustworthy observations can be detected and removed. This may be less easy in recursive estimation, where the quality of the estimates varies during an iteration, starting relatively poor but improving as more observations are processed. The significance of a given error value correspondingly increases in the course of the iteration. The effect is most pronounced in early iterations and with short records.

The high cost of recomputing the estimate at short intervals by batch methods makes recursive methods preferable for time-varying systems.

The flexibility and computational economy of recursive algorithms has led to great practical and theoretical emphasis on them over the past decade. We should not forget batch methods, however, as apart from their occasional advantages they provide an introduction and motivation for recursive methods.

7.2.2 Iterative Generalised Least-Squares†

Model (7.1.3) is used with C zero and $1 + D$ of the form $(1 + A)(1 + D')$, giving

$$Y = -AY + Bz^{-k}U + (1/(1 + D'))V \equiv -AY + Bz^{-k}U + E \quad (7.2.1)$$

Unless D' is zero, which is improbable, $\{e\}$ is autocorrelated. An equation-error method therefore must take action to avoid bias arising as in Section 5.2.4. The algorithm filters $\{y\}$ and $\{u\}$ with a transfer function $1 + D'$, producing $\{y^*\}$ and $\{u^*\}$ related by

$$Y^* \triangleq (1 + D')Y = -AY^* + Bz^{-k}U^* + V \quad (7.2.2)$$

With $\{v\}$ white, there is no correlation between v_t and regressors y^*_{t-1} to y^*_{t-n} by way of v_{t-1} and earlier samples. There is no correlation between $\{v\}$ and $\{u^*\}$ either, so minimum-covariance linear unbiased estimates of the coefficients in A and B can be found by o.l.s. The problem is how to find D', not normally known in advance. It is estimated iteratively, alternately with A and B, as follows.

Batch Iterative g.l.s. Iteration i is

(i) estimate $\hat{A}^{(i)}$ and $\hat{B}^{(i)}$ by o.l.s. using filtered $Y^{*(i-1)}$ and $U^{*(i-1)}$ produced in the previous iteration;

(ii) form residuals

$$\hat{E}^{(i)} = (1 + \hat{A}^{(i)})Y^{*(i-1)} - \hat{B}^{(i)}z^{-k}U^{*(i-1)}$$

(iii) fit an autoregressive model

$$\hat{E}^{(i)} = -\hat{D}^{(i)}\hat{E}^{(i)} + \hat{V}^{(i)}$$

to $\hat{E}^{(i)}$ by o.l.s., yielding $\hat{D}^{(i)}$;

(iv) filter $Y^{*(i-1)}$ and $U^{*(i-1)}$ to form

$$\hat{Y}^{*(i)} = (1 + \hat{D}^{(i)})\hat{Y}^{*(i-1)}, \qquad \hat{U}^{*(i)} = (1 + \hat{D}^{(i)})\hat{U}^{*(i-1)}$$

Step (i) of the first iteration uses Y and U as $Y^{*(0)}$ and $U^{*(0)}$. Subsequent iterations gradually build up a noise-whitening filter as a cascade of $1 + \hat{D}^{(i)}$, $i = 1, 2, \ldots$. Each $1 + \hat{D}^{(i)}$ is of low order, up to order 3 or so. The convergence rate is markedly influenced by the structure of E, but is usually quite rapid, with insignificant reduction of the sum of squares of residuals from step (iii) beyond about five to ten iterations.

† Clarke, 1967.

7.2.3 Maximum-Likelihood Algorithm†

The model (7.1.3) is rewritten to look like a regression equation, by taking D as identical to A then multiplying by $1 + A$:

$$Y = -AY + Bz^{-k}U + (1 + C)V \tag{7.2.3}$$

or in difference-equation form

$$\begin{aligned} y_t &= -a_1 y_{t-1} - \cdots - a_n y_{t-n} + b_1 u_{t-k-1} + \cdots + b_m u_{t-k-m} \\ &\quad + c_1 v_{t-1} + \cdots + c_q v_{t-q} + v_t \\ &\equiv \mathbf{u}_t^{\mathrm{T}} \boldsymbol{\theta} + v_t, \qquad t = 1, 2, \ldots, N \end{aligned} \tag{7.2.4}$$

where

$$\begin{aligned} \mathbf{u}_t^{\mathrm{T}} &= [y_{t-1} \ \cdots \ y_{t-n} \ u_{t-k-1} \ \cdots \ u_{t-k-m} \ v_{t-1} \ \cdots \ v_{t-q}] \\ \boldsymbol{\theta}^{\mathrm{T}} &= [-a_1 \ \cdots \ -a_n \ b_1 \ \cdots \ b_m \ c_1 \ \cdots \ c_q] \end{aligned} \tag{7.2.5}$$

The "regression-equation noise" v_t is white, but we cannot apply o.l.s. to (7.2.4) because regressors v_{t-1} to v_{t-q} are unknown. The idea of replacing them by estimates from some auxiliary model is pursued in Section 7.4.1. Instead, Åström and Bohlin recognised that, with $\{u\}$ and $\{y\}$ given, the conditional p.d.f. $p(Y \mid \boldsymbol{\theta})$ of the observations given $\boldsymbol{\theta}$ is determined solely by the joint p.d.f. of $\{v\}$. We looked in Section 6.4.2 at the special case for which $\{v\}$ is Gaussian and all the regressors in each $\mathbf{u}_t$ are known (deterministic). Writing (7.2.4) as

$$\mathbf{y} = U\boldsymbol{\theta} + \mathbf{v}, \qquad \operatorname{cov} \mathbf{v} = R \tag{7.2.6}$$

we found that the log-likelihood function is

$$L(\boldsymbol{\theta}) = -\tfrac{1}{2} \ln\{(2\pi)^N |R|\} - \tfrac{1}{2}(\mathbf{y} - U\boldsymbol{\theta})^{\mathrm{T}} R^{-1} (\mathbf{y} - U\boldsymbol{\theta}) \tag{7.2.7}$$

so if R is known, we are left with the minimisation of the Markov cost function $(\mathbf{y} - U\boldsymbol{\theta})^{\mathrm{T}} R^{-1} (\mathbf{y} - U\boldsymbol{\theta})$. The present situation differs in that U is not completely known but depends on $\{v\}$. Accordingly, $\mathbf{v}^{\mathrm{T}} R^{-1} \mathbf{v}$ must be used in place of $(\mathbf{y} - U\boldsymbol{\theta})^{\mathrm{T}} R^{-1} (\mathbf{y} - U\boldsymbol{\theta})$, with $\mathbf{v}$ estimated iteratively along with $\boldsymbol{\theta}$ and any unknown parameters of R. Assuming $\{v\}$ is zero-mean, white and of constant but unknown variance σ^2, R is $\sigma^2 I$ and σ can be adjoined to the parameters to be estimated. To sum up, we are faced with the non-linear programming problem of finding $\boldsymbol{\theta}$ and σ to maximise

$$L(\boldsymbol{\theta}, \sigma) = -\frac{N}{2} \ln(2\pi) - N \ln \sigma - \frac{1}{2\sigma^2} \sum_{t=1}^{N} v_t^2 \tag{7.2.8}$$

† Åström and Bohlin, 1966.

Each $\partial L/\partial\theta_i$ is found via $\{\partial v_t/\partial\theta_i\}$, which in turn come from a recurrence relation obtained by differentiating (7.2.4). The algorithm takes m equal to $n+1$ and q equal to n, then from (7.2.4)

$$\frac{\partial v_t}{\partial\theta_i}+c_1\frac{\partial v_{t-n}}{\partial\theta_i}+\cdots+c_n\frac{\partial v_{t-n}}{\partial\theta_i}=\begin{cases}-y_{t-i} & \text{for } 1\leq i\leq n\\ -u_{t-k-i+n+1} & \text{for } n+1\leq i\leq 2n+1\\ -v_{t-i+2n+1} & \text{for } 2n+2\leq i\leq 3n+1\end{cases} \tag{7.2.9}$$

The second derivatives of L follow from (7.2.8):

$$\frac{\partial^2 L}{\partial\theta_i\partial\theta_j}=-\frac{1}{\sigma^2}\sum_{t=1}^{N}\left\{v_t\frac{\partial^2 v_t}{\partial\theta_i\partial\theta_j}+\frac{\partial v_t}{\partial\theta_i}\frac{\partial v_t}{\partial\theta_j}\right\} \tag{7.2.10}$$

in which all $\partial^2 v_t/\partial\theta_i\partial\theta_j$ are zero except those with i and j between $2n+2$ and $3n+1$ inclusive, which are obtained by differentiating (7.2.9). Only the derivatives involving $\{v\}$ have to be recalculated at each iteration, the rest being zero or fixed by $\{u\}$ and $\{y\}$. Lastly $\partial L/\partial\sigma$ and $\partial^2 L/\partial\sigma^2$ are found by differentiating (7.2.8).

Batch iterative m.l. Iteration i for the parameter vector $\boldsymbol{\theta}'^{\mathrm{T}} \triangleq [\boldsymbol{\theta}^{\mathrm{T}}\sigma]$ is

(i) recalculate all first and second derivatives which depend on $\boldsymbol{\theta}'$, at $\hat{\boldsymbol{\theta}}'^{(i-1)}$, and form $\partial L/\partial\boldsymbol{\theta}'$ and the Hessian matrix $[H_{\theta'\theta'}L]$;

(ii) update $\hat{\theta}'$ by a Newton–Raphson step

$$\hat{\boldsymbol{\theta}}'^{(i)}=\hat{\boldsymbol{\theta}}'^{(i-1)}-[H_{\theta'\theta'}L]^{-1}\partial L/\partial\boldsymbol{\theta}'$$

(iii) check whether the estimates and/or likelihood function have settled enough to stop.

Applications of this widely used algorithm appear in Åström (1967) and Gustavsson (1969).

7.3 RECURSIVE ESTIMATION

Recursive estimation consists of repeatedly updating the estimates, each update processing only one output observation. Its most obvious application is to real-time control or prediction. An example is hourly updating of a flow-prediction model of a river catchment on receipt of a new sample of input (an hour's rainfall) and output (present river flow). Less obviously, recursive estimators are valuable off line because their structure is simple and they are able, with simple modifications at most, to track time-varying dynamics as discussed in Section 8.1. They also allow prior information in the form of an existing estimate to be exploited as the starting value for the recursion.

Some recursive estimators update a covariance to indicate the reliability of the estimates. A starting value for the covariance is provided by the user, along with the initial estimate itself, and tells the estimator how good the initial estimate is. This sort of estimator can be viewed as a pared-down Bayes estimator. Each update produces in place of the posterior p.d.f. just the posterior mean and covariance, which provide prior values for the next update. For a p.d.f. which is completely determined by its mean and covariance, as a Gaussian p.d.f. is, the updating amounts to full Bayes estimation. We shall not normally take this view of the estimators, partly to avoid thinking of the mean and covariance as necessarily the whole story, and partly to retain the option of regarding the parameters as unknown but deterministic.

Recursive updating is popular in state estimation (Jazwinski, 1970; Maybeck, 1979) for the same reasons as it is in identification, and some identification algorithms differ from state-estimation algorithms only in detailed interpretation. The next two sections will emphasize the similarity by

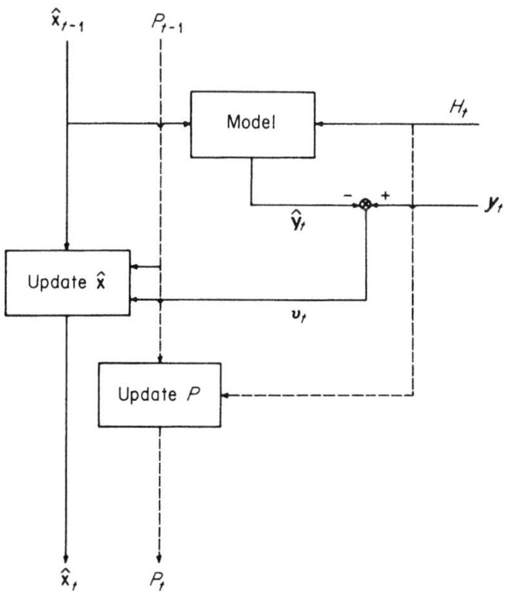

Fig. 7.3.1 One step of recursive estimator, where P_{t-1}, is estimated covariance of $\hat{\mathbf{x}}_{t-1}$, H_t is observation (regressor) matrix, y_t is observed output, $\hat{y}_t$ is predicted output and v_t is prediction error (innovation).

discussing estimation rather than specifically identification, and employing a more general notation. Figure 7.3.1 introduces the notation and shows an updating step. The broken lines indicate parts not present in all recursive estimators. For generality, a vector output $\mathbf{y}$ is considered. Our exploration of recursive estimators starts by seeing how the updating mechanism of Fig. 7.3.1 arises if we demand that the estimator is linear and has good finite-record statistical properties.

7.3.1 Linear Unbiased Updating

Suppose we have at time t an old unbiased estimate $\hat{\mathbf{x}}_{t-1}$ of a vector $\mathbf{x}$ (parameter or state) with a covariance P_{t-1}. We receive new noisy observations making up $\mathbf{y}_t$, related linearly to $\mathbf{x}$ by

$$\mathbf{y}_t = H_t\mathbf{x} + \mathbf{v}_t \tag{7.3.1}$$

The observation noise $\mathbf{v}_t$ has zero mean and covariance R, and is assumed uncorrelated with the error in $\hat{\mathbf{x}}_{t-1}$. We wish to combine $\hat{\mathbf{x}}_{t-1}$ and $\mathbf{y}_t$ linearly (to keep the computation and analysis simple), forming a new estimate $\hat{\mathbf{x}}_t$. In other words, we want

$$\hat{\mathbf{x}}_t = J_t\hat{\mathbf{x}}_{t-1} + K_t\mathbf{y}_t \tag{7.3.2}$$

with matrices J_t and K_t chosen to make $\hat{\mathbf{x}}_t$ a good estimate. If we ask for $\hat{\mathbf{x}}_t$ to be unbiased, it means that for any $\mathbf{x}$ and any given H_t

$$E\hat{\mathbf{x}}_t = J_tE\hat{\mathbf{x}}_{t-1} + K_tE\mathbf{y}_t = J_t\mathbf{x} + K_tH_t\mathbf{x} = \mathbf{x} \tag{7.3.3}$$

Hence

$$J_t + K_tH_t = I \tag{7.3.4}$$

and so

$$\hat{\mathbf{x}}_t = (I - K_tH_t)\hat{\mathbf{x}}_{t-1} + K_t\mathbf{y}_t = \hat{\mathbf{x}}_{t-1} + K_t(\mathbf{y}_t - H_t\hat{\mathbf{x}}_{t-1}) \tag{7.3.5}$$

Our new linear and unbiased estimate must therefore add to the old estimate a correction proportional to the prediction error between the new observation and its value predicted by the old estimate. The prediction error is often called the *innovation*. Already the structure of Fig. 7.3.1 is partly explained.

Example 7.3.1 Old unbiased estimates $\hat{x}_{t-1}^{(1)} = 5, \hat{x}_{t-1}^{(2)} = -2$ of two parameters $x^{(1)}$ and $x^{(2)}$ are to be updated using a new observation

$$y_t = x^{(1)} + x^{(2)} + v_t = 4.5$$

in which the noise is zero-mean. The new estimate is to be linear and unbiased, so it has to be of the form

$$\hat{\mathbf{x}}_t = \begin{bmatrix} 5 \\ -2 \end{bmatrix} + \begin{bmatrix} k_t^{(1)} \\ k_t^{(2)} \end{bmatrix} \left(4.5 - [1 \quad 1] \begin{bmatrix} 5 \\ -2 \end{bmatrix} \right) = \begin{bmatrix} 5 + 1.5k_t^{(1)} \\ -2 + 1.5k_t^{(2)} \end{bmatrix} \qquad \triangle$$

We have yet to fix K_t.

7.3.2 Minimum-Covariance Linear Unbiased Updating

We next ask for $\hat{\mathbf{x}}_t$ to have the smallest possible covariance. Its covariance is

$$\begin{aligned} P_t &= E[(\hat{\mathbf{x}}_t - E\hat{\mathbf{x}}_t)(\cdot)^{\mathrm{T}}] = E[((I - K_t H_t)\hat{\mathbf{x}}_{t-1} + K_t \mathbf{y}_t - \mathbf{x})(\cdot)^{\mathrm{T}}] \\ &= E[((I - K_t H_t)(\hat{\mathbf{x}}_{t-1} - \mathbf{x}) + K_t(\mathbf{y}_t - H_t \mathbf{x}))(\cdot)^{\mathrm{T}}] \\ &= (I - K_t H_t) P_{t-1} (I - H_t^{\mathrm{T}} K_t^{\mathrm{T}}) + K_t R K_t^{\mathrm{T}} \end{aligned} \tag{7.3.6}$$

where $(\cdot)$ denotes a repeat of the previous bracketed expression. The last step in (7.3.6) relies on the noise $\mathbf{y}_t - H_t\mathbf{x}$, that is $\mathbf{v}_t$, being uncorrelated with the error $\hat{\mathbf{x}}_{t-1} - \mathbf{x}$. We can find the K_t that minimises P_t by writing down the change ΔP_t due to a small change ΔK_t and choosing K_t to make the rate of change of P_t with K_t zero:

$$\begin{aligned} \Delta P_t &= (I - (K_t + \Delta K_t) H_t) P_{t-1} (I - H_t^{\mathrm{T}} (K_t^{\mathrm{T}} + \Delta K_t^{\mathrm{T}})) \\ &\quad + (K_t + \Delta K_t) R (K_t^{\mathrm{T}} + \Delta K_t^{\mathrm{T}}) - (I - K_t H_t) P_{t-1} (I - H_t^{\mathrm{T}} K_t^{\mathrm{T}}) - K_t R K_t^{\mathrm{T}} \\ &\simeq \Delta K_t (-H_t P_{t-1} (I - H_t^{\mathrm{T}} K_t^{\mathrm{T}}) + R K_t^{\mathrm{T}}) + (-(I - K_t H_t) P_{t-1} H_t^{\mathrm{T}} + K_t R) \Delta K_t^{\mathrm{T}} \end{aligned} \tag{7.3.7}$$

Here the second-degree terms in ΔK_t have been neglected as we are about to make ΔK_t tend to zero. For ΔP_t to have zero rate of change with ΔK_t whatever the relative sizes of the elements of ΔK_t, the expressions multiplying ΔK_t and ΔK_t^{T} in (7.3.7) must both be zero, requiring

$$-(I - K_t H_t) P_{t-1} H_t^{\mathrm{T}} + K_t R = 0 \tag{7.3.8}$$

The optimal gain matrix for the updating is therefore

$$K_t = P_{t-1} H_t^{\mathrm{T}} (H_t P_{t-1} H_t^{\mathrm{T}} + R)^{-1} \tag{7.3.9}$$

The inverse in this expression exists unless $\mathbf{y}_t$ is both noisefree, making R zero, and part-redundant, making $H_t P_{t-1} H_t^{\mathrm{T}}$ singular. The neglected second-degree terms in ΔK_t are readily seen to be positive-definite with the same exceptions, confirming that a minimum of P_t has been achieved.

Example 7.3.2 We are told the covariance of the old estimate $\hat{\mathbf{x}}_{t-1}$ in Example 7.3.1 and the noise variance of the new observation y_t, respectively

$$P_{t-1} = \begin{bmatrix} 4 & 0 \\ 0 & 1 \end{bmatrix} \quad \text{and} \quad R = 5$$

The minimum-covariance linear unbiased new estimate of $\mathbf{x}$ is obtained using the correction gain matrix

$$K \equiv \begin{bmatrix} k_t^{(1)} \\ k_t^{(2)} \end{bmatrix} = \begin{bmatrix} 4 & 0 \\ 0 & 1 \end{bmatrix} \begin{bmatrix} 1 \\ 1 \end{bmatrix} \left([1 \quad 1] \begin{bmatrix} 4 & 0 \\ 0 & 1 \end{bmatrix} \begin{bmatrix} 1 \\ 1 \end{bmatrix} + 5 \right)^{-1} = \begin{bmatrix} 4 \\ 1 \end{bmatrix} \Big/ (5+5) = \begin{bmatrix} 0.4 \\ 0.1 \end{bmatrix}$$

so

$$\hat{\mathbf{x}}_t = \begin{bmatrix} 5 + 1.5k_t^{(1)} \\ -2 + 1.5k_t^{(2)} \end{bmatrix} = \begin{bmatrix} 5.6 \\ -1.85 \end{bmatrix}$$

Notice that

(i) We can see in this small example why $k_t^{(1)}$ is larger than $k_t^{(2)}$. According to P_{t-1}, $\hat{x}_{t-1}^{(1)}$ is less reliable than $\hat{x}_{t-1}^{(2)}$ (it has a larger variance) and $\hat{x}_{t-1}^{(1)}$ is not correlated with $\hat{x}_{t-1}^{(2)}$; as in addition H_t shows that $\hat{x}_{t-1}^{(1)}$ and $\hat{x}_{t-1}^{(2)}$ affect $\hat{y}_t$ equally, the error between y_t and $\hat{y}_t$ should induce a larger correction in $\hat{x}^{(1)}$ than in $\hat{x}^{(2)}$.

(ii) The dimensions of the matrix inverted in the computation of K_t are fixed by the dimensions of $\mathbf{y}_t$. Here $\mathbf{y}_t$ is a scalar and the matrix is 1×1. △

We can find the minimal covariance of $\hat{\mathbf{x}}_t$ by substituting (7.3.9) into (7.3.6):

$$P_t = P_{t-1} - K_t H_t P_{t-1} - P_{t-1} H_{t-1}^{\mathrm{T}} K_t^{\mathrm{T}} + K_t (H_t P_{t-1} H_t^{\mathrm{T}} + R) K_t^{\mathrm{T}}$$

$$= P_{t-1} - P_{t-1} H_t^{\mathrm{T}} (H_t P_{t-1} H_t^{\mathrm{T}} + R)^{-1} H_t P_{t-1} \tag{7.3.10}$$

$$= (I - K_t H_t) P_{t-1} \tag{7.3.11}$$

A simpler expression for K_t can be found from (7.3.10) and (7.3.9):

$$P_t H_t^{\mathrm{T}} R^{-1} = P_{t-1} H_t^{\mathrm{T}} (I - (H_t P_{t-1} H_t^{\mathrm{T}} + R)^{-1} H_t P_{t-1} H_t^{\mathrm{T}}) R^{-1}$$

$$= P_{t-1} H_t^{\mathrm{T}} (H_t P_{t-1} H_t^{\mathrm{T}} + R)^{-1} (H_t P_{t-1} H_t^{\mathrm{T}} + R - H_t P_{t-1} H_t^{\mathrm{T}}) R^{-1}$$

$$= P_{t-1} H_t^{\mathrm{T}} (H_t P_{t-1} H_t^{\mathrm{T}} + R)^{-1} = K_t \tag{7.3.12}$$

Equations (7.3.11) and (7.3.12) for P_t and K_t simplify the algebra but are not necessarily good recipes for computation, where sensitivity to round-off error may have to be considered. They could not be used together, in any case, since P_t would require K_t and vice versa.

Example 7.3.3 The covariance of $\hat{\mathbf{x}}_t$ in Example 7.3.2 is given by (7.3.11) as

$$P_t = \left(\begin{bmatrix}1 & 0\\0 & 1\end{bmatrix} - \begin{bmatrix}0.4\\0.1\end{bmatrix}[1 \quad 1]\right)\begin{bmatrix}4 & 0\\0 & 1\end{bmatrix} = \begin{bmatrix}2.4 & -0.4\\-0.4 & 0.9\end{bmatrix}$$

so the estimated standard deviation of $\hat{x}_t^{(1)}$ is $\sqrt{2.4} \simeq 1.55$, that of $\hat{x}_t^{(2)}$ is $\sqrt{0.9} \simeq 0.95$, and $\hat{x}_t^{(1)}$ and $\hat{x}_t^{(2)}$ are negatively correlated.

Sub-optimal correction gains $k_t^{(1)} = k_t^{(2)} = 0.2$, say, would make the covariance, given by (7.3.6),

$$\begin{aligned}P_t &= \begin{bmatrix}0.8 & -0.2\\-0.2 & 0.8\end{bmatrix}\begin{bmatrix}4 & 0\\0 & 1\end{bmatrix}\begin{bmatrix}0.8 & -0.2\\-0.2 & 0.8\end{bmatrix} + \begin{bmatrix}0.2\\0.2\end{bmatrix}5[0.2 \quad 0.2]\\ &= \begin{bmatrix}2.8 & -0.6\\-0.6 & 1.0\end{bmatrix}\end{aligned}$$

The variances of $\hat{x}_t^{(1)}$ and $\hat{x}_t^{(2)}$ are larger, and the difference between this P_t and the optimal one is easily seen to be non-negative-definite, e.g. by Sylvester's criterion. △

At this juncture we can look back at Fig. 7.3.1 and recognise the whole mechanism in our equations. The only other things required are initial conditions $\hat{\mathbf{x}}_0$ and P_0 to start the recursion when $\mathbf{y}_1$ arrives. If $\hat{\mathbf{x}}_0$ is poor and P_0 very large, the effect is that

$$\begin{aligned}H_1\hat{\mathbf{x}}_1 &= H_1(\hat{\mathbf{x}}_0 + K_1(\mathbf{y}_1 - H_1\hat{\mathbf{x}}_0))\\ &= H_1(\hat{\mathbf{x}}_0 + H_1P_0H_1^{\mathrm{T}}(H_1P_0H_1^{\mathrm{T}} + R)^{-1}(\mathbf{y}_1 - H_1\hat{\mathbf{x}}_0))\\ &\simeq H_1\hat{\mathbf{x}}_0 + (\mathbf{y}_1 - H_1\hat{\mathbf{x}}_0) = \mathbf{y}_1 \qquad (7.3.13)\end{aligned}$$

That is to say, the correction of $\hat{\mathbf{x}}_0$ to $\hat{\mathbf{x}}_1$ is almost enough to make $H_1\hat{\mathbf{x}}_1$ fit $\mathbf{y}_1$ exactly. The influence of $\hat{\mathbf{x}}_0$ is negligible because its uncertainty, specified by P_0, is much larger than that in $\mathbf{y}_1$, specified by the noise covariance R. Generally, the larger P_0 the smaller the influence of $\hat{\mathbf{x}}_0$.

In state estimation, (7.3.9), (7.3.5) and (7.3.6) or their alternatives listed below are a large part of the famous *Kalman filter* (Kalman, 1960; Jazwinski, 1970; Maybeck, 1979). The only items missing account for time variation of the state, as described by a state equation. We are not primarily concerned with state estimation, but we shall bring in a state equation to describe evolution of the parameters of time-varying systems in Section 8.1.5. The resemblance between state and parameter estimation was pointed out not long after the Kalman filter was devised (Mayne, 1963; Lee, 1964).

Let us summarise what we have found so far.

Recursive Minimum-Covariance Linear Unbiased Estimator In order of computation,

$$\left.\begin{array}{lr} K_t = P_{t-1}H_t^{\mathrm{T}}(H_tP_{t-1}H_t^{\mathrm{T}} + R)^{-1} & (7.3.9) \\ \hat{\mathbf{x}}_t = \hat{\mathbf{x}}_{t-1} + K_t(\mathbf{y}_t - H_t\hat{\mathbf{x}}_{t-1}) & (7.3.5) \\ P_t = (I - K_tH_t)P_{t-1}(I - H_t^{\mathrm{T}}K_t^{\mathrm{T}}) + K_tRK_t^{\mathrm{T}} & (7.3.6) \\ & (\text{or } (7.3.11)) \end{array}\right\} \text{ or } \left\{\begin{array}{l} (7.3.10) \\ (7.3.12) \\ (7.3.5) \end{array}\right.$$

for $t = 1, 2, \ldots, N$; $\hat{\mathbf{x}}_0, P_0$ given.

The updating equations can be written in several other ways, differing in numerical properties such as sensitivity to round-off error.

It is important to remember that we assumed $\mathbf{v}_t$ to be uncorrelated with $\hat{\mathbf{x}}_{t-1} - \mathbf{x}$. Since $\hat{\mathbf{x}}_{t-1}$ depends on $\mathbf{y}_{t-1}$ and through $\hat{\mathbf{x}}_{t-2}$ on all earlier observations, it depends on the noise present in those observations. Our assumption therefore implies that $\{\mathbf{v}\}$ is white. The covariance R is between noise variables all at one time, and does not describe correlation between successive samples.

We have treated H_t as deterministic in deriving this algorithm. The observation equation (7.3.1) is a vector-output generalisation of our usual regression-type model, with H_t made up of regressors. When we rewrite a transfer-function model like (7.1.3) in regression-equation form, (7.1.9) or (7.2.4), the regressors are partly stochastic since they include earlier output samples, and they are also generally correlated with the regression-equation noise. A vector-output version would have H_t stochastic and correlated with $\mathbf{v}_t$ in (7.3.1). Section 7.4 is concerned with finding minimum-covariance linear recursive estimates in those more complicated circumstances.

Having just derived the estimator from first principles, by direct minimisation of the covariance subject to linearity and unbiasedness, we trace an alternative derivation in the next section.

7.3.3 Recursive Minimum-Covariance Estimator Derived from Least Squares

The idea behind this derivation is to apply the batch Markov (g.l.s.) estimate given by (5.3.25) to all the observations, and pick out the effect of the new observation and corresponding regressor samples at time t. Again we do the vector-output case. The regression equation (7.3.1) from 1 to time t gives altogether

$$\mathcal{Y}_t = \mathcal{H}_t \mathbf{x} + \mathcal{V}_t \tag{7.3.14}$$

where

$$\mathcal{Y}_t = \begin{bmatrix} \mathbf{y}_1 \\ \mathbf{y}_2 \\ \vdots \\ \mathbf{y}_t \end{bmatrix} \quad \mathcal{H}_t = \begin{bmatrix} H_1 \\ H_2 \\ \vdots \\ H_t \end{bmatrix} \quad \mathcal{V}_t = \begin{bmatrix} \mathbf{v}_1 \\ \mathbf{v}_2 \\ \vdots \\ \mathbf{v}_t \end{bmatrix} \tag{7.3.15}$$

The Markov estimate based on (7.3.14) is

$$\hat{\mathbf{x}}_t = (\mathcal{H}_t^{\mathrm{T}} \mathcal{R}_t^{-1} \mathcal{H}_t)^{-1} \mathcal{H}_t^{\mathrm{T}} \mathcal{R}_t^{-1} \mathcal{Y}_t \tag{7.3.16}$$

where $\mathcal{R}_t$ is the covariance matrix of $\mathcal{V}_t$. As in the previous section, we assume that $\{\mathbf{v}\}$ is white, with cov $\mathbf{v}_t$ given by R_t. Therefore, cov $\mathcal{V}_t$ is a block-diagonal matrix and has a block-diagonal inverse:

$$\mathcal{R}_t = \begin{bmatrix} R_1 & 0 & & \cdots & 0 \\ 0 & R_2 & 0 & \cdots & 0 \\ \vdots & & & & \\ 0 & \cdots & 0 & & R_t \end{bmatrix} \qquad \mathcal{R}_t^{-1} = \begin{bmatrix} R_1^{-1} & 0 & & \cdots & 0 \\ 0 & R_2^{-1} & 0 & \cdots & 0 \\ \vdots & & & & \\ 0 & \cdots & 0 & & R_t^{-1} \end{bmatrix} \tag{7.3.17}$$

Hence

$$\mathcal{H}_t^{\mathrm{T}} \mathcal{R}_t^{-1} = [H_1^{\mathrm{T}} R_1^{-1} \quad H_2^{\mathrm{T}} R_2^{-1} \quad \cdots \quad H_t^{\mathrm{T}} R_t^{-1}] \tag{7.3.18}$$

and defining $(\mathcal{H}_t^{\mathrm{T}} \mathcal{R}_t^{-1} \mathcal{H}_t)^{-1}$ as P_t, which we know from the end of Section 5.3.5 is the covariance of $\hat{\mathbf{x}}_t$, we have

$$\begin{aligned} P_t^{-1} &= \mathcal{H}_t^{\mathrm{T}} \mathcal{R}_t^{-1} \mathcal{H}_t = \sum_{i=1}^{t} H_i^{\mathrm{T}} R_i^{-1} H_i = \mathcal{H}_{t-1}^{\mathrm{T}} \mathcal{R}_{t-1}^{-1} \mathcal{H}_{t-1} + H_t^{\mathrm{T}} R_t^{-1} H_t \\ &= P_{t-1}^{-1} + H_t^{\mathrm{T}} R_t^{-1} H_t \end{aligned} \tag{7.3.19}$$

Similarly,

$$\mathcal{H}_t^{\mathrm{T}} \mathcal{R}_t^{-1} \mathcal{Y}_t = \mathcal{H}_{t-1}^{\mathrm{T}} \mathcal{R}_{t-1}^{-1} \mathcal{Y}_{t-1} + H_t^{\mathrm{T}} R_t^{-1} \mathbf{y}_t \tag{7.3.20}$$

and so

$$\begin{aligned} \hat{\mathbf{x}}_t &= P_t(\mathcal{H}_{t-1}^{\mathrm{T}} \mathcal{R}_{t-1}^{-1} \mathcal{Y}_{t-1} + H_t^{\mathrm{T}} R_t^{-1} \mathbf{y}_t) = P_t(P_{t-1}^{-1} \hat{\mathbf{x}}_{t-1} + H_t^{\mathrm{T}} R_t^{-1} \mathbf{y}_t) \\ &= P_t((P_t^{-1} - H_t^{\mathrm{T}} R_t^{-1} H_t)\hat{\mathbf{x}}_{t-1} + H_t^{\mathrm{T}} R_t^{-1} \mathbf{y}_t) \\ &= \hat{\mathbf{x}}_{t-1} + P_t H_t^{\mathrm{T}} R_t^{-1} (\mathbf{y}_t - H_t \hat{\mathbf{x}}_{t-1}) \end{aligned} \tag{7.3.21}$$

We recognise this updating equation as identical to (7.3.5) with K_t replaced by $P_t H_t^{\mathrm{T}} R_t^{-1}$ as in (7.3.12). The covariance updating equation (7.3.19) is less familiar. We can twist it into a recognisable form by writing

$$\begin{aligned} P_{t-1} &= P_{t-1} P_t^{-1} P_t = P_{t-1}(P_{t-1}^{-1} + H_t^{\mathrm{T}} R_t^{-1} H_t) P_t \\ &= P_t + P_{t-1} H_t^{\mathrm{T}} R_t^{-1} H_t P_t \end{aligned} \tag{7.3.22}$$

which gives

$$H_t P_{t-1} = (I + H_t P_{t-1} H_t^{\mathrm{T}} R_t^{-1}) H_t P_t \tag{7.3.23}$$

On substituting (7.3.23) into (7.3.22), we find

$$P_{t-1} = P_t + P_{t-1} H_t^{\mathrm{T}} R_t^{-1} (I + H_t P_{t-1} H_t^{\mathrm{T}} R_t^{-1})^{-1} H_t P_{t-1} \tag{7.3.24}$$

so

$$P_t = P_{t-1} - P_{t-1} H_t^{\mathrm{T}} (R_t + H_t P_{t-1} H_t^{\mathrm{T}})^{-1} H_t P_{t-1} \tag{7.3.25}$$

This is (7.3.10). The equivalence of (7.3.19) and (7.3.25) is called *the matrix-inversion lemma*. It is a special case of

$$(A + BC)^{-1} = A^{-1} - A^{-1} B (I + C A^{-1} B)^{-1} C A^{-1} \tag{7.3.26}$$

and is often helpful in obtaining alternative ways of writing the updating equations. Although (7.3.25) looks more complicated than (7.3.19), it uses only one matrix inversion to go from P_{t-1} to P_t, whereas (7.3.19) takes two. Moreover, the matrix inverted in (7.3.25) has as many rows or columns as the dimension of $\mathbf{y}_t$, usually smaller than the dimension of $\mathbf{x}$, so the matrix is smaller than P_{t-1} or P_t. If the covariance is updated on a short-word-length computer over many steps, (7.3.25) is risky, since round-off error may inflate the last term and ultimately cause P_t to be indefinite rather than positive-definite. Safer alternatives exist (Bierman, 1977; Maybeck, 1979), but floating-point computation with longer word length, say 30 bits, seems to avoid trouble.

The scalar-output algorithm has $\mathbf{h}_t^{\mathrm{T}}$ for H_t and σ_t^2 for R_t.

Scalar-Output Recursive Minimum-Covariance Linear Unbiased Estimator

$$P_t = P_{t-1} - P_{t-1} \mathbf{h}_t \mathbf{h}_t^{\mathrm{T}} P_{t-1} / (\sigma_t^2 + \mathbf{h}_t^{\mathrm{T}} P_{t-1} \mathbf{h}_t) \tag{7.3.27}$$

$$\hat{\mathbf{x}}_t = \hat{\mathbf{x}}_{t-1} + P_t \mathbf{h}_t (y_t - \mathbf{h}_t^{\mathrm{T}} \hat{\mathbf{x}}_{t-1}) / \sigma_t^2 \tag{7.3.28}$$

for $t = 1, 2, \ldots, N$; $\hat{\mathbf{x}}_0, P_0$ given.

The noise variance σ_t^2 may well be constant but unknown. If so, we can define a normalised covariance $\bar{P}_t$ as P_t/σ^2 and write

$$\bar{P}_t = \bar{P}_{t-1} - \bar{P}_{t-1}\mathbf{h}_t\mathbf{h}_t^{\mathrm{T}}\bar{P}_{t-1}/(1+\mathbf{h}_t^{\mathrm{T}}\bar{P}_{t-1}\mathbf{h}_t) \tag{7.3.29}$$

$$\hat{\mathbf{x}}_t = \hat{\mathbf{x}}_{t-1} + \bar{P}_t\mathbf{h}_t(y_t - \mathbf{h}_t^{\mathrm{T}}\hat{\mathbf{x}}_{t-1}) \tag{7.3.30}$$

This algorithm is the cornerstone of s.i.s.o. recursive identification. Although the algorithm was generalised and brought to prominence by Plackett (1950), its essentials were derived, without the benefit of matrix algebra, by Gauss (Young, 1984).

Example 7.3.4 In example 4.1.4, a weighted l.s. estimate of the parameter vector [initial position, initial velocity, acceleration]^{T} was computed for the tracking problem of Example 4.1.1. Observations at six sampling instants were processed using the batch w.l.s. estimator

$$\hat{\boldsymbol{\theta}}_W = [U^{\mathrm{T}}WU]^{-1}U^{\mathrm{T}}W\mathbf{y}$$

The weighting matrix W was diagonal with principal diagonal

$$[4 \quad 4 \quad 1 \quad 1 \quad 1 \quad 4]$$

$\hat{\boldsymbol{\theta}}_w$ is equal to the Markov estimate for a system with white noise of covariance $R = \mathrm{diag}[\frac{1}{4} \quad \frac{1}{4} \quad 1 \quad 1 \quad 1 \quad \frac{1}{4}]$, i.e. of time-varying variance $\frac{1}{4}, \frac{1}{4}, 1, 1, 1, \frac{1}{4}$, since the Markov and w.l.s. estimates differ only by R^{-1} replacing W. We use our recursive Markov estimator instead. The model was

$$x(\tau) = x_0 + v_0\tau + a\tau^2/2$$

and the data were

τ	0	0.2	0.4	0.6	0.8	1
$x(\tau) \equiv x_t$	3	59	98	151	218	264
t	1	2	3	4	5	6

In our present notation,

$$y_t \equiv x_t, \qquad \mathbf{x} \equiv [x_0 \quad v_0 \quad a]^{\mathrm{T}}$$

$$\mathbf{h}_t^{\mathrm{T}} \equiv [1 \quad \tau \quad \tau^2/2] \equiv [1 \quad 0.2(t-1) \quad 0.02(t-1)^2]$$

Let us go through two recursion steps, starting with a guess $\hat{\mathbf{x}}_0 = \mathbf{0}$ and $P_0 = 10^4 I$, which states correctly that $\hat{\mathbf{x}}_0$ is very poor. Calculation to about eight figures is necessary because of ill-conditioning, but results will be quoted to four figures for conciseness.

At $\tau = 0$, t is 1 and $\mathbf{h}_1^T = [1 \quad 0 \quad 0]$, $\mathbf{h}_1^T P_0 = [10^4 \quad 0 \quad 0]$. Equations (7.3.27) and (7.3.28) then give

$$P_1 = 10^4 I - \begin{bmatrix} 10^8 & 0 & 0 \\ 0 & 0 & 0 \\ 0 & 0 & 0 \end{bmatrix} \Bigg/ \left(\frac{1}{4} + 10^4\right) \simeq \begin{bmatrix} 0.25 & 0 & 0 \\ 0 & 10^4 & 0 \\ 0 & 0 & 10^4 \end{bmatrix}$$

$$\hat{\mathbf{x}}_1 \simeq \mathbf{0} + \begin{bmatrix} 0.25 & 0 & 0 \\ 0 & 10^4 & 0 \\ 0 & 0 & 10^4 \end{bmatrix} \begin{bmatrix} 1 \\ 0 \\ 0 \end{bmatrix} \left(\frac{3 - \mathbf{h}_1^T \mathbf{0}}{0.25} \right) = \begin{bmatrix} 3 \\ 0 \\ 0 \end{bmatrix}$$

At $\tau = 0.2$, $t = 2$, $\mathbf{h}_2^T = [1 \quad 0.2 \quad 0.02]$, $\mathbf{h}_2^T P_1 \simeq [0.25 \quad 2000 \quad 200]$, $\sigma_2^2 + \mathbf{h}_2^T P_1 \mathbf{h}_2 \simeq 404.5$,

$$P_2 \simeq \begin{bmatrix} 0.2498 & -1.236 & -0.1236 \\ -1.236 & 111.2 & -988.9 \\ -0.1236 & -988.9 & 9901 \end{bmatrix}$$

$$P_2 \mathbf{h}_2 \simeq \begin{bmatrix} 1.545 \times 10^{-4} \\ 1.236 \\ 0.1236 \end{bmatrix}$$

$$y_2 - \mathbf{h}_2^T \hat{\mathbf{x}}_1 \simeq 56, \qquad \hat{\mathbf{x}}_2 \simeq \begin{bmatrix} 3.035 \\ 276.9 \\ 27.69 \end{bmatrix}$$

Figure 7.3.2 shows the evolution of $\hat{\mathbf{x}}_t$ and P_t over six updates. Convergence is rapid, and $\hat{\mathbf{x}}_6$ is $[4.613 \quad 250.1 \quad 19.30]^T$, close to the batch estimate $[4.592 \quad 250.2 \quad 18.97]^T$. As the initial error in v_0 is about 250, its initial estimated variance 10^4 used as element (2, 2) of P_0 is rather small. A guide is that each principal-diagonal element of P_0 should not be smaller than the square of the largest initial error in that parameter which would be unremarkable. For a Gaussian random variable, the $\pm\sigma$ limits encompass about $\frac{2}{3}$ of all samples, so the guide is reasonable.

The main numerical difficulty in this example is in calculating the gain $P_t \mathbf{h}_t$. △

7.3.4 Information Updating

The updating equation (7.3.19) for P^{-1} is worth closer inspection. An illuminating interpretation is that P^{-1} measures information, and (7.3.19) says how much information about $\mathbf{x}$ is supplied by the new observation $\mathbf{y}_t$. For a Gaussian $\hat{\mathbf{x}}_t$, we can show that P_t^{-1} is in fact the Fisher information matrix.

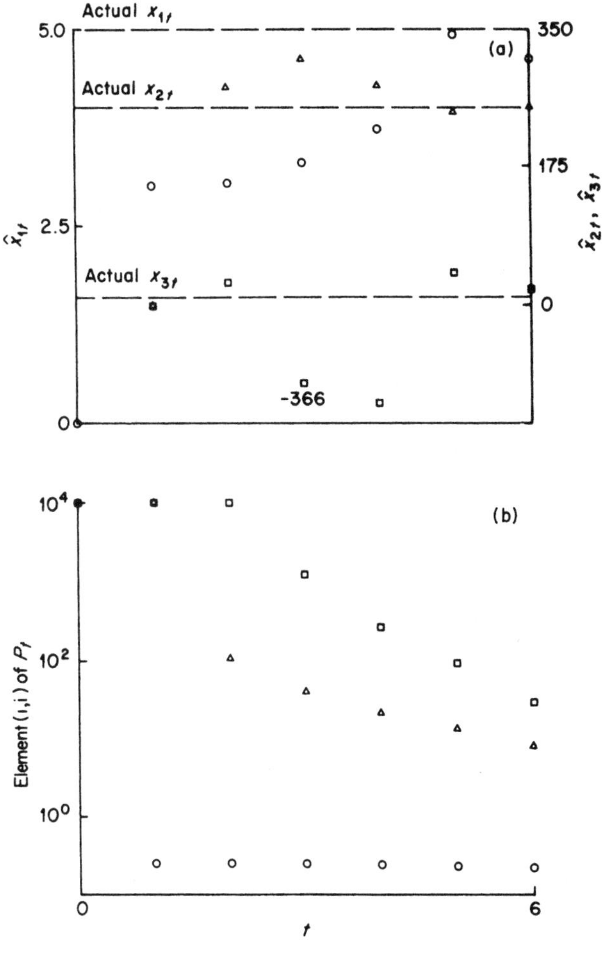

Fig. 7.3.2 Recursive minimum-covariance unbiased estimates and their computed variances, Example 7.3.4, where ○: $\hat{x}_{1t}$; △: $\hat{x}_{2t}$; □: $\hat{x}_{3t}$.

First we must realise that the p.d.f. $p(\hat{\mathbf{x}}_t)$ can act the part of the p.d.f. of the observations given the parameters, in the definition (5.4.1) of the information matrix. We can think of $\hat{\mathbf{x}}_t$ as a vector of processed observations, containing the information in all the original observations up to time t. Omitting the explicit dependence of $\hat{\mathbf{x}}_t$ on $\mathbf{x}$ from our notation for brevity, we have

$$p(\hat{\mathbf{x}}_t) = \text{const} \times \exp(-\tfrac{1}{2}(\hat{\mathbf{x}}_t - \mathbf{x}_t)^{\mathrm{T}} P_t^{-1}(\hat{\mathbf{x}}_t - \mathbf{x}_t)) \qquad (7.3.31)$$

since $\mathbf{x}_t$ is the mean of the unbiased $\hat{\mathbf{x}}_t$ and P_t its covariance. Hence

$$\frac{\partial}{\partial \hat{\mathbf{x}}_t}((\ln p(\hat{\mathbf{x}}_t)) = \frac{\partial}{\partial \hat{\mathbf{x}}_t}(\text{const} - \tfrac{1}{2}(\hat{\mathbf{x}}_t - \mathbf{x}_t)^{\mathrm{T}} P_t^{-1}(\hat{\mathbf{x}}_t - \mathbf{x}_t))$$

$$= -P_t^{-1}(\hat{\mathbf{x}}_t - \mathbf{x}_t) \tag{7.3.32}$$

and the Fisher information matrix for $\hat{\mathbf{x}}_t$ is

$$F_t \triangleq E\left[\frac{\partial}{\partial \hat{\mathbf{x}}_t}(\ln p(\hat{\mathbf{x}}_t))\left(\frac{\partial}{\partial \hat{\mathbf{x}}_t}(\ln p(\hat{\mathbf{x}}_t))\right)^{\mathrm{T}}\right]$$

$$= P_t^{-1} E[(\hat{\mathbf{x}}_t - \mathbf{x}_t)(\hat{\mathbf{x}}_t - \mathbf{x}_t)^{\mathrm{T}}] P_t^{-1} = P_t^{-1} \tag{7.3.33}$$

With this interpretation, the role of $H_t^{\mathrm{T}} R_t^{-1} H_t$ in (7.3.19) is clear. A larger observation-noise covariance implies that $\mathbf{y}_t$ brings less information about $\mathbf{x}$. A large P_0 is seen to signal that $\hat{\mathbf{x}}_0$ contains little information.

The idea of updating P^{-1} rather than P has received a lot of attention in state estimation (Bierman, 1977) but not apparently in identification. The relative dimensions of $\mathbf{y}$ and $\mathbf{x}$ determine whether there is any computational saving.

7.4 RECURSIVE IDENTIFICATION INCLUDING A NOISE-STRUCTURE MODEL

We now examine several techniques which approximate the minimum-covariance estimate when the noise is not white but can be modelled as linearly filtered white noise. The notation that $\mathbf{x}$ is the parameter vector and $\mathbf{h}$ the "regressor" vector will be retained, to allow easy comparison with the preceding recursive algorithms and avoid confusion between a regressor vector $\mathbf{u}$ and inputs u. It also helps when, in Chapter 8, we borrow further state-estimation methods.

7.4.1 Extended Least Squares

The aim of this algorithm is to modify (7.3.27) and (7.3.28) as little as possible yet attain acceptably small covariance and bias in the presence of autocorrelated noise. The algorithm takes D equal to A in the standard model (7.1.3) like the batch m.l. algorithm. The result is (7.2.4), or in our present notation

$$y_t = \mathbf{h}_t^{\mathrm{T}} \mathbf{x}_t + v_t \tag{7.4.1}$$

where

$$\mathbf{h}_t^{\mathrm{T}} = [y_{t-1} \quad \cdots \quad y_{t-n} \quad u_{t-k-1} \quad \cdots \quad u_{t-k-m} \quad v_{t-1} \quad \cdots \quad v_{t-q}]$$

$$\mathbf{x} = [-a_1 \quad \cdots \quad -a_n \quad b_1 \quad \cdots \quad b_m \quad c_1 \quad \cdots \quad c_q] \tag{7.4.2}$$

with $\{v\}$ zero-mean, white and usually of constant variance σ^2. If we knew v_{t-1} to v_{t-q} at time t, (7.3.27) and (7.3.28) would give minimum-covariance estimates, unbiased as v_t is uncorrelated with $\mathbf{h}_t$. As we step through the records updating $\hat{\mathbf{x}}$, we can generate residuals

$$\hat{v}_{t-i} = y_{t-i} - \mathbf{h}_{t-i}^{\mathrm{T}}\hat{\mathbf{x}}_{t-i}, \qquad i = q, q-1, \ldots, 1 \tag{7.4.3}$$

to stand in for the unknown v_{t-1} to v_{t-q} in $\mathbf{h}_t$. By doing so, we introduce an indirect link between the earlier estimates $\hat{\mathbf{x}}_{t-i}$ and $\hat{\mathbf{x}}_t$. The asymptotic and finite-sample behaviour of the estimator is thereby altered, and we can no longer be sure even that it is consistent. These worries are postponed to Section 7.5. For now, it is enough to know that misbehaviour is possible but is very seldom seen in practice.

The extended least squares (e.l.s.) algorithm became popular (Panuska, 1968; Young, 1968) under a variety of other names such as *Panuska's method*, *r.m.l.* 1 (recursive maximum-likelihood 1; r.m.l. 2 is discussed later) and, when used for noise-structure estimation as in Section 7.4.5, *a.m.l.* (approximate maximum-likelihood). A common variant of e.l.s. uses the innovation

$$\nu_{t-i} = y_{t-i} - \mathbf{h}_{t-i}^{\mathrm{T}}\hat{\mathbf{x}}_{t-i-1} \tag{7.4.4}$$

in place of $\hat{v}_{t-i}$. As $\hat{\mathbf{x}}_{t-i-1}$ is one step out of date, ν_{t-i} approximates v_{t-i} less well than does $\hat{v}_{t-i}$, and the performance of the algorithm suffers. There is rarely any computational saving since the residuals are needed for model validation anyway. Some authors reserve the names r.m.l. 1 for the version using (7.4.4) and a.m.l. for that using (7.4.3).

Extended Least-Squares Algorithm

For $t = 1, 2, \ldots, N$

(i) Update $\mathbf{h}_{t-1}$ to $\mathbf{h}_t$ as in (7.4.2), using $\hat{v}_{t-1}$ for v_{t-1}.

(ii) $P_t = P_{t-1} - P_{t-1}\mathbf{h}_t\mathbf{h}_t^{\mathrm{T}}P_{t-1}/(\sigma_t^2 + \mathbf{h}_t^{\mathrm{T}}P_{t-1}\mathbf{h}_t)$. (7.4.5)

(iii) Calculate innovation $\nu_t = y_t - \mathbf{h}_t^{\mathrm{T}}\hat{\mathbf{x}}_{t-1}$. (7.4.6)

(iv) $\hat{\mathbf{x}}_t = \hat{\mathbf{x}}_{t-1} + P_t\mathbf{h}_t\nu_t/\sigma_t^2$; $\hat{\mathbf{x}}_0$, P_0 given. (7.4.7)

Unknown early samples of v and perhaps u and y in early $\mathbf{h}_t$'s are taken as zero. Alternatively, batch o.l.s. can be performed on the first few sets of samples to get $\hat{\mathbf{x}}_0$, P_0 and the earliest residuals, but the extra programming effort is scarcely worthwhile.

A constant term can be added to the model by adjoining 1 as an extra element of $\mathbf{h}_t$, to go with the unknown constant as an element of $\mathbf{x}$. The extra term copes automatically with unknown means of $\{u\}$, $\{y\}$ and $\{v\}$ providing they do not change too quickly. Deterministic trends, e.g. ramps, could also be accommodated by extra regression terms, but we shall find more flexible and elegant methods in Section 8.1.

Most of the results in Chapter 10 were obtained by e.l.s.

7.4.2 Extended Matrix Method

A weakness of e.l.s. is that the noise-model order may have to be high. Coefficients c_1 to c_q are the u.p.r. ordinates of the filter which shapes the noise. If this transfer function, more generally $(1+C)(1+A)/(1+D)$, includes a z-plane pole just inside the unit circle, the u.p.r. has a long tail and q has to be large. Estimating an a.r.m.a. noise model avoids this problem (Talmon and van den Boom, 1973). With

$$E = \frac{(1+C)(1+A)}{1+D} V \triangleq \frac{1+C'}{1+D} V \tag{7.4.8}$$

a regression equation can be written in terms of both E and V which is linear in all the plant- and noise-model coefficients:

$$\begin{aligned} Y &= -AY + Bz^{-k}U + (1+C')V/(1+D) \\ &= -AY + Bz^{-k}U + C'V - DE + V \end{aligned} \tag{7.4.9}$$

The e.l.s. updating equations are used to estimate

$$\mathbf{x} = [-a_1 \quad \cdots \quad -a_n \quad b_1 \quad \cdots \quad b_m \quad c'_1 \quad \cdots \quad c'_q \quad -d_1 \quad \cdots \quad -d_r]^{\mathrm{T}} \tag{7.4.10}$$

Assuming enough previous samples are available, approximate values of v_{t-1} and e_{t-1} to update $\mathbf{h}_{t-1}$ can be provided by

$$\hat{e}_{t-1} = y_{t-1} + \hat{a}_1 y_{t-2} + \cdots + \hat{a}_n y_{t-n-1} - \hat{b}_1 u_{t-k-2} - \cdots - \hat{b}_m u_{t-k-m-1} \tag{7.4.11}$$

$$\hat{v}_{t-1} = -\hat{c}_1 \hat{v}_{t-2} - \cdots - \hat{c}_q \hat{v}_{t-q-1} + \hat{e}_{t-1} + \cdots + \hat{d}_r \hat{e}_{t-r-1} \tag{7.4.12}$$

where the coefficient estimates come from $\hat{\mathbf{x}}_{t-1}$. Initially zeros are used for $\hat{e}$ and $\hat{v}$ in $\mathbf{h}$, until $\hat{e}_{t-r-1}$ to $\hat{e}_{t-1}$ have all been found from (7.4.11). From that point on, $\mathbf{h}$ fills with values calculated by (7.4.11) and (7.4.12).

Talmon and van den Boom advise exponential weighting-out of old residuals, as in Section 8.1.3, to weaken the influence of poor initial estimates.

7.4.3 Extended Least-Squares as Approximate Maximum-Likelihood; Recursive Maximum-Likelihood 2

Extended least-squares appears in a new light if we think of (7.4.7) as a step of a numerical search routine to minimise the sum of squares of residuals

$$S_t = \sum_{l=1}^{t} \hat{v}_l^2 = \sum_{l=1}^{t} (y_l - \mathbf{h}_l^{\mathrm{T}}\hat{\mathbf{x}})^2 \tag{7.4.13}$$

The gradient is

$$\frac{\partial S_t}{\partial \hat{\mathbf{x}}} = 2\sum_{l=1}^{t} \frac{\partial \hat{v}_l}{\partial \hat{\mathbf{x}}}\hat{v}_l = -2\sum_{l=1}^{t} (\mathbf{h}_l + [J_{\hat{x}}\mathbf{h}_l]^{\mathrm{T}}\hat{\mathbf{x}})\hat{v}_l \tag{7.4.14}$$

where element (i, j) of the Jacobian matrix $[J_{\hat{x}}\mathbf{h}_l]$ is $\partial h_{l,i}/\partial \hat{x}_j$. The Jacobian is non-zero because $\hat{\mathbf{x}}$ affects $\hat{v}_{l-1}$ to $\hat{v}_{l-q}$ in $\mathbf{h}_l$, which are computed as $y_{l-i} - \mathbf{h}_{l-i}^{\mathrm{T}}\hat{\mathbf{x}}$, $1 \le i \le q$. Precise evaluation of $\partial S_t/\partial \hat{\mathbf{x}}$ would mean going back over the entire record to recalculate every $\hat{v}$ and $\mathbf{h}$ at each new value of $\hat{\mathbf{x}}$; the algorithm would not be recursive. To keep the computation recursive and simple, we might ignore at time t the influence of $\hat{\mathbf{x}}$ on all earlier v and $\mathbf{h}$ values, neglecting even $[J_{\hat{x}}\mathbf{h}_t]$. Only $-2\mathbf{h}_t\hat{v}_t$ would remain in $\partial S_t/\partial \hat{\mathbf{x}}$, and we could regard $\mathbf{h}_t v_t$ in (7.4.7) as approximating $-\frac{1}{2}\partial S_t/\partial \hat{\mathbf{x}}$ evaluated at $\hat{\mathbf{x}}_{t-1}$.

Element (i, j) of the Hessian matrix of S_t with respect to $\hat{\mathbf{x}}$ is

$$[H_{\hat{x}\hat{x}}S_t]_{ij} = \frac{\partial^2 S_t}{\partial \hat{x}_i \partial \hat{x}_j} \simeq -2\sum_{l=1}^{t} h_{li}\frac{\partial \hat{v}_l}{\partial \hat{x}_j} \simeq 2\sum_{l=1}^{t} h_{li}h_{lj} \tag{7.4.15}$$

neglecting the Jacobian again. The whole Hessian is then roughly

$$[H_{\hat{x}\hat{x}}S_t] \simeq 2\sum_{l=1}^{t} \mathbf{h}_l\mathbf{h}_l^{\mathrm{T}} \tag{7.4.16}$$

It relates easily to P_t in e.l.s., since from the information equation (7.3.19) specialised to scalar output

$$P_t^{-1} = P_{t-1}^{-1} + \mathbf{h}_t\mathbf{h}_t^{\mathrm{T}}/\sigma_t^2 \tag{7.4.17}$$

so that if the noise variance σ^2 is constant and P_0 is large,

$$P_t^{-1} = P_0^{-1} + \frac{\sum_{l=1}^{t} \mathbf{h}_l \mathbf{h}_l^{\mathrm{T}}}{\sigma^2} \simeq \frac{\sum_{l=1}^{t} \mathbf{h}_l \mathbf{h}_l^{\mathrm{T}}}{\sigma^2} \simeq [H_{\hat{x}\hat{x}} S_t]/2\sigma^2 \tag{7.4.18}$$

The e.l.s. updating of $\hat{\mathbf{x}}$ is therefore

$$\begin{aligned} \hat{\mathbf{x}} &= \hat{\mathbf{x}}_{t-1} + P_t \mathbf{h}_t v_t/\sigma^2 \\ &\simeq \hat{\mathbf{x}}_{t-1} - \{[H_{\hat{x}\hat{x}} S_t]^{-1} \partial S_t/\partial \hat{\mathbf{x}}\}_{\hat{\mathbf{x}} = \hat{\mathbf{x}}_{t-1}} \end{aligned} \tag{7.4.19}$$

We recognise this as a Newton–Raphson step (Adby and Dempster, 1974) towards minimising S_t.

Extended least-squares can be modified relatively cheaply to include the Jacobian in the gradient calculation (7.4.14) by finding an expression for $\partial v_l/\partial \hat{\mathbf{x}}$ which does not necessitate going back over the whole record (Söderström, 1973). Since

$$\begin{aligned} \hat{v}_l = y_l - h_l^{\mathrm{T}} \hat{\mathbf{x}} \equiv y_l + \hat{a}_1 y_{l-1} + \cdots + \hat{a}_n y_{l-n} - \hat{b}_1 u_{l-k-1} - \cdots - \hat{b}_m u_{l-k-m} \\ - \hat{c}_1 \hat{v}_{l-1} - \cdots - \hat{c}_q \hat{v}_{l-q} \end{aligned} \tag{7.4.20}$$

we have for each element $\hat{x}_i$ of $\hat{\mathbf{x}}$

$$\frac{\partial \hat{v}_l}{\partial \hat{x}_i} = -\begin{pmatrix} \text{element } i \\ \text{of } \mathbf{h}_l \end{pmatrix} - \hat{c}_1 \frac{\partial \hat{v}_{l-1}}{\partial \hat{x}_i} - \cdots - \hat{c}_q \frac{\partial \hat{v}_{l-q}}{\partial \hat{x}_i} \tag{7.4.21}$$

Consequently $-\mathbf{h}_l$ is $\partial \hat{v}_l/\partial \hat{\mathbf{x}}$ filtered by $1 + \hat{C}$, and $\partial \hat{v}_l/\partial \hat{\mathbf{x}}$ is obtainable by filtering $-\mathbf{h}_l$ by $1/(1 + \hat{C})$, instead of using $-\mathbf{h}_l$ for $\partial \hat{v}_l/\partial \hat{\mathbf{x}}$ as we did formerly. The modification improves the asymptotic properties of e.l.s., ensuring consistency for all parameter values, which e.l.s. lacks (Ljung, Söderström and Gustavsson, 1975). The modified algorithm is often called r.m.l. 2 to distinguish it from r.m.l. 1, which is e.l.s.

The actual performance of r.m.l. 2 at realistic s.n.r. is an object lesson in the dangers of relying on asymptotic results. The filtering by $1/(1 + \hat{C})$ is often unstable, as $\hat{C}$ is initially poor (Norton, 1977). Even when it is not unstable, the filtering is liable to be counterproductive, because it uses (7.4.21) with old values from q previous steps as $\partial \hat{v}_{l-1}/\partial \hat{x}_i$ to $\partial \hat{v}_{l-q}/\partial \hat{x}_i$. Hence $\partial \hat{v}_l/\partial \hat{x}_i$ is still not accurate, especially when $\hat{\mathbf{x}}$ is changing rapidly from step to step, and errors persist and accumulate whenever $1/(1 + \hat{C})$ has one or more poles close to the unit circle. A stability check is therefore no guarantee of good performance.

7.4.4 Stochastic Approximation

The heaviest computational task in e.l.s. and related methods is updating the covariance. We could dispense with the covariance updating and calculate the correction gain some other way, while keeping the corrector form of updating for $\hat{\mathbf{x}}$, but only at the price of a larger actual covariance for $\hat{\mathbf{x}}$. Unbiasedness, by contrast, does not depend on the choice of correction-gain matrix, K_t in (7.3.5), providing the regressors are not correlated with the noise.

The simplest scheme would be to replace P_t/σ_t^2 in (7.3.28) or (7.4.7) by a predetermined positive scalar γ_t, giving a *stochastic approximation* (s.a.) algorithm.

Stochastic Approximation Algorithm

(i) Update $\mathbf{h}_{t-1}$ to $\mathbf{h}_t$.
(ii) Calculate innovation $v_t = y_t - \mathbf{h}_t^T \hat{\mathbf{x}}_{t-1}$.
(iii) $\hat{\mathbf{x}}_t = \hat{\mathbf{x}}_{t-1} + \gamma_t \mathbf{h}_t v_t$ (7.4.22)
for $t = 1, 2, \ldots, N$; $\{\gamma\}$ given.

Stochastic approximation (Robbins and Monro, 1951) originated as a trial-and-error technique for finding the root x of a scalar equation

$$m(x) = y \tag{7.4.23}$$

from noisy observations $\hat{m}(\hat{x}_t)$, $t = 1, 2, \ldots$, where the forms of $m(\cdot)$ and the noise p.d.f. are not necessarily known, but the noise has zero mean. Successive trial values $\hat{x}_t$ are found by the scalar version of (7.4.22) with $y - \hat{m}(\hat{x}_{t-1})$ for $\mathbf{h}_t v_t$. A distinctive feature of s.a. is that $m(\cdot)$ need not be parameterised, so its adoption for parameter estimation is ironical.

As the correction to $\hat{\mathbf{x}}$ is still proportional to $\mathbf{h}_t v_t$, we can interpret s.a. as another gradient-based method for minimising S_t. The step-size factors $\{\gamma\}$ have to be both large enough for $\hat{\mathbf{x}}$ to reach the correct value and small enough finally for $\hat{\mathbf{x}}$ to stay there. As v_t depends on the noise, $\mathbf{h}_t v_t$ is a random variable with, in general, a finite covariance, so $\{\hat{\mathbf{x}}\}$ can settle to a zero-covariance limit as $t \to \infty$ only if $\gamma_t \to 0$. In e.l.s. the corrections decrease automatically as P_t decreases. A sufficient condition for $\{\hat{\mathbf{x}}\}$ to converge w.p.1 (and in mean square) is that $\sum_{l=1}^{t} \gamma_l^2$ stays finite as $t \to \infty$. The commonest choice for γ_t, $Ct^{-\alpha}$ with $\frac{1}{2} < \alpha \leq 1$ and C a positive constant, satisfies this condition. The sequence $\{\gamma\}$ can decrease more slowly without losing convergence if the noise is bounded, which it is in actual observations; $0 < \alpha \leq 1$ is allowed (Ljung, 1977a). If $\{\gamma\}$ is to decrease slowly enough for $\{\hat{\mathbf{x}}\}$ to reach the correct value however far away it started at $\hat{\mathbf{x}}_0$, $\sum_{l=1}^{t} \gamma_l$ must tend to infinity as $t \to \infty$.

There is a large literature on how to select $\{\gamma\}$, but s.a. incurs such a penalty in slower convergence through replacing a matrix-dependent updating gain by a gain depending on a scalar that it is useful only as a last resort, when computing power is very limited. One of the original attractions of s.a., the relative ease with which convergence can be guaranteed, has lost its significance now that the convergence properties of more ambitious algorithms are better understood (Ljung, 1977a; Goodwin and Sin, 1984; Solo, 1979).

Example 7.4.1 We shall repeat Example 7.3.4, but using the s.a. updating equation (7.4.22) with

(i) $\gamma_t = 1/t$ so that the update at $t = 1$ is the same as in Example 7.3.4;
(ii) $\gamma_t = 1/10t$;
(iii) $\gamma_t = 10/\mathrm{t}$.

Figure 7.4.1 shows the resulting estimates.

Case (i) shows clearly the ill effects of restricting the correction to be a scalar times $\mathbf{h}_t$; the initial-position estimate rises far too rapidly and the initial velocity far too slowly. The results also underline the fact that consistency (ensured by this choice of $\{\gamma\}$) is no guarantee of good, or even tolerable, small-sample behaviour. Case (ii) shows the effects of observation noise less, but is even slower to converge than case (i). Case (iii) illustrates the serious consequences of making γ_t larger. We can explain them by examining the estimation-error recursion resulting from (7.4.22),

$$\tilde{\mathbf{x}}_t = (I - \gamma_t \mathbf{h}_t \mathbf{h}_t^{\mathrm{T}})\tilde{\mathbf{x}}_{t-1} + \gamma_t \mathbf{h}_t v_t$$

where $\tilde{\mathbf{x}}$ is $\mathbf{x} - \hat{\mathbf{x}}$. The sum of the eigenvalues of the transition matrix $I - \gamma_t \mathbf{h}_t \mathbf{h}_t^{\mathrm{T}}$ is its trace $p - \gamma_t \mathbf{h}_t^{\mathrm{T}} \mathbf{h}_t$. If γ_t is large enough and $\mathbf{h}_t^{\mathrm{T}} \mathbf{h}_t$ not too small, at least one eigenvalue must be below -1. The recursion step is then unstable, and contributes to divergence of $\tilde{\mathbf{x}}$. △

7.4.5 Recursive Instrumental-Variable Algorithm†

The e.l.s., e.m.m. and r.m.l. 2 algorithms circumvented the bias due to correlation between regressors and regression-equation noise by integrating a noise model into the regression equation. The remaining noise-generating term v_t was white and uncorrelated with the input, and as a result uncorrelated with the regressors. An alternative stratagem, seen in batch form in Section 5.3.6, is to replace the noise-correlated regressors by instrumental variables, which are totally uncorrelated with the noise. They must also be strongly correlated with the information-bearing regressors to yield low-covariance estimates. The idea can be implemented recursively, but it shares a weakness with the alternative recursive methods. The instrumental variables and the

† Söderström and Stoica, 1983; Young, 1974, 1984.

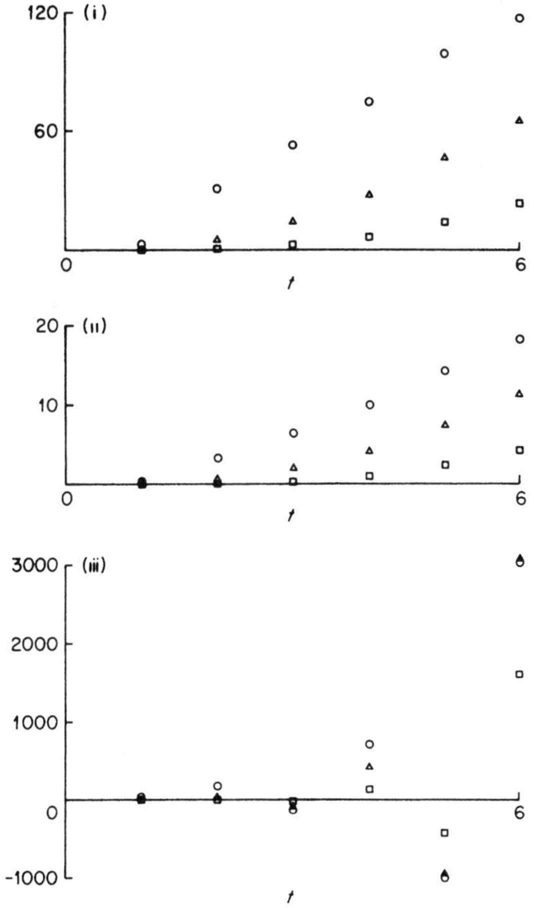

Fig. 7.4.1 Stochastic approximation estimates, Example 7.4.1, where ○: x_{1t}; △: x_{2t}; □: x_{3t}. (i) $\gamma_t = 1/t$, (ii) $\gamma_t = 1/10t$ and (iii) $\gamma_t = 10/t$.

noise models of the alternatives are only as good as the parameter estimates they depend on. Early estimates are poor, so bias occurs initially, declining, one hopes, as more observations are processed. It is not at all obvious that any of the methods will indeed converge to unbiased parameter estimates with acceptable covariance. We pursue convergence further in Section 7.5.

The recursive instrumental-variable (r.i.v.) algorithm to be described has a distinctive feature, separation of plant-parameter from noise-parameter estimation. The separation allows a free choice of noise-parameter estimation method, but the best-tested is a.m.l., the a.r.m.a. version of e.l.s. More significantly, splitting the estimation of a $(m+n+q+r)$ vector of parameters into separate estimation of an $(m+n)$ vector and a $(q+r)$ vector saves computing, as the computing demand rises more than linearly with the number of parameters.

The mechanics of r.i.v. plant-parameter estimation are straightforward. From Section (5.3.6), the batch i.v. estimate is

$$\hat{\boldsymbol{\theta}}_z = [Z^{\mathrm{T}}U]^{-1}Z^{\mathrm{T}}\mathbf{y} \tag{7.4.24}$$

with Z the matrix of instrumental variables. We need only replace $\mathscr{H}_t^{\mathrm{T}}\mathscr{R}_t^{-1}$ in the algebra leading to (7.3.21) and (7.3.25), (7.3.27) and (7.3.28) by Z^{T} to obtain the algorithm.

Recursive Instrumental-Variable Algorithm

For $t = 1, 2, \ldots, N$

(i) Update $\mathbf{h}_{t-1}$ and $\mathbf{z}_{t-1}$ to $\mathbf{h}_t$ and $\mathbf{z}_t$. (7.4.25)
(ii) $M_t = M_{t-1} - M_{t-1}\mathbf{z}_t\mathbf{h}_t^{\mathrm{T}}M_{t-1}/(1+\mathbf{h}_t^{\mathrm{T}}M_{t-1}\mathbf{z}_t)$.
(iii) Calculate innovation $v_t = y_t - \mathbf{h}_t^{\mathrm{T}}\hat{\mathbf{x}}_{t-1}$.
(iv) $\hat{\mathbf{x}}_t = \hat{\mathbf{x}}_{t-1} + M_t\mathbf{z}_t v_t$; (7.4.26)

mechanism for generating $\{\mathbf{z}\}$ given; $\hat{\mathbf{x}}_0$, M_0 given;

$$\mathbf{h}_t \triangleq [y_1 \quad \cdots \quad y_n \quad u_{t-k-1} \quad \cdots \quad u_{t-k-m}]^{\mathrm{T}}$$

$$\mathbf{x}_t \triangleq [-a_1 \quad \cdots \quad a_n \quad b_1 \quad \cdots \quad b_m]^{\mathrm{T}}$$

Here $\mathbf{z}_t$ is column t of Z^{T}. We see from (7.4.25) that M_t is not symmetric, and of course it is no longer the error covariance of $\hat{\mathbf{x}}_t$.

We have yet to specify $\mathbf{z}_t$. Intuitively, the ideal thing would be to replace the noisy output samples in $\mathbf{h}_t$ which cause all the trouble by "clean" values y_{t-1}^{c} to y_{t-n}^{c} computed, as in the output-error approach of Section 7.1.3, by

$$y_t^{\mathrm{c}} = -a_1 y_{t-1}^{\mathrm{c}} - \cdots - a_n y_{t-n}^{\mathrm{c}} + b_1 u_{t-k-1} + \cdots + b_m u_{t-k-m} \tag{7.4.27}$$

We do not know the plant parameters, and the best we can do is use the current estimates. More generally, we can generate an i.v. sequence as the output of an auxiliary model.

We have some freedom in generating the instrumental variables, since $\hat{\boldsymbol{\theta}}_z$ in (7.4.24) is unaffected by any non-singular linear transformation of Z, i.e. by any linear reversible filtering of the sequences of samples forming Z, for if we premultiply Z^{T} by G,

$$\hat{\boldsymbol{\theta}}_{z'} = [GZ^{\mathrm{T}}U]^{-1}GZ^{\mathrm{T}}\mathbf{y} = [Z^{\mathrm{T}}U]^{-1}G^{-1}GZ^{\mathrm{T}}\mathbf{y} = \hat{\boldsymbol{\theta}}_z \tag{7.4.28}$$

A surprising consequence is that the numerator $\hat{B}$ of the auxiliary model has no effect on the i.v. estimates. To see this, consider the sequence $\{u'\}$ produced by filtering $\{u\}$ with $1/(1+\hat{A})$. Since

$$\hat{y}_t^c = \hat{b}_1 u'_{t-k-1} + \cdots + \hat{b}_m u'_{t-k-m}, \qquad u_t = u'_t + \hat{a}_1 u'_{t-1} + \cdots + \hat{a}_n u'_{t-n} \tag{7.4.29}$$

we can write column t of Z^T as

$$\mathbf{z}_t = \begin{bmatrix} \hat{y}_{t-1}^c \\ \vdots \\ \hat{y}_{t-n}^c \\ u_{t-k-1} \\ \vdots \\ u_{t-k-m} \end{bmatrix} = \begin{bmatrix} 0 & \hat{b}_1 & \cdots & \hat{b}_m & 0 & \cdots & & 0 \\ 0 & 0 & \hat{b}_1 & \cdots & \hat{b}_m & 0 & \cdots & 0 \\ \vdots & & & & & & & \vdots \\ 0 & \cdots & & & & & & \hat{b}_m \\ 1 & \hat{a}_1 & \cdots & \hat{a}_n & 0 & \cdots & & 0 \\ 0 & 1 & \cdots & & \hat{a}_n & 0 & \cdots & 0 \\ \vdots & & & & & & & \vdots \\ 0 & \cdots & & & & & & \hat{a}_n \end{bmatrix} \begin{bmatrix} u'_{t-k-1} \\ \vdots \\ u'_{t-k-m-n} \end{bmatrix}$$

$$\equiv G\mathbf{u}'_t \tag{7.4.30}$$

where $\mathbf{u}'_t$ is column t of an alternative i.v. matrix entirely composed of samples of $\{u'\}$. It turns out that G in (7.4.30) (a *Sylvester matrix*) is invertible unless $\hat{B}$ and $1+\hat{A}$ have a zero in common, so $\mathbf{u}'_t$ gives precisely the same parameter estimates as $\mathbf{z}_t$, algebraically. There may, however, be numerical differences, especially before the effects of initial unknown values subside in the recursion giving $\{u'\}$ from $\{u\}$.

The auxiliary model may be updated after every plant-parameter update, or less often. For instance, the algorithm may be iterated, performing repeated passes through the records with the auxiliary model updated at the end of each. Iteration of a recursive algorithm is clumsy and may be too slow for real-time identification, but it may help in getting usable results from short records. If the final M or P from one iteration is carried over as the initial value for the next, the records are being treated arbitrarily as periodic with period equal to their original length. Instead, it has been suggested (Young, 1984) that in r.i.v. each iteration be initialised with M from $(1/l)$th of the way through the previous iteration, where l iterations in all are performed. This seems equally arbitrary and results in heavy dependence of the final M on the first $(1/l)$th of the record, but appears to yield good results.

Approximate maximum-likelihood estimates of the noise parameters are obtained from the model (7.1.5), written as the a.r.m.a. model

$$w_t = -d_1 w_{t-1} = \cdots - d_r w_{t-r} + c_1 v_{t-1} + \cdots + c_q v_{t-q} + v_t$$

$$\equiv \mathbf{h}_t^{n\mathrm{T}} \mathbf{x}^n + v_t \tag{7.4.31}$$

where superscript n denotes noise, and

$$\mathbf{h}_t^n \triangleq [w_{t-1} \quad \cdots \quad w_{t-r} \quad v_{t-1} \quad \cdots \quad v_{t-q}]^{\mathrm{T}},$$
$$\mathbf{x}^n \triangleq [-d_1 \quad \cdots \quad -d_r \quad c_1 \quad \cdots \quad c_q]^{\mathrm{T}} \tag{7.4.32}$$

Both $\{w\}$ and $\{v\}$ are unobservable, of course, and have to be estimated by

$$\hat{w}_{t-i} = y_{t-i} - \hat{y}^{\mathrm{c}}_{t-i} \qquad \hat{v}_{t-j} = \hat{w}_{t-j} - \hat{\mathbf{h}}^{n\mathrm{T}}_{t-j}\hat{\mathbf{x}}^n_{t-j}$$
$$i = 1, 2, \ldots, r, \qquad j = 1, 2, \ldots, q \tag{7.4.33}$$

where $\{\hat{y}^{\mathrm{c}}\}$ is found as in (7.4.27) with estimated plant parameters, and $\hat{\mathbf{h}}^n$ is defined analogously to $\mathbf{h}^n$.

A long series of refinements and extensions to the basic r.i.v./a.m.l. combination has been reported (Young and Jakeman, 1979, 1980; Jakeman and Young, 1979), and supported by extensive results from real and simulated records.

7.4.6 Prediction-Error Algorithms

With a few exceptions, we have examined algorithms which give parameter estimates with zero bias and minimum covariance. As noted in Section 5.3.2, the attraction of minimum parameter-estimate covariance is that it guarantees, for any scalar quantity linear in the parameters, the smallest variance and m.s. error obtainable with zero bias. It thereby kills two birds with one stone; it gives a minimum m.s. error both in each individual parameter and in the output at fixed values of the explanatory variables.

To find the parameter-estimate covariance, we must specify not only the estimation algorithm but also the form of the model and the system actually generating the records. However, when the model is intended for prediction, we are not interested in the parameter estimates or model structure for their own sake, and can concentrate on prediction performance. In comparing the performance of a predictor estimated by a particular algorithm with the best attainable by that type of predictor, we need not specify either the original model from which the predictor arose or the actual plant and noise dynamics. In these circumstances, it is simplest to regard the predictor itself as the model, and think of identification as predictor-building. The adequacy of the parameter estimates for purposes other than prediction can then be left as a separate issue.

Example 7.4.1 A one-step-ahead predictor of the output can be developed from the regression equation model (7.2.3):

$$Y = -AY + Bz^{-k}U + (1 + C)V$$

by noting that in the corresponding difference equation

$$y_t = -a_1 y_{t-1} - \cdots - a_n y_{t-n} + b_1 u_{t-k-1} + \cdots + b_m u_{t-k-m} + c_1 v_{t-1} + \cdots + c_q v_{t-q} + v_t$$

the only quantity on the right-hand side not known at instant $t-1$ is v_t if A, B, C and k are known. Since Ev_t is zero, the conditional-mean predictor $\hat{y}_t$ is $y_t - v_t$, obtained by setting $\hat{v}_t$ to zero and using the known values for all the other terms making up y_t. We can avoid working out the terms in v_{t-q} to v_{t-1} explicitly by expressing $\hat{y}_t$ in terms of u's, y's and $\hat{y}$'s up to $t-1$. In z-transforms,

$$\hat{Y} = Y - V = -AY + Bz^{-k}U + CV = (C - A)Y + Bz^{-k}U - C\hat{Y}$$

so the one-step-ahead predictor is a recursion

$$\hat{y}_t = c_1' y_{t-1} + \cdots + c_s' y_{t-s} + b_1 u_{t-k-1} + \cdots + b_m u_{t-k-m} - c_1 \hat{y}_{t-1} - \cdots - c_q \hat{y}_{t-q}$$

where c_i' is $c_i - a_i$ and s is $\max(q, n)$. Although the predictor came from the model (7.2.3), it could equally have been devised empirically or simply chosen as a convenient structure. For prediction, the parameter vector which must be identified is $[c_1' \;\cdots\; c_s' \;\; b_1 \;\cdots\; b_m \;\; -c_1 \;\cdots\; -c_q]$. However, the original model implies some redundancy in these parameters if q is greater than n. △

The way the simplified view of identification as predictor-building allows us to defer consideration of the underlying model and the process generating the records has led to strong emphasis on *prediction-error algorithms* recently (Ljung and Söderström, 1983). These algorithms aim to minimise a scalar risk function

$$r_t(\mathbf{x}) \triangleq E[V_t(\mathbf{x})] \triangleq E\left[\frac{1}{t}\sum_{i=1}^{t} l(\mathbf{y}_i - \hat{\mathbf{y}}_i, \mathbf{x})\right] \qquad (7.4.34)$$

The loss function l measures the error in predicting output $\mathbf{y}_i$ one sample ahead. The expectation is over the observations up to t, with the parameter vector $\mathbf{x}$ fixed. The algorithms actually minimise $V_t(\mathbf{x})$ in lieu of $r_t(\mathbf{x})$ by adjusting $\mathbf{x}$ according to the derivatives of $V_r(\mathbf{x})$ with respect to $\mathbf{x}$, just as r.m.l. 2 did in Section 7.4.3. They consequently require each output prediction to be a known and sufficiently differentiable function of $\mathbf{x}$. The algorithms' inability to compute $r_t(\mathbf{x})$ without making assumptions about the p.d.f. of the observations does not matter asymptotically (Ljung, 1981), for under weak assumptions the $\hat{\mathbf{x}}_t$ that minimises $V_t(\mathbf{x})$ tends w.p.1 to the $\hat{\mathbf{x}}$ that minimises $\lim_{t\to\infty} r_t(\mathbf{x})$, as t tends to infinity.

Minimisation of $r_t(\mathbf{x})$ for a given predictor structure provides initial motivation for a prediction-error algorithm, but as soon as we enquire whether a different predictor structure might be better, we have to discuss the process actually generating the records. Statements can be made about the asymptotic covariance of $\hat{\mathbf{x}}_t$, i.e. the efficiency of the algorithm, at the price of assuming that the model underlying the predictor can explain the observations, in the sense that the observations could have been generated by the model with some true value of $\mathbf{x}$.

A recursive prediction-error algorithm can be developed in quite a general form (Ljung and Söderström, 1983) and then specialised by choice of loss function, model structure and detailed implementation. Many familiar algorithms can be obtained in this way as prediction-error or approximate prediction-error algorithms. This reinterpretation may suggest modifications to improve asymptotic properties, as when e.l.s. was modified into r.m.l. 2. The prediction-error approach is direct and elegant when the problem really is prediction, but it is also valuable because it allows degree of specialisation to be traded against power of asymptotic convergence and efficiency results when the algorithms are analysed.

*7.5 CONVERGENCE ANALYSIS FOR RECURSIVE IDENTIFICATION ALGORITHMS

Since the early 1970's, several methods have been developed for convergence analysis of a wide range of recursive identification algorithms. Their development was stimulated by two trends, a proliferation of recursive algorithms with similar updating structure (Åström and Eykhoff, 1971) and a rebirth of adaptive control in the shape of self-tuning controllers (Åström and Wittenmark, 1973). Self-tuning controllers combine a recursive identifier and a control law based on the updated model, as will be shown in Fig. 8.3.2 and covered in Section 8.3.3. A convergence analysis was needed for such closed-loop systems as well as systems with input independent of earlier output. The analysis outlined below (Ljung, 1977a) is general enough to cover both.

As even an informal account of the analysis is quite complicated, and its implications are much easier to see in specific examples, a careful reading of Examples 7.5.1–7.5.3 is strongly advised.

7.5.1 Formulation of an Archetype Algorithm

Our first aim is to fit a broad class of recursive algorithms into a single framework, the archetype algorithm. It has to encompass a variety of recursive schemes, including o.l.s., e.l.s., s.a. and the commonest self-tuning controllers. It consists of two difference equations, one the parameter-updating equation and the other relating current samples of the output and regressors to earlier samples. Covariance updating will be added later.

Parameter-updating equations (7.3.28) for recursive minimum-covariance estimation, (7.4.7) for e.l.s. and e.m.m., (7.4.22) for s.a. and (7.4.26) for r.i.v. are all of the form

$$\hat{\mathbf{x}}_t = \hat{\mathbf{x}}_{t-1} + \gamma_t \mathbf{s}_t(y_t - \mathbf{h}_t^T \hat{\mathbf{x}}_{t-1}) \tag{7.5.1}$$

where γ_t is a scalar gain, sometimes 1. Let us collect the regressor vector $\mathbf{h}_t$ and output y_t into

$$\boldsymbol{\phi}_t = [y_t \ \mathbf{h}_t^T]^T \tag{7.5.2}$$

and write an equation stating how $\boldsymbol{\phi}_t$ evolves from $\boldsymbol{\phi}_{t-1}$. To obtain $\boldsymbol{\phi}_t$ from $\boldsymbol{\phi}_{t-1}$ we delete the oldest samples of input, output and, if present, noise estimate, shift the rest down one place and insert the new values. Unless the new input sample can be treated as deterministic, it is modelled as a function of earlier input samples and a white driving sequence and, if feedback is present, earlier output samples. In self-tuning controllers the control input at time $t-1$ depends on the parameter estimates $\hat{\mathbf{x}}_{t-1}$ used in the control computation, and this relation must be included in the equation for $\boldsymbol{\phi}_t$. The new output y_t depends on the rest of $\boldsymbol{\phi}_t$ via the system dynamics, and the new noise estimate $\hat{v}_{t-1}$ depends on $\hat{\mathbf{x}}_{t-1}$ and part of $\boldsymbol{\phi}_{t-1}$ through the model, e.g.

$$\hat{v}_{t-1} = y_{t-1} - \mathbf{h}_{t-1}^T \hat{\mathbf{x}}_{t-1} \tag{7.5.3}$$

The evolution of $\boldsymbol{\phi}$ in a linear system is altogether described by

$$\boldsymbol{\phi}_t = K(\hat{\mathbf{x}}_{t-1})\boldsymbol{\phi}_{t-1} + L(\hat{\mathbf{x}}_{t-1})\mathbf{e}_t \tag{7.5.4}$$

where K and L depend on $\hat{\mathbf{x}}_{t-1}$ in general but not in some of the simpler algorithms, and $\{\mathbf{e}\}$ is a white sequence.

Example 7.5.1 Extended least squares is based on

$$\begin{aligned} y_t = &-a_1 y_{t-1} - \cdots - a_n y_{t-n} + b_1 u_{t-k-1} + \cdots + b_m u_{t-k-m} \\ &+ c_1 v_{t-1} + \cdots + c_q v_{t-q} + v_t \end{aligned}$$

so y_t and the regressors $\mathbf{h}_t$ form

$$\begin{aligned} \boldsymbol{\phi}_t &= [y_t \ \cdots \ y_{t-n} \,|\, u_{t-k-1} \ \cdots \ u_{t-k-m} \,|\, \hat{v}_{t-1} \ \cdots \ \hat{v}_{t-q}]^T \\ &= \left[0 \ \begin{matrix}\text{elements 2 to } n \\ \text{of } \boldsymbol{\phi}_{t-1}\end{matrix} \,\middle|\, 0 \ \begin{matrix}\text{elements } n+2 \text{ to} \\ n+m \text{ of } \boldsymbol{\phi}_{t-1}\end{matrix} \,\middle|\, 0 \ \begin{matrix}\text{elements } m+n+2 \\ \text{to } m+n+q \text{ of } \boldsymbol{\phi}_{t-1}\end{matrix}\right]^T \\ &\quad + [\mathbf{h}_t^T \mathbf{x} + v_t \ \ \mathbf{0}^T \,|\, u_{t-k-1} \ \mathbf{0}^T \,|\, y_{t-1} - \mathbf{h}_{t-1}^T \hat{\mathbf{x}}_{t-1} \ \ \mathbf{0}^T]^T \end{aligned}$$

If $\{u\}$ is, for instance, an autoregression of order $l<m$ so that

$$u_{t-k-1}=g_1u_{t-k-2}+\cdots+g_lu_{t-k-l-1}+w_{t-k-1}$$

then $\boldsymbol{\phi}_t$ can be written as

$$\boldsymbol{\phi}_t=\Psi\boldsymbol{\phi}_{t-1}+\begin{bmatrix}0 & \mathbf{x}^{\mathrm{T}}\\ & \mathbf{0}\\ \hdashline & \mathbf{0}\\ \hdashline & \mathbf{0}\end{bmatrix}\boldsymbol{\phi}_t+\begin{bmatrix} & \mathbf{0} & \\ \hdashline \mathbf{0}^{\mathrm{T}} & \mathbf{g}^{\mathrm{T}} & \mathbf{0}^{\mathrm{T}}\\ & \mathbf{0} & \\ \hdashline 1 & \hat{\mathbf{x}}_{t-1}^{\mathrm{T}} & \\ & \mathbf{0} & \end{bmatrix}\boldsymbol{\phi}_{t-1}+\begin{bmatrix}1 & 0\\ 0 & 0\\ \hdashline 0 & 1\\ 0 & 0\\ \hdashline 0 & 0\\ 0 & 0\end{bmatrix}\mathbf{e}_t$$

$$\equiv\Psi\boldsymbol{\phi}_{t-1}+F\boldsymbol{\phi}_t+H\boldsymbol{\phi}_{t-1}+J\mathbf{e}_t\quad\text{say}$$

Here the partitioning separates samples of y, u and $\hat{v}$, the forcing $\mathbf{e}_t$ is $[v_t\ w_{t-k-1}]^{\mathrm{T}}$ and Ψ accounts for the downward shift of most elements from $\boldsymbol{\phi}_{t-1}$ to $\boldsymbol{\phi}_t$. The partitions of Ψ are

$$\Psi_{ii}=\begin{bmatrix}0 & & & \cdots & 0\\ 1 & 0 & & \cdots & 0\\ 0 & 1 & 0 & \cdots & 0\\ \vdots & & & & \\ 0 & \cdots & & 1 & 0\end{bmatrix}\qquad \Psi_{ij}=0,\quad i\neq j,\quad i=1,2,3,\quad j=1,2,3$$

Hence (7.5.4) for this algorithm and input is

$$\boldsymbol{\phi}_t=[I-F]^{-1}[(\Psi+H)\boldsymbol{\phi}_{t-1}+J\mathbf{e}_t]\equiv K(\hat{\mathbf{x}}_{t-1})\boldsymbol{\phi}_{t-1}+L\mathbf{e}_t$$

where L does not depend on $\hat{\mathbf{x}}_{t-1}$ but K does. Both depend on $\mathbf{x}$. △

In (7.5.1), $\mathbf{s}_t$ depends on $\mathbf{h}_t$ directly, and in most algorithms also indirectly through P_t or M_t. The correction term in the parameter-updating equation is consequently a complicated function of $\boldsymbol{\phi}_t$, so we write

$$\hat{\mathbf{x}}_t=\hat{\mathbf{x}}_{t-1}+\gamma_t\mathbf{q}(\boldsymbol{\phi}_t,\hat{\mathbf{x}}_{t-1})\tag{7.5.5}$$

where $\mathbf{q}(\cdot,\cdot)$ is a vector of functions we must specify, along with the exact contents of $\boldsymbol{\phi}_t$, for each algorithm analysed.

The convergence analysis applies to the archetype algorithm (7.5.4) and (7.5.5). Covariance updating is adjoined to (7.5.5), as we see shortly. The analysis is difficult because (7.5.4) and (7.5.5) are coupled; in particular, $\hat{\mathbf{x}}_t$ depends not only on $\hat{\mathbf{x}}_{t-1}$ but also, through $\boldsymbol{\phi}_t$, on all earlier $\hat{\mathbf{x}}$'s in general.

Let us now see how l.s. covariance and parameter updating fit into (7.5.5). As usual it is easiest to consider the information-updating equation

$$P_t^{-1}=P_{t-1}^{-1}+\mathbf{h}_t\mathbf{h}_t^{\mathrm{T}}/\sigma_t^2\tag{7.5.6}$$

We can make this look very like (7.5.1) by writing it as

$$\frac{P_t^{-1}}{t} = \frac{P_{t-1}^{-1}}{t-1} + \frac{1}{t}\left(\frac{\mathbf{h}_t\mathbf{h}_t^{\mathrm{T}}}{\sigma_t^2} - \frac{P_{t-1}^{-1}}{t-1}\right) \tag{7.5.7}$$

For each column $\mathbf{r}_t^{(i)}$ of P_t^{-1}/t, defining γ_t as $1/t$ we have

$$\mathbf{r}_t^{(i)} = \mathbf{r}_{t-1}^{(i)} + \gamma_t(\mathbf{h}_t^{(i)}\mathbf{h}_t/\sigma_t^2 - \mathbf{r}_{t-1}^{(i)}), \qquad i = 1, 2, \ldots, p \tag{7.5.8}$$

and

$$\hat{\mathbf{x}}_t = \hat{\mathbf{x}}_{t-1} + \frac{P_t\mathbf{h}_t}{\sigma_t^2}(y_t - \mathbf{h}_t^{\mathrm{T}}\hat{\mathbf{x}}_{t-1})$$

$$\equiv \hat{\mathbf{x}}_{t-1} + \frac{\gamma_t(P_t^{-1}/t)^{-1}\mathbf{h}_t}{\sigma_t^2}(y_t - \mathbf{h}_t^{\mathrm{T}}\hat{\mathbf{x}}_{t-1}) \tag{7.5.9}$$

We then need only define an augmented "parameter" vector consisting of $\hat{\mathbf{x}}_t$ followed by columns $\mathbf{r}_t^{(i)}$ to $\mathbf{r}_t^{(p)}$ to make (7.5.8) and (7.5.9) into (7.5.5). Example 7.5.2 illustrates the procedure for e.l.s.

7.5.2 Asymptotic Behaviour of Archetype Algorithm

Rigorous analysis of the archetype algorithm is complicated and difficult, but its spirit is conveyed by the following loose and heuristic argument.

In (7.5.4), the influence of the forcing $\mathbf{e}$ at one instant on later $\boldsymbol{\phi}$'s will eventually fade until negligible in almost any conceivable practical circumstances, and so will the effect of the initial value $\boldsymbol{\phi}_0$. Also, if the parameter estimates converge, $\hat{\mathbf{x}}_t - \hat{\mathbf{x}}_{t-1}$ gets very small as t increases. (We are not embarking on a circular argument along the lines "if it converges then . . . it converges", as we shall be enquiring into the value to which $\hat{\mathbf{x}}$ converges and its manner of convergence, not *whether* it converges.) We should not then be too far out if, in calculating $\boldsymbol{\phi}_i$ at any time i close to a large time t, we replace all earlier $\hat{\mathbf{x}}$ values by $\hat{\mathbf{x}}_t$ and ignore the effect of $\boldsymbol{\phi}_0$, giving

$$\boldsymbol{\phi}_i = \prod_{j=1}^{i} K(\hat{\mathbf{x}}_{j-1})\boldsymbol{\phi}_0 + \sum_{k=1}^{i}\prod_{j=k+1}^{i} K(\hat{\mathbf{x}}_{j-1})L(\hat{\mathbf{x}}_{k-1})\mathbf{e}_k$$

$$\simeq \sum_{k=1}^{i}\{K(\hat{\mathbf{x}}_t)\}^{i-k}L(\hat{\mathbf{x}}_t)\mathbf{e}_k \equiv \tilde{\boldsymbol{\phi}}_i(\hat{\mathbf{x}}_t) \text{ say} \tag{7.5.10}$$

Furthermore, if $\{\mathbf{e}\}$ is stationary and $K(\hat{\mathbf{x}}_t)$ has only stable eigenvalues, $\tilde{\boldsymbol{\phi}}_i(\hat{\mathbf{x}}_t)$ will become a stationary random vector as i increases. We can now approximate the change in $\hat{\mathbf{x}}$ over s steps from time t using (7.5.5):

$$\hat{\mathbf{x}}_{t+s} - \hat{\mathbf{x}}_t = \sum_{i=t+1}^{t+s} \gamma_i \mathbf{q}(\boldsymbol{\phi}_i, \mathbf{x}_{i-1}) \simeq \sum_{i=t+1}^{t+s} \gamma_i \mathbf{q}(\tilde{\boldsymbol{\phi}}_i(\hat{\mathbf{x}}_t), \hat{\mathbf{x}}_t) \quad (7.5.11)$$

At large enough i, $\mathbf{q}(\tilde{\boldsymbol{\phi}}_i(\hat{\mathbf{x}}_t), \hat{\mathbf{x}}_t)$ also becomes a stationary random variable by virtue of its dependence on $\tilde{\boldsymbol{\phi}}_i$ (with $\hat{\mathbf{x}}_t$ regarded strictly as a parameter, not a variable, at the moment). We can thus express $\mathbf{q}$ in terms of its deviation $\mathbf{p}_i$ from its mean $\mathbf{f}(\hat{\mathbf{x}}_t)$:

$$\mathbf{q}(\tilde{\boldsymbol{\phi}}_i(\hat{\mathbf{x}}_t), \hat{\mathbf{x}}_t) = \underset{\mathbf{e}}{E}\,\mathbf{q}(\tilde{\boldsymbol{\phi}}_i(\hat{\mathbf{x}}_t), \hat{\mathbf{x}}_t) + \mathbf{p}_i \equiv \mathbf{f}(\hat{\mathbf{x}}_t) + \mathbf{p}_i \quad (7.5.12)$$

and write (7.5.11) as

$$\hat{\mathbf{x}}_{t+s} - \hat{\mathbf{x}}_t = \mathbf{f}(\hat{\mathbf{x}}_t) \sum_{i=t+1}^{t+s} \gamma_i + \sum_{i=t+1}^{t+s} \gamma_i \mathbf{p}_i \simeq \mathbf{f}(\hat{\mathbf{x}}_t) \sum_{i=t+1}^{t+s} \gamma_i \quad (7.5.13)$$

This last approximation is clearly acceptable if we are interested in the mean behaviour of $\hat{\mathbf{x}}$, since the term neglected has zero mean. It is not clear without a more careful analysis how far (7.5.13) is true for individual realisations of $\{\hat{\mathbf{x}}\}$. We can tidy up (7.5.13) without essentially altering it by the trick of treating $\sum_{i=t+1}^{t+s} \gamma_i$ as the increment of a transformed time τ, giving

$$\hat{\mathbf{x}}_{t+s} - \hat{\mathbf{x}}_t \equiv \hat{\mathbf{x}}_{\tau+\Delta\tau} - \hat{\mathbf{x}}_\tau \simeq \mathbf{f}(\hat{\mathbf{x}}_\tau) \sum_{i=t+1}^{t+s} \gamma_i \equiv \mathbf{f}(\hat{\mathbf{x}}_\tau)\,\Delta\tau \quad (7.5.14)$$

This can be viewed as the discrete-time version of a time-invariant differential equation

$$\dot{\mathbf{x}}(\tau) \simeq \mathbf{f}(\mathbf{x}(\tau)) \quad (7.5.15)$$

The asymptotic behaviour of $\{\hat{\mathbf{x}}\}$ can then be investigated by solving the differential equation numerically or, with luck, analytically.

Example 7.5.2 We continue with the e.l.s. algorithm, defining $\hat{\mathbf{x}}'_t$ as $\hat{\mathbf{x}}_t$ augmented by the stacked columns $\mathbf{r}_t^{(1)}$ to $\mathbf{r}_t^{(p)}$ of P_t^{-1}/t, and R_t as P_t^{-1}/t. (R_t is Ljung's notation, unrelated to regression-equation noise covariance). The joint parameter and covariance updating equation (7.5.5) for $\hat{\mathbf{x}}'_t$ then gives

$$\mathbf{q}(\tilde{\boldsymbol{\phi}}_i(\hat{\mathbf{x}}'_t), \hat{\mathbf{x}}'_t) = \begin{bmatrix} R_t^{-1}\bar{\mathbf{h}}_t(y_i - \bar{\mathbf{h}}_i^{\mathrm{T}}\hat{\mathbf{x}}_t)/\sigma^2 \\ \text{stacked columns of } \bar{\mathbf{h}}_i\bar{\mathbf{h}}_i^{\mathrm{T}}/\sigma^2 - R_t \end{bmatrix}$$

where σ^2 is constant as $\{\mathbf{e}\}$ has been assumed stationary, and $\bar{\mathbf{h}}_i$ is what the regressor vector would be if the fixed value $\hat{\mathbf{x}}_t$ were used for evaluating K and L at every update from the indefinite past to time i. Only $\hat{v}_{i-1}$ to $\hat{v}_{i-q}$ are affected, in fact.

Denoting $E[\bar{\mathbf{h}}_i(y_i - \bar{\mathbf{h}}_i^{\mathrm{T}}\hat{\mathbf{x}}_t)]$ by $\mathbf{d}(\hat{\mathbf{x}}_t)$ and $E[\bar{\mathbf{h}}_i\bar{\mathbf{h}}_i^{\mathrm{T}}]$ by $G(\hat{\mathbf{x}}_t)$, we obtain

$$\mathbf{f}(\hat{\mathbf{x}}_t') = E\mathbf{q}(\bar{\boldsymbol{\phi}}_i(\hat{\mathbf{x}}_t'), \hat{\mathbf{x}}_t') = \begin{bmatrix} R_t^{-1}\mathbf{d}(\hat{\mathbf{x}}_t)/\sigma^2 \\ \text{stacked columns of } G(\hat{\mathbf{x}}_t)/\sigma^2 - R_t \end{bmatrix}$$

The differential equation (7.5.15) is then

$$\dot{\hat{\mathbf{x}}}(\tau) = R^{-1}(\tau)\mathbf{d}(\hat{\mathbf{x}}(\tau))/\sigma^2$$

for the $\hat{\mathbf{x}}$ part of $\hat{\mathbf{x}}'$, and for the remainder, writing the columns of R and G side by side once more,

$$\dot{R}(\tau) = G(\hat{\mathbf{x}}(\tau))/\sigma^2 - R(\tau)$$

The converged state is given by $\dot{\hat{\mathbf{x}}}$ and $\dot{R}$ both zero, i.e.

$$R = G/\sigma^2 \qquad \text{and} \qquad R^{-1}\mathbf{d}/\sigma^2 = \mathbf{0}$$

Assuming no redundancy among the regressors in $\bar{\mathbf{h}}_i$, G is non-singular and, for bounded input and output, finite. So therefore is the asymptotic R. Hence R^{-1} is finite, so $\mathbf{d}$ must generally converge to zero. That is to say, the residuals $y_i - \bar{\mathbf{h}}_i^{\mathrm{T}}\hat{\mathbf{x}}_t$ are asymptotically uncorrelated with the regressors $\bar{\mathbf{h}}_i$. We recognise this as the probabilistic counterpart of the orthogonality property of l.s.

△

For (7.5.15) to approximate the large-sample behaviour of $\{\hat{\mathbf{x}}\}$ adequately, some regularity conditions must be imposed on $\{\gamma\}$, K, L, $\{\mathbf{e}\}$ and $\mathbf{q}$. Without wishing to go into them in detail, we should look briefly at them to see that they are not very restrictive. They are:

(i) $\{\mathbf{e}\}$ must be a sequence of independent random variables, not necessarily stationary although we took them to be so in the heuristic argument.

(ii) Either (a) $\mathbf{e}_t$ must be bounded w.p.l. at all t (real-life disturbances and forcing *are* bounded), and $\mathbf{q}$ must be continuously differentiable with respect to $\boldsymbol{\phi}$ and $\hat{\mathbf{x}}$, the derivatives being bounded at all t, or (b) to permit analysis with $\mathbf{e}_t$ not bounded, other conditions on $\mathbf{e}_t$ and $\mathbf{q}$ must be assumed (Ljung, 1977a).

(iii) K and L must be Lipschitz continuous (Vidyasagar, 1978).

(iv) $\mathbf{q}$ must be such that $\lim_{t\to\infty} \mathbf{f}(\hat{\mathbf{x}}_t)$ exists.

(v) $\{\gamma\}$ must be a decreasing sequence with $\sum_{t=1}^{\infty}\gamma_t$ infinite, $\sum_{t=1}^{\infty}\gamma_t^{\alpha}$ finite for some α and $1/\gamma_t - 1/\gamma_{t-1}$ finite as $t\to\infty$.

The conditions on $\mathbf{q}$, K and L have to be met over some region of $\hat{\mathbf{x}}$, say D_R, throughout which K has only strictly stable eigenvalues.

7.5.3 Convergence Theorems for Archetype Algorithm

Three theorems (Ljung, 1977a) state how individual sequences $\{\hat{\mathbf{x}}\}$ of estimates relate to the differential equation (7.5.15).

Theorem 1 says that the distance between $\hat{\mathbf{x}}_t$ and the nearest point of a region D_c, which contains only trajectories of (7.5.15) which have been and will remain in it forever, tends to zero w.p.1 as $t \to \infty$. At its simplest, D_c (called an invariant set of (7.5.15)) is a single equilibrium point where $\mathbf{f}$ is zero. The theorem applies when, w.p.1 and at an infinite number of instants, $|\tilde{\boldsymbol{\phi}}_t|$ is below some value and $\hat{\mathbf{x}}_t$ is in a region from which no trajectory leaves D_R and all trajectories converge to D_c as $t \to \infty$.

The usefulness of Theorem 1 depends on how easily the domain of attraction of D_c, the region from which trajectories of (7.5.15) converge to D_c, can be identified. Lyapunov stability analysis (Vidyasagar, 1978) is required for all but trivial cases, and is notoriously more an art than a science. Alternative approaches, some not relying on Lyapunov theory, have more recently been developed (Goodwin *et al.*, 1980; Solo, 1980; Fogel, 1981; Goodwin *et al.*, 1984).

Theorem 2 says that if $\hat{\mathbf{x}}_t$ has non-zero probability of converging to within a distance ρ of some point $\mathbf{x}^*$ however small ρ is, then $\mathbf{x}^*$ must be a stable equilibrium point of (7.5.15). It applies so long as cov $\mathbf{q}(\tilde{\boldsymbol{\phi}}(\mathbf{x}^*), \mathbf{x}^*)$ is positive-definite and $\mathbf{f}(\hat{\mathbf{x}}_t)$ has continuous derivatives throughout a neighbourhood of $\mathbf{x}^*$, the derivatives converging uniformly as $t \to \infty$.

An algorithm therefore cannot give consistent estimates unless the true value $\mathbf{x}$ is a stable equilibrium point of (7.5.15).

Theorem 3 says how closely $\{\hat{\mathbf{x}}\}$ follows a trajectory of (7.5.15). It states that if the solution of (7.5.15) from $\hat{\mathbf{x}}_{t_0}$ onwards is compared at instants t_1 to t_N with $\{\hat{\mathbf{x}}\}$ obtained from the same $\hat{\mathbf{x}}_{t_0}$, the probability of the difference exceeding ε at any of the instants is not more than $K\sum_{j=0}^{N}(\gamma_{t_j}/\varepsilon^4)^\alpha$ for any $\alpha \geq 1$ and any ε up to ε_0 and t_0 beyond T_0, where ε_0, T_0 and constant K depend on α and on the minimum spacing between instants.

Now if we increase t_0 but nothing else, so that K stays the same, then with $\{\gamma\}$ decreasing and $\sum_{t=1}^{\infty}\gamma_t^\alpha$ finite for some α (two of our earlier assumptions (v)), $K\sum_{j=0}^{n}(\gamma_{tj}/\varepsilon^4)^\alpha$ gets smaller and smaller. Hence the estimates from $\hat{\mathbf{x}}_{t0}$ onwards stay close to the solution of (7.5.15) more and more certainly as we increase t_0; the later the section of $\{\hat{\mathbf{x}}\}$ considered, the more certainly the trajectory starting at the first $\hat{\mathbf{x}}_t$ of the section approximates the rest of $\{\hat{\mathbf{x}}\}$ well.

Example 7.5.3 completes the analysis of e.l.s. and introduces the important property of positive realness.

Example 7.5.3 (Ljung, 1977b). In Example 7.5.2 we found that for e.l.s. $\mathbf{d}(\hat{\mathbf{x}}_t)$ is zero in the converged state. Let us check whether the actual $\mathbf{x}$ is an equilibrium point by examining $\mathbf{d}(\hat{\mathbf{x}}_t)$. By definition

$$\mathbf{d}(\hat{\mathbf{x}}_t) = E[\bar{\mathbf{h}}_i(y_i - \bar{\mathbf{h}}_i^{\mathrm{T}}\mathbf{x}_t)] = E[\bar{\mathbf{h}}_i\bar{v}_i]$$

where

$$\bar{v}_i = y_i - \bar{\mathbf{h}}_i^{\mathrm{T}}\hat{\mathbf{x}}_t = \mathbf{h}_i^{\mathrm{T}}\mathbf{x} + v_i - \bar{\mathbf{h}}_i^{\mathrm{T}}\hat{\mathbf{x}}_t = \mathbf{x}^{\mathrm{T}}(\mathbf{h}_i - \bar{\mathbf{h}}_i) + \bar{\mathbf{h}}_i^{\mathrm{T}}(\mathbf{x} - \hat{\mathbf{x}}_t) + v_i$$

Now in e.l.s. $\mathbf{h}_i - \bar{\mathbf{h}}_i$ is zero except for the last q elements $v_{i-1} - \bar{v}_{i-1}$ to $v_{i-q} - \bar{v}_{i-q}$, so

$$\bar{v}_i - v_i + \mathbf{x}^{\mathrm{T}}(\bar{\mathbf{h}}_i - \mathbf{h}_i) = \bar{v}_i - v_i + c_1(\bar{v}_{i-1} - v_{i-1}) + \cdots + c_q(\bar{v}_{i-q} - v_{i-q})$$

We therefore get $\bar{v}_i - v_i$ if we pass $\bar{\mathbf{h}}_i^{\mathrm{T}}(\mathbf{x} - \hat{\mathbf{x}}_t)$ through the filter $1/(1 + C(z^{-1}))$, i.e.

$$\bar{v}_i - v_i = \tilde{\mathbf{h}}_i^{\mathrm{T}}(\mathbf{x} - \hat{\mathbf{x}}_t)$$

where $\tilde{\mathbf{h}}_i$ is $\bar{\mathbf{h}}_i$ filtered by $1/(1 + C(z^{-1}))$. Consequently,

$$\mathbf{d}(\hat{\mathbf{x}}_t) = E[\bar{\mathbf{h}}_i(\tilde{\mathbf{h}}_i^{\mathrm{T}}(\mathbf{x} - \hat{\mathbf{x}}_t) + v_i)] = E[\bar{\mathbf{h}}_i\tilde{\mathbf{h}}_i^{\mathrm{T}}](\mathbf{x} - \hat{\mathbf{x}}_t) = \tilde{G}(\mathbf{x} - \hat{\mathbf{x}}_t)$$

where $\tilde{G}$ is $E[\bar{\mathbf{h}}_i\tilde{\mathbf{h}}_i^{\mathrm{T}}]$, itself a function of $\hat{\mathbf{x}}_t$. Here we are treating $\hat{\mathbf{x}}_t$ as a parameter, not a random variable, and $E[\bar{\mathbf{h}}_i v_i]$ is zero since $\{v\}$ is white. Any converged value of $\hat{\mathbf{x}}_t$ has to make $\mathbf{d}$ and hence $\tilde{G}(\mathbf{x} - \hat{\mathbf{x}}_t)$ zero. One such value is $\mathbf{x}$, but there may be others.

We can discover how any other $\hat{\mathbf{x}}_t$ might make $\mathbf{d}$ zero by looking at $(\mathbf{x} - \hat{\mathbf{x}}_t)^{\mathrm{T}}\tilde{G}(\mathbf{x} - \hat{\mathbf{x}}_t)$, which is zero whenever $\mathbf{d}$ is zero. Defining $\bar{\xi}_i$ and $\tilde{\xi}_i$ through

$$(\mathbf{x} - \hat{\mathbf{x}}_t)^{\mathrm{T}}\tilde{G}(\mathbf{x} - \hat{\mathbf{x}}_t) = E[(\mathbf{x} - \hat{\mathbf{x}}_t)^{\mathrm{T}}\bar{\mathbf{h}}_i\tilde{\mathbf{h}}_i^{\mathrm{T}}(\mathbf{x} - \hat{\mathbf{x}}_t)] = E[\bar{\xi}_i\tilde{\xi}_i]$$

we see that when $\mathbf{d}$ is zero, the cross-correlation at lag zero between $\bar{\xi}_i$ and $\tilde{\xi}_i$ is zero. As $\tilde{\xi}_i$ is $\bar{\xi}_i$ filtered by $1/(1 + C(z^{-1}))$, the cross-correlation is easily written in terms of the power spectral density $\phi_{\bar{\xi}}(j\omega)$ of $\bar{\xi}_i$ by Parseval's theorem:

$$E[\bar{\xi}_i\tilde{\xi}_i] = \frac{T}{2\pi}\int_{-\pi/T}^{\pi/T} \phi_{\bar{\xi}}(j\omega)/(1 + C(e^{-j\omega T}))\, d\omega$$

Like any auto-spectral density, $\phi_{\bar{\xi}}(j\omega)$ is real and non-negative at all ω, so the integrand is negative only when $\mathrm{Re}\{1/(1 + C(e^{-j\omega T}))\}$ is. We can say more about the $\hat{\mathbf{x}}_t$ values which make $\mathbf{d}$ zero if we assume $\mathrm{Re}\{1/(1 + C(e^{-j\omega T}))\}$ to be positive for all ωT between $-\pi$ and π, i.e. $1/(1 + C(z^{-1}))$ to be *strictly positive real*. When it is, $E[\bar{\xi}_i\tilde{\xi}_i]$ is zero only if $\phi_{\bar{\xi}}(j\omega)$ is zero over this range of ω, implying that $\{\bar{\xi}\}$ has zero power. That is, e.l.s. can converge at some $\hat{\mathbf{x}}_t$ other than $\mathbf{x}$ only if $(\mathbf{x} - \hat{\mathbf{x}}_t)^{\mathrm{T}}\bar{\mathbf{h}}_i$ is zero at all instants i w.p.1. Since $(\mathbf{x} - \hat{\mathbf{x}}_t)^{\mathrm{T}}\bar{\mathbf{h}}_i$ is $\bar{v}_i - v_i$ filtered by $1 + C(z^{-1})$, we conclude that the residuals $\{\bar{v}\}$ of any converged model are identical to $\{v\}$. The model output thus coincides with that given by the true parameter and regressor values. If the model order is minimal (no pole-zero cancellation) $\hat{\mathbf{x}}_t$ actually equals $\mathbf{x}$, as the parameters are transfer-function coefficients.

Without the strictly-positive-real assumption, no such conclusions can be drawn; the converged estimates may differ from the true values. A stable and unremarkable example which is not strictly positive real is

$$1/(1+C(z^{-1})) = 1/(1+1.6z^{-1}+0.8z^{-2})$$

which has a negative real part for any z between the 98.4 and 148.6° or -98.4 and $-148.6°$ radii on the unit circle. A positive-real condition turns up in many recursive algorithms (Ljung, 1977b).

We complete our check whether $\{\hat{x}\}$ converges w.p.1 to $\mathbf{x}$ by testing the local stability of the solution of (7.5.15) about the equilibrium point $\mathbf{x}$, as required by Theorem 2. From Example 7.5.2,

$$\frac{d\hat{\mathbf{x}}}{d\tau} = \frac{R^{-1}\mathbf{d}}{\sigma^2} = R^{-1}\tilde{G}(\mathbf{x}-\hat{\mathbf{x}}), \qquad \frac{dR}{d\tau} = \frac{G}{\sigma^2} - R$$

and so at $\hat{\mathbf{x}} = \mathbf{x}$, R is G/σ^2 and the linearised equations are

$$\frac{d(\hat{\mathbf{x}}-\mathbf{x})}{d\tau} = -\sigma^2 G^{-1}\tilde{G}(\hat{\mathbf{x}}-\mathbf{x}), \qquad \frac{d(R-G/\sigma^2)}{d\tau} = \frac{G}{\sigma^2} - R + \frac{\partial G}{\partial \hat{\mathbf{x}}}\frac{\hat{\mathbf{x}}-x}{\sigma^2}$$

with G, $\tilde{G}$ and $\partial G/d\hat{\mathbf{x}}$ evaluated at $\mathbf{x}$. The second equation is stable if the first is, so that the term in $\hat{\mathbf{x}}-\mathbf{x}$ decays. Stability of the first follows from G^{-1} being positive-definite and $\tilde{G}+\tilde{G}^{\mathrm{T}}$ positive-semi-definite, but only under the strictly positive-real assumption. Ljung (1977b) gives the details. △

Example 7.5.4 The motivation for the r.m.l. 2 modification of e.l.s. (Section 7.4.3) is clear in the present context. Recursive maximum-likelihood 2 filters $\mathbf{h}_i$ by $1/(1+C(z^{-1}))$, with the effect asymptotically of replacing $\tilde{\mathbf{h}}_i$ by $\bar{\mathbf{h}}_i$ in $\tilde{G}$, making $\tilde{G}$ and G identical. The linearised differential equation for $\hat{\mathbf{x}}$ about $\mathbf{x}$ is then

$$\frac{d(\hat{\mathbf{x}}-\mathbf{x})}{d\tau} = -\sigma^2(\hat{\mathbf{x}}-\mathbf{x})$$

which is stable without the strictly positive-real condition on $1/(1+C(z^{-1}))$. △

FURTHER READING

Several excellent books on recursive identification have appeared recently. Ljung and Söderström (1983) give broad coverage and are strong on the analysis of algorithms and the connections between them. Young (1984) introduces recursive identification via least-squares and pays close practical attention to time-varying models (which we discuss in Section 8.1) and instrumental-variable algorithms. Söderström and Stoica (1983) analyse instrumental-variable algorithms at a fairly advanced level, but point out the practical implications of their results and present a number of case studies. A detailed account of stochastic approximation is given in an older book by Wasan (1969).

We took only a cursory glance, in Section 7.1.2, at the connection between an a.r.m.a. model of a stochastic process and its autocorrelation function, for two reasons. First, the inputs and noise in actual dynamic systems are at best only approximately modelled by low-order a.r.m.a. models. Second, identification is most conveniently carried out by trying successive models, each estimated straight from the input and output records, until an adequate one is found. It is not normally feasible to choose even the noise-model structure mainly by reference to observed correlation functions. However, correlation functions might on occasion give some guidance, and an acquaintance with representation theory for stochastic processes might help, so two widely respected textbooks are recommended: Åström (1970) and Whittle (1963, extended and reprinted 1984).

REFERENCES

Adby, P. R., and Dempster, M. A. H. (1974). "Introduction to Optimization Methods". Chapman & Hall, London.

Åström, K. J. (1967). Computer control of a paper machine—an application of linear stochastic control theory. *IBM J. Res. Dev.* **11**, 389–405.

Åström, K. J. (1970). "Introduction to Stochastic Control Theory". Academic Press, New York and London.

Åström, K. J., and Bohlin, T. (1966). Numerical identification of linear dynamic systems from normal operating records. *In* "Theory of Self-Adaptive Control Systems" (P. H. Hammond, ed.). Plenum, New York.

Åström, K. J., and Eykhoff, P. (1971). System identification—a survey. *Automatica* **7**, 123–164.

Åström, K. J., and Wittenmark, B. (1973). On self-tuning regulators. *Automatica* **9**, 185–199.

Åström, K. J., Borisson, U., Ljung, L., and Wittenmark, B. (1977). Theory and applications of self-tuning regulators. *Automatica* **13**, 457–476.

Automatica Special Issue (1981). On Identification and System Parameter Estimation, *Automatica*, **17**.

Bekey, G. A., and Saridis, G. N. (eds.) (1983). "Identification and System Parameter Estimation 1982" (2 vols.). *Proc. IFAC Symp., 6th, Washington, D.C., 7–11 June 1982.* Pergamon, Oxford.

Bierman, G. J. (1977). "Factorization Methods for Discrete Sequential Estimation". Academic Press, New York and London.

Clarke, D. W. (1967). Generalized-least-squares estimation of the parameters of a dynamic model. *IFAC SYMP. Identification Autom. Control, Prague, Czechoslovakia*, paper 3.17.

Dugard, L., and Landau, I. D. (1980). Recursive output error identification algorithms theory and evaluation. *Automatica* **16,** 443–462.

Eykhoff, P. (1974). "System Identification". Wiley, New York.

Eykhoff, P. (ed.) (1981). "Trends and Progress in System Identification". Pergamon, Oxford.

Fogel, E. (1981). A fundamental approach to the convergence analysis of least squares algorithms. *IEEE Trans. Autom. Control* **AC–26,** 646–655.

Goodwin, G. C., and Sin, K. S. (1984). "Adaptive Filtering Prediction and Control". Prentice-Hall, Englewood Cliffs, New Jersey.

Goodwin, G. C., Ramadge, P. J., and Caines, P. E. (1980). Discrete-time multivariable adaptive control. *IEEE Trans. Autom. Control* **AC–25**, 449–456.

Goodwin, G. C., Hill, D. J., and Palaniswami, M. (1984). A perspective on convergence of adaptive control algorithms. *Automatica* **20**, 519–531.

Gustavsson, I. (1969). Maximum likelihood identification of dynamics of the Agesta reactor and comparison with results of spectral analysis. Rep. 6903. Division of Automatic Control, Lund Institute of Technology, Sweden.

Isermann, R. (ed.) (1980). "Identification and System Parameter Estimation" (2 vols.). *Proc. IFAC Symp. 5th, Darmstadt, FRG, September 1979.* Pergamon, Oxford.

Jakeman, A. J., and Young, P. C. (1979). Refined instrumental variable methods of recursive time-series analysis, Part II: Multivariable systems. *Int. J. Control* **29**, 621–644.

Jazwinski, A. H. (1970). "Stochastic Processes and Filtering Theory". Academic Press, New York and London.

Kalman, R. E. (1960). A new approach to linear filtering and prediction. *Trans. ASME. Ser. D. J. Basic Eng.* **82**, 35–45.

Kurz, H., and Goedecke, W. (1981). Digital parameter-adaptive control of processes with unknown dead time. *Automatica* **17**, 245–252.

Lee, R. C. K. (1964). "Optimal Estimation, Identification and Control". MIT Press, Cambridge, Massachusetts.

Ljung, L. (1977a). Analysis of recursive stochastic algorithms. *IEEE Trans. Autom. Control* **AC–22**, 551–575.

Ljung, L. (1977b). On positive real transfer functions and the convergence of some recursive schemes. *IEEE Trans. Autom. Control* **AC-22**, 539–551.

Ljung, L. (1981). Analysis of a general recursive prediction error identification algorithm. *Automatica* **17**, 89–99.

Ljung, L., and Soderstrom, T. (1983). "Theory and Practice of Recursive Identification". MIT Press, Cambridge, Massachusetts, and London.

Ljung, L., Soderstrom, T., and Gustavsson, I. (1975). Counterexample to the general convergence of a commonly used recursive identification method. *IEEE Trans. Autom. Control* **AC–20**, 643–652.

Maybeck, P. S. (1979). "Stochastic Models, Estimation, and Control", Vol. 1. Academic Press, New York and London.

Mayne, D. Q. (1963). Optimal non-stationary estimation of the parameters of a linear system with Gaussian inputs. *J. Electron. Control* **14**, 101–112.

Norton, J. P. (1977). Initial convergence of recursive identification algorithms. *Electron. Lett.* **13**, 621–622.

Panuska, V. (1968). A stochastic approximation method for identification of linear systems using adaptive filtering. *Proc. Joint Autom. Control Conf. Ann Arbor, Michigan*, 1014–1021.

Plackett, R. L. (1950). Some theorems in least squares. *Biometrika* **371**, 149–157.

Robbins, H., and Monro, S. (1951). A stochastic approximation method. *Ann. Math. Stat.* **22**, 400–407.

Söderström, T. (1973). An on-line algorithm for approximate maximum-likelihood identification of linear dynamic systems. Rep. 7308. Division Automatic Control, Lund Institute of Technology, Sweden.

Söderström, T., and Stoica, P. G. (1983). "Instrumental Variable Methods for System Identification". Springer-Verlag, Berlin and New York.

Solo, V. (1979). The convergence of AML. *IEEE Trans. Autom. Control* **AC–24**, 958–962.

Solo, V. (1980). Some aspects of recursive parameter estimation. *Int. J. Control* **32**, 395–410.

Talmon, J. L., and van den Boom, A. J. W. (1973). On the estimation of the transfer function parameters of process and noise dynamics using a single-stage estimation. *In* "Identification and System Parameter Estimation" (P. Eykhoff, ed.), *Proc. IFAC Symp., 3rd, Part 2, The Hague/Delft, The Netherlands, June 1973*. North-Holland, Amsterdam and American Elsevier, New York.

Vidyasagar, M. (1978). "Nonlinear System Analysis". Prentice-Hall, Englewood Cliffs, New Jersey.

Wasan, M. T. (1969). "Stochastic Approximation". Cambridge Univ. Press, London and New York.

Wellstead, P. E., Edmunds, J. M., Prager, D., and Zanker, P. (1979). Self-tuning pole/zero assignment regulators. *Int. J. Control* **30**, 1–26.

Whittle, P. (1984). "Prediction and Regulation", 2nd ed. Blackwell, Oxford.

Young, P. C. (1968). The use of linear regression and related procedures for the identification of dynamic processes. *Proc. IEEE Symp. Adaptive Processes, 7th, Univ. of California, Los Angeles*. IEEE, New York.

Young, P. C. (1974). Recursive approaches to time series analysis. *Bull. IMA* **10**, 209–224.

Young, P. C. (1984). "Recursive Estimation and Time-Series Analysis". Springer-Verlag, Berlin and New York.

Young, P. C., and Jakeman, A. J. (1979). Refined instrumental variable methods of recursive time-series analysis, Part I: Single input, single output systems. *Int. J. Control* **29**, 1–30.

Young, P. C., and Jakeman, A. J. (1980). Refined instrumental variable methods of recursive time-series analysis, Part III: Extensions. *Int. J. Control* **31**, 741–764.

PROBLEMS

7.1 An autocorrelated sequence is represented as the result of passing a white sequence though a filter with transfer function $(1+0.5z^{-1})/(1-0.8z^{-1})$. Find the a.c.f. of the autocorrelated sequence, normalised by the m.s. value. Show that the autocorrelated sequence has m.s. value 5.694 times that of the white sequence

7.2 In the recursive, minimum-covariance algorithm of Section 7.3.2, $H_tP_{t-1}H_t^{\mathrm{T}}+R$ must be inverted. If the noise and modelling error contributions to the observed variables are negligible, R is zero. If also there is some redundancy in the observation vector, the rows of H_t are linearly dependent, so $H_tP_{t-1}H_t^{\mathrm{T}}$ is singular and the inversion is not feasible. Can we make it feasible by deleting observed variables until the redundancy vanishes? Does the answer change if only one of the observed variables is free of noise and modelling error? What is special about P when one of the observed variables contains no noise or modelling error? Can the information-updating form of the algorithm be employed in such a case?

7.3 Repeat Example 7.3.4 retaining only five figures at the end of each stage of calculation, to see the effects of ill-conditioning.

7.4 Section 7.3.2 mentioned state estimation, in which $\mathbf{x}$ is modelled as evolving in accordance with a state equation $\mathbf{x}_t = \mathbf{\Phi}_{t-1}\mathbf{x}_{t-1} + \Gamma_{t-1}\mathbf{w}_{t-1}$ where $\{\mathbf{w}\}$ is a zero-mean, white sequence. As we shall see in Section 8.1, it sometimes makes sense to model a parameter vector $\mathbf{x}$ in this way. With that in mind, show that the only unbiased estimator of $\mathbf{x}_t$ in the form $A\hat{\mathbf{x}}_{t-1}$, where $\hat{\mathbf{x}}_{t-1}$ is unbiased, is $\Phi_{t-1}\hat{\mathbf{x}}_{t-1}$.

7.5 Two possible choices for the step-size factors $\{\gamma\}$ in stochastic approximation (in addition to those in Example 7.4.1) are $\gamma_t = 1/t^{1/2}$ and $\gamma_t = 10/t^{1/2}$. Try them out over six steps on the problem considered in Examples 7.3.4 and 7.4.1.

7.6 This problem traces the steps proving stability of the differential equation for $\hat{\mathbf{x}} - \mathbf{x}$ at the end of Example 7.5.3, and thus showing that the true value $\mathbf{x}$ is a possible convergence point for e.l.s.

(i) Show that if $\tilde{G} + \tilde{G}^{\mathrm{T}}$ is positive-semi-definite and G^{-1} positive-definite, a positive-definite matrix P can be found such that $P(-G^{-1}\tilde{G}) + (-G^{-1}\tilde{G})^{\mathrm{T}}P \leq 0$. [The easiest way to show it is to produce a suitable P.]

(ii) By considering $\mathbf{m}^{\mathrm{T}}(PA + A^{\mathrm{T}}P)\mathbf{m}$, where $\mathbf{m}$ is any eigenvector of A and P is positive-definite, and then $\mathbf{n}^{\mathrm{T}}(A + A^{\mathrm{T}})\mathbf{n}$, where $\mathbf{n}$ is any eigenvector of A, show that negative-semi-definiteness of $PA + A^{\mathrm{T}}P$ ensures that all eigenvalues of A are in the left-hand half plane.

(iii) Notice that with $-\sigma^2 G^{-1}\tilde{G}$ as A, (ii) proves that the d.e. in $\hat{\mathbf{x}} - \mathbf{x}$ in Example 7.5.3 is stable when the assumptions on G^{-1} and $\tilde{G}$ are met.

Chapter 8

Specialised Topics in Identification

8.1 RECURSIVE IDENTIFICATION OF LINEAR, TIME-VARYING MODELS

8.1.1 Role of Time-Varying Models

Up to this point we have treated the dynamics, and hence the model structure and parameters, as constant. We could argue that we have no alternative, since a parameter is essentially not a variable; a sufficiently comprehensive *constant-parameter* model should be able to represent the dynamics throughout. Practical factors force us to take a less dogmatic view, however. We may know or care too little about the detailed dynamics to propose a comprehensive model structure. The model may have to be linear and low-order because modelling effort is limited or the end use requires a simple model, even at the price of the parameters having to vary to accommodate non-linear or higher-order behaviour. In those circumstances the distinction between parameters and variables is blurred, so we adopt the ad hoc definition that a parameter is anything we want to regard as such, usually because of its physical interpretation or its place in a standard model structure.

Example 8.1.1 We need a model, for control design, of an industrial boiler in which water passes through tubes heated from outside by an oil burner. In time the heat transfer through the tube walls slows as soot from incomplete combustion builds up on the outside and minerals accrete on the inside of the tubes. With enough instrumentation, time, access and skill a constant-parameter model could be built, no doubt non-linear, relating the heat-transfer coefficient to the history of burner air/fuel ratio, fuel rate, water or steam flow rate, and fuel and water composition. In practice, the coefficient would at most be treated as a parameter and measured, directly or by estimation from operating records, periodically or more probably once and for all. More probably still, its influence would be lumped with the rest of the boiler dynamics into a simplified overall model relating fuel, steam and feedwater flow rates to steam temperature and pressure. △

A less obvious justification for a time-varying model is that systematic time-variation of parameters in a tentative model, induced by unmodelled behaviour, can be very effective as a guide to how the model should be extended or modified.

Example 8.1.2 Figure 8.1.1 shows time variation of the estimated steady-state gain of a river catchment, the Mackintosh in Tasmania, with hourly rainfall as input and river flow as output. The gain was estimated by an extension of the e.l.s. algorithm as described in Section 8.1.5.

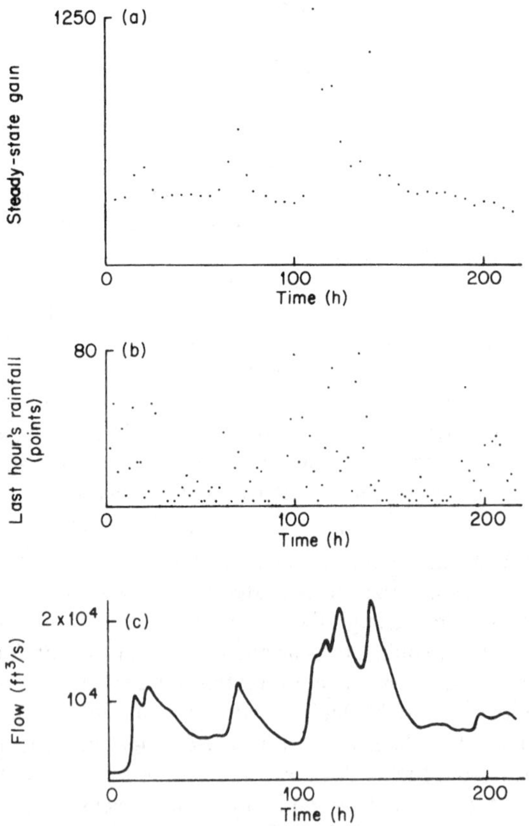

Fig. 8.1.1 Time-variation of river-catchment model, Example 8.1.2. (a) Steady-state gain, (b) last hour's rainfall, and (c) flow.

As the highest gains coincide with flow peaks and the lowest with relatively dry spells, the variation is plainly due to changing soil dryness, with consequent variation in the proportion of rainfall running off rapidly. A model allowing for storage and saturation would be worth investigating, it seems. △

8.1.2 Modification of Recursive Algorithms to Track Time Variation

Our task is to modify the recursive algorithms of Sections 7.3 and 7.4 to track time-varying parameters. The observation (or regression) equation

$$y_t = \mathbf{h}_t^{\mathrm{T}} \mathbf{x}_t + v_t \tag{8.1.1}$$

with parameter vector $\mathbf{x}_t$ constant gave rise to updating equations (7.3.28), (7.4.7), (7.4.22) and (7.4.26), all of the form

$$\hat{\mathbf{x}}_t = \hat{\mathbf{x}}_{t-1} + \mathbf{k}_t(y_t - \mathbf{h}_t^{\mathrm{T}} \hat{\mathbf{x}}_{t-1}) \tag{8.1.2}$$

which can track parameter variation provided the correction gain $\mathbf{k}_t$ is not too small. The problem is that the gain decreases as t increases, in any algorithm which for time-invariant dynamics yields ever-increasing accuracy for $\hat{\mathbf{x}}_{t-1}$. If the prediction error $y_t - \mathbf{h}_t^{\mathrm{T}} \hat{\mathbf{x}}_{t-1}$ is due less and less to error in $\hat{\mathbf{x}}_{t-1}$ and is ultimately due mostly to observation noise, a small correction gain is appropriate. On the other hand, with time-varying dynamics $\hat{\mathbf{x}}_{t-1}$ may not be a good estimator of $\mathbf{x}_t$ even at large t, and a larger gain is necessary. To improve tracking ability, the gain $\mathbf{k}_t$ has to be increased in some systematic way. In the Markov, e.l.s. and other recursive l.s. algorithms, $\mathbf{k}_t$ is $P_t \mathbf{h}_t / \sigma_t^2$, so $\mathbf{k}_t$ may be increased by increasing P_t, that is by making the covariance reflect less confidence in the updated $\hat{\mathbf{x}}_t$. We shall examine three ways of doing so.

8.1.3 Recursive Weighted-Least-Squares with a Forgetting Factor

In recursive o.l.s., $\hat{\mathbf{x}}_t$ differs little from $\hat{\mathbf{x}}_{t-1}$ at large t because $\hat{\mathbf{x}}_{t-1}$ minimises the sum of squared residuals from time 1 to $t-1$, and the new residual $\hat{v}_t$ has a proportionately small influence on the sum from 1 to t. If we think the dynamics are time-varying, we can attach more weight to recent than to earlier residuals, reasoning that $\hat{\mathbf{x}}_t$ should not have to give small residuals at past times when $\mathbf{x}$ was very different from $\mathbf{x}_t$. Thus we are led to minimise a weighted sum

$$S_t = \sum_{i=1}^{t} w_i \hat{v}_i^2 \equiv \hat{\mathbf{v}}_t^{\mathrm{T}} W_t \hat{\mathbf{v}}_t \tag{8.1.3}$$

where W_t is a diagonal matrix and the weights w_t increase with time. The scalar-output Markov estimator in Section 7.3.3 attaches weight $1/\sigma_i^2$ to $\hat{v}_i^2$, giving (7.3.27) and (7.3.28). If we put w_i for $1/\sigma_i^2$ we obtain (8.1.2) with

$$\mathbf{k}_t = w_t P_t \mathbf{h}_t \tag{8.1.4}$$

and

$$P_t = P_{t-1} - \frac{w_t P_{t-1} \mathbf{h}_t \mathbf{h}_t^T P_{t-1}}{1 + w_t \mathbf{h}_t^T P_{t-1} \mathbf{h}_t} \tag{8.1.5}$$

These equations are often written in terms of $w_t P_t$ and w_{t-1}/w_t, denoted by P'_t and β_t:

$$P'_t = \frac{1}{\beta_t}\left(P'_{t-1} - \frac{P'_{t-1}\mathbf{h}_t\mathbf{h}_t^T P'_{t-1}}{\beta_t + \mathbf{h}_t^T P'_{t-1}\mathbf{h}_t}\right), \qquad \mathbf{k}_t = P'_t \mathbf{h}_t \tag{8.1.6}$$

The commonest choice of β_t is a constant *forgetting factor* β just below 1, generating an exponentially increasing sequence of weights $w_t = \beta^{-t}$. The value of β is adjusted until credible parameter variation and acceptable residuals are obtained. A typical value is between 0.95 and 0.99.

8.1.4 Covariance Resetting

Alternatively, P_t can be prevented from becoming too small as t increases by being reset to a fixed large value whenever its size, measured for instance by $\operatorname{tr} P_t$, falls below a certain value. By doing so we express disbelief that $\hat{\mathbf{x}}_t$ is really as good as P_t says, and dismiss the confidence in $\hat{\mathbf{x}}_t$ derived from earlier observations.

The method has the technical virtue of allowing convergence to be proved relatively easily for time-invariant parameters (Goodwin and Sin, 1984). On the other hand, it gives much more influence to observations immediately after a covariance resetting than to those just before, generally for no good reason.

8.1.5 Explicit Modelling of Parameter Variation

We cannot easily pick a forgetting factor or resetting threshold in advance even if we know roughly what parameter variation to expect. The best forgetting factor may in any case be a poor compromise when some parameters vary much more rapidly than others. Greater flexibility and simpler incorporation of prior knowledge can be achieved by basing the estimator on an explicit model of the parameter variation. If we are to avoid a substantial extra identification problem, we must keep the model of the variation very simple.

One simple yet flexible model is a random walk for each parameter:

$$\mathbf{x}_t = \mathbf{x}_{t-1} + \mathbf{w}_{t-1}, \qquad \operatorname{cov} \mathbf{w}_{t-1} = Q_{t-1} \tag{8.1.7}$$

Here $\mathbf{w}_{t-1}$ is independent of $\mathbf{x}_{t-1}$, zero-mean and white, i.e. $E[\mathbf{w}_s \mathbf{w}_t^{\mathrm{T}}]$ is zero for $s \neq t$. It can usually be taken as wide-sense stationary, so Q_{t-1} can be written as just Q. In the absence of special background knowledge we take Q as diagonal, implying that the parameters vary independently, and thus we need only specify the mean-square variation of each parameter. This *simple random walk (s.r.w.) model* is a special case of the more general parameter-evolution model

$$\mathbf{x}_t = \Phi_{t-1}\mathbf{x}_{t-1} + \Gamma_{t-1}\mathbf{w}_{t-1}, \qquad \operatorname{cov} \mathbf{w}_{t-1} = Q_{t-1} \tag{8.1.8}$$

We recognise (8.1.8) as a state equation (D'Azzo and Houpis, 1981; Gabel and Roberts, 1980) to accompany the observation equation (8.1.1). Estimation of $\mathbf{x}_t$ is therefore a sort of state estimation problem, with a parameter vector as the state. We should not find this too paradoxical, given the haziness of the distinction between time-varying parameters and variables, discussed earlier. The view of parameter estimation as state estimation is enormously helpful, as it opens the door to a great armoury of state-estimation technique, as we shall soon see. There is one difference between parameter and state estimation. The observation vector $\mathbf{h}_t$ (matrix H_t for vector observations) is taken as known and deterministic in state estimation, but in parameter estimation it is usually stochastic, containing noisy previous output samples and/or samples of the noise-generating variable, as well as input samples which may be viewed as stochastic. We saw in Section 7.5 that the stochastic and often complicated nature of $\mathbf{h}_t$ makes analysis of recursive parameter estimators difficult.

We normally know too little to specify Φ and Γ in the full parameter-variation model (8.1.8), but we can see its effects on recursive least-squares algorithms with very little more algebraic effort than considering (8.1.7), and end up with an algorithm recognisable as a standard state estimator. State equation (8.1.8) adds a new stage to each recursion step. It is used to project $\hat{\mathbf{x}}_{t-1}$ forward in time to a new prior estimate $\hat{\mathbf{x}}_{t|t-1}$ of $\mathbf{x}_t$. The second subscript indicates that $\hat{\mathbf{x}}_{t|t-1}$ is based on observations up to $t-1$. For $\hat{\mathbf{x}}_{t|t-1}$ to be linear and unbiased, it must be of the form $A\hat{\mathbf{x}}_{t-1} + \mathbf{b}$, with

$$E\hat{\mathbf{x}}_{t|t-1} = AE\hat{\mathbf{x}}_{t-1} + \mathbf{b} = AE\mathbf{x}_{t-1} + \mathbf{b} = E\mathbf{x}_t = \Phi_{t-1}E\mathbf{x}_{t-1} \tag{8.1.9}$$

since $\hat{\mathbf{x}}_{t-1}$ is unbiased and $E\mathbf{w}_{t-1}$ is zero. To satisfy (8.1.9), whatever the value of $\mathbf{x}_{t-1}$, we must have A equal to Φ_{t-1} and $\mathbf{b}$ zero, so

$$\hat{\mathbf{x}}_{t|t-1} = \Phi_{t-1}\hat{\mathbf{x}}_{t-1} \tag{8.1.10}$$

The covariance $P_{t|t-1}$ of $\hat{\mathbf{x}}_{t|t-1}$ is found by noting that neither $\mathbf{x}_{t-1}$ nor $\hat{\mathbf{x}}_{t-1}$ is correlated with $\mathbf{w}_{t-1}$, as both depend only on earlier values of $\mathbf{w}$. Hence, denoting $\mathbf{x}_{t-1} - \hat{\mathbf{x}}_{t-1}$ by $\tilde{\mathbf{x}}_{t-1}$,

$$\begin{aligned} P_{t|t-1} &= E[(-\Phi_{t-1}\tilde{\mathbf{x}}_{t-1} - \Gamma_{t-1}\mathbf{w}_{t-1})(-\Phi_{t-1}\tilde{\mathbf{x}}_{t-1} - \Gamma_{t-1}\mathbf{w}_{t-1})^{\mathrm{T}}] \\ &= \Phi_{t-1}E[\tilde{\mathbf{x}}_{t-1}\tilde{\mathbf{x}}_{t-1}^{\mathrm{T}}]\Phi_{t-1}^{\mathrm{T}} + \Gamma_{t-1}E[\mathbf{w}_{t-1}\mathbf{w}_{t-1}^{\mathrm{T}}]\Gamma_{t-1}^{\mathrm{T}} \\ &= \Phi_{t-1}P_{t-1}\Phi_{t-1}^{\mathrm{T}} + \Gamma_{t-1}Q_{t-1}\Gamma_{t-1}^{\mathrm{T}} \end{aligned} \tag{8.1.11}$$

For the s.r.w. model (8.1.7) with constant-covariance $\{\mathbf{w}\}$, this reduces to

$$P_{t|t-1} = P_{t-1} + Q \tag{8.1.12}$$

and $\hat{\mathbf{x}}_{t|t-1}$ is just $\hat{\mathbf{x}}_{t-1}$.

The rest of the recursion step is exactly as for time-invariant parameters, with $\hat{\mathbf{x}}_{t|t-1}$ and $P_{t|t-1}$ replacing $\hat{\mathbf{x}}_{t-1}$ and P_{t-1} in (8.1.2) and in the covariance equation

$$P_t = P_{t-1} - \frac{P_{t-1}\mathbf{h}_t\mathbf{h}_t^{\mathrm{T}}P_{t-1}}{\sigma^2 + \mathbf{h}_t^{\mathrm{T}}P_{t-1}\mathbf{h}_t} \tag{8.1.13}$$

or its w.l.s. counterpart.

For vector observations, (8.1.10) and (8.1.11) are unchanged. Together with (7.3.9), (7.3.5) and (7.3.6) they form the complete algorithm. The algorithm is identical to the Kalman filter for state estimation. The notation $\hat{\mathbf{x}}_{t|t}$ and $P_{t|t}$ is usual for what we have been calling $\hat{\mathbf{x}}_t$ and P_t, to emphasise that the latest observation $\mathbf{y}_t$ has been processed.

Parameter tracking with an s.r.w. model adds only (8.1.12) to the recursion. The choice of Q must avoid over-inflation of P and consequent inefficiency of $\hat{\mathbf{x}}$ on the one hand, and inability to follow rapid variations on the other. Prior knowledge is not usually enough to fix Q, so we must adjust Q with reference to some performance measure. A simple and effective technique (Norton, 1975) is to compare the m.s. innovations and residuals at trial values of Q with those obtained when Q is zero.

Background knowledge rarely provides Φ_{t-1} and Γ_{t-1} in the more general model (8.1.8) either. It is not easy to estimate them along with $\mathbf{x}$, partly because too many unknowns are to be estimated and partly because products of unknown elements of Φ_{t-1} with the unknown elements of $\mathbf{x}_{t-1}$ make the overall estimation problem non-linear. This problem is pursued further in Section 8.9. There is, however, one versatile model more elaborate than the s.r.w. but actually easier to use, the *integrated random walk* (*i.r.w.*) *model* (Norton, 1976) for a parameter vector $\mathbf{x}'$:

$$\left.\begin{aligned} \mathbf{s}_t &= \mathbf{s}_{t-1} + \mathbf{w}_{t-1} \\ \mathbf{x}'_t &= \mathbf{x}'_{t-1} + \mathbf{s}_{t-1} \end{aligned}\right\} \qquad \operatorname{cov} \mathbf{w}_{t-1} = Q \tag{8.1.14}$$

The parameter increments $\mathbf{s}$ are now random walks and Q dictates the smoothness of the parameter variation rather than the m.s. variation itself.

Both $\mathbf{x}'$ and $\mathbf{s}$ are estimated, so

$$\mathbf{x}_t \triangleq \begin{bmatrix} \mathbf{x}'_t \\ \mathbf{s}_t \end{bmatrix}, \qquad \Phi_{t-1} = \begin{bmatrix} I_p & I_p \\ 0 & I_p \end{bmatrix}, \qquad \Gamma_{t-1} = \begin{bmatrix} 0 \\ I_p \end{bmatrix} \tag{8.1.15}$$

In the observation equation (8.1.1), $\mathbf{h}_t$ is padded with zeros to multiply $\mathbf{s}_t$. We partition $P_{t|t-1}$ into $(p \times p)$ blocks to match $\mathbf{x}'$ and $\mathbf{s}$:

$$P_{t|t-1} \equiv \begin{bmatrix} P^{(11)} & P^{(12)} \\ P^{(21)} & P^{(22)} \end{bmatrix}_{t|t-1} \tag{8.1.16}$$

and similarly for P_t. Defining

$$\mathbf{f}_t \equiv \begin{bmatrix} \mathbf{f}_t^{(1)} \\ \mathbf{f}_t^{(2)} \end{bmatrix} \triangleq \begin{bmatrix} P_{t|t-1}^{(11)} \\ P_{t|t-1}^{(21)} \end{bmatrix} \frac{\mathbf{h}_t}{\sigma^2} \tag{8.1.17}$$

we obtain

$$\hat{\mathbf{x}}_t = \Phi\hat{\mathbf{x}}_{t-1} + \mathbf{f}_t(y_t - \mathbf{h}_t^{\mathrm{T}}\Phi\hat{\mathbf{x}}_{t-1})/(1 + \mathbf{h}_t^{\mathrm{T}}\mathbf{f}_t^{(1)}) \tag{8.1.18}$$

$$P_t = P_{t|t-1} - \sigma^2\mathbf{f}_t\mathbf{f}_t^{\mathrm{T}}/(1 + \mathbf{h}_t^{\mathrm{T}}\mathbf{f}_t^{(1)}) \tag{8.1.19}$$

It is much easier to find, by trial and error, a satisfactory value for Q in the i.r.w. model (8.1.14) than in the s.r.w. model (8.1.7), because the effect of a less-than-ideal choice in the i.r.w. model is merely to make the parameter-estimate sequence $\{\hat{\mathbf{x}}\}$ a little too smooth or rough. The overall extent of time variation in $\{\hat{\mathbf{x}}\}$, which is usually our main interest, is not very sensitive to Q in the i.r.w. model.

8.1.6 Optimal Smoothing

When the parameters are modelled as time-varying and stochastic as in (8.1.8), the extra uncertainty introduced by $\Gamma_{t-1}\mathbf{w}_{t-1}$ adds $\Gamma_{t-1}Q_{t-1}\Gamma_{t-1}^{\mathrm{T}}$ to the covariance of the parameter estimates, as in (8.1.11). The increase in uncertainty makes it important that at every sample instant $\hat{\mathbf{x}}$ should utilise the information in as many observations as possible. Also, as the detailed time variation in $\{\hat{\mathbf{x}}\}$ may be of great interest, we should like good parameter estimates throughout the record, early as well as late. This applies especially with short records, where error in $\{\hat{\mathbf{x}}\}$ due to a poor initial guess $\hat{\mathbf{x}}_0$ may decrease slowly enough to obscure the time variation over much of the record.

The key to improved parameter estimates is the fact that $\mathbf{x}_t$ influences all later values of $\mathbf{x}$ through the state equation, and hence all later observations up to the last, y_N. Consequently $\hat{\mathbf{x}}_t$ should embody information from all later observations as well as those up to y_t. That is, we should compute $\hat{\mathbf{x}}_{t|N}$, not just $\hat{\mathbf{x}}_{t|t}$. Computation of $\hat{\mathbf{x}}_{t|N}$ is the function of *fixed-interval optimal smoothing* in state estimation (Jazwinski, 1970; Bierman, 1977; Maybeck, 1982). In the same way that recursive l.s. identification is identical to Kalman filtering except for the stochastic nature of H_t, optimal smoothing algorithms developed for state estimation can be used unchanged for parameter estimation.

We can derive a fixed-interval smoothing algorithm as the optimally weighted Markov l.s. estimator of $\mathbf{x}_t$ from $\mathbf{y}_1$ to $\mathbf{y}_N$. Vector observations do not add to the complication, so we shall consider them. We start by writing down all the available information about $\{\mathbf{x}\}$:

$$\begin{aligned}
&\mathbf{y}_t - H_t\mathbf{x}_t = \mathbf{v}_t && \operatorname{cov}\mathbf{v}_t = R_t, && t=1,2,\ldots,N\\
&\mathbf{x}_t = \Phi_{t-1}\mathbf{x}_{t-1} + \Gamma_{t-1}\mathbf{w}_{t-1}, && \operatorname{cov}\mathbf{w}_{t-1} = Q_{t-1}, && t=1,2,\ldots,N\\
&\mathbf{x}_0 = \hat{\mathbf{x}}_{0|0} + \tilde{\mathbf{x}}_{0|0}, && \operatorname{cov}\tilde{\mathbf{x}}_{0|0} = P_{0|0}, && \hat{\mathbf{x}}_{0|0}\text{ given}
\end{aligned} \tag{8.1.20}$$

The Markov estimates $\hat{\mathbf{x}}_{0|N}$ to $\hat{\mathbf{x}}_{N|N}$ and the corresponding estimates $\hat{\mathbf{w}}_0$ to $\hat{\mathbf{w}}_{N-1}$ must minimise

$$S_N = \sum_{t=1}^{N}\{(\mathbf{y}_t - H_t\hat{\mathbf{x}}_{t|N})^{\mathrm{T}}R_t^{-1}(\mathbf{y}_t - H_t\hat{\mathbf{x}}_{t|N}) + \hat{\mathbf{w}}_{t-1}^{\mathrm{T}}Q_{t-1}^{-1}\hat{\mathbf{w}}_{t-1}\}$$
$$+ (\hat{\mathbf{x}}_{0|N} - \hat{\mathbf{x}}_{0|0})^{\mathrm{T}}P_{0|0}^{-1}(\hat{\mathbf{x}}_{0|N} - \hat{\mathbf{x}}_{0|0}) \tag{8.1.21}$$

subject to equality constraints

$$\hat{\mathbf{x}}_{t|N} = \Phi_{t-1}\hat{\mathbf{x}}_{t-1|N} + \Gamma_{t-1}\hat{\mathbf{w}}_{t-1}, \qquad t=1,2,\ldots,N \tag{8.1.22}$$

We use the Lagrange multiplier method (Adby and Dempster, 1974), defining

$$L = \tfrac{1}{2}S_N - \sum_{t=1}^{N}\boldsymbol{\lambda}_{t-1}^{\mathrm{T}}(\hat{\mathbf{x}}_{t|N} - \Phi_{t-1}\hat{\mathbf{x}}_{t-1|N} - \Gamma_{t-1}\hat{\mathbf{w}}_{t-1}) \tag{8.1.23}$$

and setting $\partial L/\partial\boldsymbol{\lambda}_{t-1}$, $\partial L/\partial\hat{\mathbf{x}}_{t|N}$ and $\partial L/\partial\hat{\mathbf{w}}_{t-1}$ to zero to find the constrained minimum of S_N. The constraints (8.1.22) are satisfied by $\partial L/\partial\boldsymbol{\lambda}_{t-1}$ being zero. Also

$$\frac{\partial L}{\partial\hat{\mathbf{x}}_{t|N}} = -H_t^{\mathrm{T}}R_t^{-1}(\mathbf{y}_t - H_t\hat{\mathbf{x}}_{t|N}) - \boldsymbol{\lambda}_{t-1} + \Phi_t^{\mathrm{T}}\boldsymbol{\lambda}_t = \mathbf{0}, \qquad t=1,2,\ldots,N-1 \tag{8.1.24}$$

$$\frac{\partial L}{\partial\hat{\mathbf{x}}_{0|N}} = P_{0|0}^{-1}(\hat{\mathbf{x}}_{0|N} - \hat{\mathbf{x}}_{0|0}) + \Phi_0^{\mathrm{T}}\boldsymbol{\lambda}_0 = \mathbf{0} \tag{8.1.25}$$

$$\frac{\partial L}{\partial\hat{\mathbf{x}}_{N|N}} = -H_N^{\mathrm{T}}R_N^{-1}(\mathbf{y}_N - H_N\hat{\mathbf{x}}_{N|N}) - \boldsymbol{\lambda}_{N-1} = \mathbf{0} \tag{8.1.26}$$

$$\frac{\partial L}{\partial\hat{\mathbf{w}}_{t-1}} = Q_{t-1}^{-1}\hat{\mathbf{w}}_{t-1} + \Gamma_{t-1}^{\mathrm{T}}\boldsymbol{\lambda}_{t-1} = \mathbf{0}, \qquad t=1,2,\ldots,N \tag{8.1.27}$$

Boundary conditions for these equations are the initial guess, $\hat{\mathbf{x}}_{0|0}$, and the last estimate $\hat{\mathbf{x}}_{N|N}$ computed by the ordinary non-smoothing Markov estimator. Once $\boldsymbol{\lambda}_{N-1}$ is found from (8.1.26) using $\hat{\mathbf{x}}_{N|N}$, the backwards-in-time recursion formed by (8.1.24), (8.1.27) and (8.1.22) gives $\boldsymbol{\lambda}_{t-1}$, $\hat{\mathbf{w}}_{t-1}$ and $\hat{\mathbf{x}}_{t-1|N}$ successively, providing Φ_{t-1} is invertible. The recursion can be written more concisely as

$$\begin{aligned} \boldsymbol{\lambda}_t &= \Phi_{t+1}^{\mathrm{T}}\boldsymbol{\lambda}_{t+1} - H_{t+1}^{\mathrm{T}}R_{t+1}^{-1}(\mathbf{y}_{t+1} - H_{t+1}\hat{\mathbf{x}}_{t+1|N}), \qquad t = N-1, N-2, \ldots, 0 \\ \hat{\mathbf{x}}_{t|N} &= \Phi_t^{-1}(\hat{\mathbf{x}}_{t+1|N} + \Gamma_t Q_t \Gamma_t^{\mathrm{T}}\boldsymbol{\lambda}_t) \end{aligned} \tag{8.1.28}$$

but it has a fatal flaw: it is unstable.

Proof that Smoothing Algorithm (8.1.28) *is Unstable* Substituting $\boldsymbol{\lambda}_t$ from the first equation into the second and dropping inessential subscripts, one step of the backward recursion is

$$\begin{aligned} \begin{bmatrix} \hat{\mathbf{x}}_{t|N} \\ \boldsymbol{\lambda}_t \end{bmatrix} &= \begin{bmatrix} \Phi_t^{-1}(I + \Gamma Q \Gamma^{\mathrm{T}} H^{\mathrm{T}} R^{-1} H) & \Phi_t^{-1}\Gamma Q \Gamma^{\mathrm{T}}\Phi_{t+1}^{\mathrm{T}} \\ H^{\mathrm{T}}R^{-1}H & \Phi_{t+1}^{\mathrm{T}} \end{bmatrix} \begin{bmatrix} \hat{\mathbf{x}}_{t+1|N} \\ \boldsymbol{\lambda}_{t+1} \end{bmatrix} \\ &\quad - \begin{bmatrix} \Phi_t^{-1}\Gamma Q \Gamma^{\mathrm{T}} H^{\mathrm{T}} R^{-1} \\ H^{\mathrm{T}}R^{-1} \end{bmatrix} \mathbf{y}_{t+1} \\ &\equiv \Psi \begin{bmatrix} \hat{\mathbf{x}}_{t+1|N} \\ \boldsymbol{\lambda}_{t+1} \end{bmatrix} + \text{forcing by } \mathbf{y}_{t+1} \end{aligned} \tag{8.1.29}$$

For stability, all eigenvalues of Ψ, i.e. zeros σ of $|\sigma I + \Psi|$, must lie inside or on the unit circle. The identities (Goodwin and Payne, 1977)

$$\begin{vmatrix} A_{11} & A_{12} \\ A_{21} & A_{22} \end{vmatrix} \equiv |A_{22}||A_{11} - A_{12}A_{22}^{-1}A_{21}| \equiv |A_{11}||A_{22} - A_{21}A_{11}^{-1}A_{12}| \tag{8.1.30}$$

give

$$\begin{aligned} |\sigma I - \Psi| &= |\sigma I - \Phi_{t+1}^{\mathrm{T}}||\sigma I - \Phi_t^{-1}(I + \Gamma Q \Gamma^{\mathrm{T}} H^{\mathrm{H}} R^{-1} H) \\ &\qquad - \Phi_t^{-1}\Gamma Q \Gamma^{\mathrm{T}}\Phi_{t+1}^{\mathrm{T}}(\sigma I - \Phi_{t+1}^{\mathrm{T}})^{-1} H^{\mathrm{T}} R^{-1} H| \\ &= |\sigma I - \Phi_{t+1}^{\mathrm{T}}||\sigma I - \Phi_t^{-1} - \sigma\Phi_t^{-1}\Gamma Q \Gamma^{\mathrm{T}}(\sigma I - \Phi_{t+1}^{\mathrm{T}})^{-1} H^{\mathrm{T}} R^{-1} H| \\ &= \begin{vmatrix} \sigma I - \Phi_t^{-1} & \sigma\Phi_t^{-1}\Gamma Q \Gamma^{\mathrm{T}} \\ H^{\mathrm{T}}R^{-1}H & \sigma I - \Phi_{t+1}^{\mathrm{T}} \end{vmatrix} \\ &= |\sigma I - \Phi_t^{-1}||\sigma I - \Phi_{t+1}^{\mathrm{T}} - H^{\mathrm{T}}R^{-1}H(\sigma I - \Phi_t^{-1})^{-1}\sigma\Phi_t^{-1}\Gamma Q \Gamma^{\mathrm{T}}| = 0 \end{aligned} \tag{8.1.31}$$

Now the zeros of $|\sigma I - \Phi_t^{-1}|$ are the eigenvalues of Φ_t^{-1}, which are the reciprocals of those of Φ_t. For any stable parameter-evolution model (8.1.8), the eigenvalues of Φ_t^{-1} are unstable, so from (8.1.31) Ψ has some unstable eigenvalues. △ △

The instability of (8.1.28) is not an essential property of the $\boldsymbol{\lambda}$'s or $\hat{\mathbf{x}}$'s, but is due to the way they are calculated. It can be avoided by computing $\hat{\mathbf{x}}_{t|N}$ from $\hat{\mathbf{x}}_{t|t-1}$ or $\hat{\mathbf{x}}_{t|t}$ rather than $\hat{\mathbf{x}}_{t+1|N}$. The mechanics of establishing a suitable relation are boring, and will only be sketched briefly. From the second equation of (8.1.28), (8.1.10) and (8.1.11),

$$\hat{\mathbf{x}}_{t+1|t} - \hat{\mathbf{x}}_{t+1|N} = \Phi_t(\hat{\mathbf{x}}_{t|t} - \hat{\mathbf{x}}_{t|N}) + (P_{t+1|t} - \Phi_t P_{t|t} \Phi_t^{\mathrm{T}})\boldsymbol{\lambda}_t \qquad (8.1.32)$$

and from the observation-updating equations giving $\hat{\mathbf{x}}_{t+1|t+1}$ from $\hat{\mathbf{x}}_{t+1|t}$ and $P_{t+1|t+1}^{-1}$ from $P_{t+1|t}^{-1}$ as in (7.3.21) and (7.3.19), substituted into the first equation of (8.1.28),

$$\boldsymbol{\lambda}_t = \Phi_{t+1}^{\mathrm{T}} \boldsymbol{\lambda}_{t+1} + P_{t+1|t+1}^{-1}(\hat{\mathbf{x}}_{t+1|N} - \hat{\mathbf{x}}_{t+1|t+1}) + P_{t+1|t}^{-1}(\hat{\mathbf{x}}_{t+1|t} - \hat{\mathbf{x}}_{t+1|N}) \quad (8.1.33)$$

Premultiplying (8.1.33) by $\Phi_{t+1} P_{t+1|t+1}^{-1}$ and then putting t for $t+1$ and substituting the result into (8.1.32) yields a recursion for $\hat{\mathbf{x}}_{t+1|t} - \hat{\mathbf{x}}_{t+1|N} - P_{t+1|t}\boldsymbol{\lambda}_t$:

$$\hat{\mathbf{x}}_{t+1|t} - \hat{\mathbf{x}}_{t+1|N} - P_{t+1|t}\boldsymbol{\lambda}_t = \Phi_t P_{t|t} P_{t|t-1}^{-1}(\hat{\mathbf{x}}_{t|t-1} - \hat{\mathbf{x}}_{t|N} - P_{t|t-1}\boldsymbol{\lambda}_{t-1}) \qquad (8.1.34)$$

An initial condition is found by adding $\Phi_0 P_{0|0}$ times (8.1.25) to (8.1.32) with t zero:

$$\hat{\mathbf{x}}_{1|0} - \hat{\mathbf{x}}_{1|N} - P_{1|0}\boldsymbol{\lambda}_0 = \mathbf{0} \qquad (8.1.35)$$

Running forwards in time, the left-hand side of (8.1.34) is therefore zero, i.e.

$$\hat{\mathbf{x}}_{t+1|N} = \hat{\mathbf{x}}_{t+1|t} - P_{t+1|t}\boldsymbol{\lambda}_t, \qquad t = 0, 1, \ldots, N-1 \qquad (8.1.36)$$

A recursion for $\boldsymbol{\lambda}_t$ in terms of $\hat{\mathbf{x}}_{t+1|t}$ can be obtained by writing the first equation of (8.1.28) as

$$\begin{aligned}\boldsymbol{\lambda}_t = {} & \Phi_{t+1}^{\mathrm{T}}\boldsymbol{\lambda}_{t+1} - H_{t+1}^{\mathrm{T}} R_{t+1}^{-1}(\mathbf{y}_{t+1} - H_{t+1}\hat{\mathbf{x}}_{t+1|t}) \\ & + H_{t+1}^{\mathrm{T}} R_{t+1}^{-1} H_{t+1}(\hat{\mathbf{x}}_{t+1|N} - \hat{\mathbf{x}}_{t+1|t})\end{aligned} \qquad (8.1.37)$$

then substituting $\hat{\mathbf{x}}_{t+1|N} - \hat{\mathbf{x}}_{t+1|t}$ from (8.1.36) and using the updating equation (7.3.19) from $P_{t+1|t}^{-1}$ to $P_{t+1|t+1}^{-1}$, $\boldsymbol{\lambda}_t$ is

$$\boldsymbol{\lambda}_t = (I - H_{t+1}^{\mathrm{T}} R_{t+1}^{-1} H_{t+1} P_{t+1|t+1})(\Phi_{t+1}^{\mathrm{T}}\boldsymbol{\lambda}_{t+1} - H_{t+1}^{\mathrm{T}} R_{t+1}^{-1}(\mathbf{y}_{t+1} - H_{t+1}\hat{\mathbf{x}}_{t+1|t}))$$
$$t = N-1, N-2, \ldots, 0 \qquad (8.1.38)$$

The rearranged optimal-smoothing algorithm uses (8.1.38) and (8.1.36) to find $\hat{\mathbf{x}}_{t+1|N}$ in a backward run, using the results $\hat{\mathbf{x}}_{t+1|t}$ and $P_{t+1|t}$ of the ordinary, non-smoothing forward recursion.

Alternatively, $\hat{\mathbf{x}}_{t+1|N}$ can be found in terms of $\hat{\mathbf{x}}_{t+1|t+1}$ by substituting (8.1.36) into (8.1.33), which gives

$$\hat{\mathbf{x}}_{t+1|N} = \hat{\mathbf{x}}_{t+1|t+1} - P_{t+1|t+1}\Phi_{t+1}^{\mathrm{T}}\boldsymbol{\lambda}_{t+1}, \qquad t = N-1, N-2, \ldots, 0 \qquad (8.1.39)$$

The matching recursion for $\boldsymbol{\lambda}_t$ follows by substituting (8.1.39) into the first equation of (8.1.28) rewritten as (8.1.37) but with $\hat{\mathbf{x}}_{t+1|t+1}$ for $\hat{\mathbf{x}}_{t+1|t}$:

$$\boldsymbol{\lambda}_t = (I - H_{t+1}^{\mathrm{T}} R_{t+1}^{-1} H_{t+1} P_{t+1|t+1})\Phi_{t+1}^{\mathrm{T}}\boldsymbol{\lambda}_{t+1} - H_{t+1}^{\mathrm{T}} R_{t+1}^{-1}(\mathbf{y}_{t+1} - H_{t+1}\hat{\mathbf{x}}_{t+1|t+1})$$
$$t = N-1, N-2, \ldots, 0 \qquad (8.1.40)$$

Summary *A fixed-interval optimal smoothing algorithm* for time-varying parameters or states consists of either (8.1.36) with (8.1.38), or (8.1.39) with (8.1.40).

The question of stability does not arise in (8.1.36) or (8.1.39) as they are not recursions, but we must prove that (8.1.38) and (8.1.40) are stable.

Proof that (8.1.38) *and* (8.1.40) *are stable* The transition matrix of either recursion is $(I - H_{t+1}^{\mathrm{T}} R_{t+1}^{-1} H_{t+1} P_{t+1|t+1})\Phi_{t+1}^{\mathrm{T}}$. The updating from $\hat{\mathbf{x}}_{t+1|t}$ to $\hat{\mathbf{x}}_{t+2|t+1}$ is altogether

$$\begin{aligned}\hat{\mathbf{x}}_{t+2|t+1} &= \Phi_{t+1}\hat{\mathbf{x}}_{t+1|t+1} \\ &= \Phi_{t+1}(\hat{\mathbf{x}}_{t+1|t} + P_{t+1|t+1}H_{t+1}^{\mathrm{T}} R_{t+1}^{-1}(\mathbf{y}_{t+1} - H_{t+1}\hat{\mathbf{x}}_{t+1|t})) \\ &= \Phi_{t+1}(I - P_{t+1|t+1}H_{t+1}^{\mathrm{T}} R_{t+1}^{-1} H_{t+1})\hat{\mathbf{x}}_{t+1|t} \\ &\quad + \text{forcing not dependent on } \hat{\mathbf{x}}_{t+1|t} \end{aligned} \qquad (8.1.41)$$

As R_{t+1}^{-1} and $P_{t+1|t+1}$ are symmetric, the transition matrix in (8.1.38) or (8.1.40) is the transpose of that of (8.1.41). Hence, it has the same eigenvalues, all stable since the forward updating (8.1.41) is stable (Jazwinski, 1970; McGarty, 1974). △ △

A noteworthy feature of these algorithms is that the covariance $P_{t+1|N}$ of $\hat{\mathbf{x}}_{t+1|N}$ need not be computed unless we want it. Of course we may wish to assess the improvement due to smoothing, particularly off line when extra computing is no problem. Despite the simplicity of (8.1.36) and (8.1.39), $P_{t+1|N}$ is very complicated to derive algebraically. We can find it much more easily from the orthogonality conditions on the smoothed estimate. Much as in Section 4.1.3, they say that the error in $\hat{\mathbf{x}}$ is orthogonal to the contribution of each observation to $\hat{\mathbf{x}}$, and hence orthogonal to the contribution of any set of observations. Denoting $\mathbf{x} - \hat{\mathbf{x}}$ by $\tilde{\mathbf{x}}$ as usual, we conclude that

$$E[\tilde{\mathbf{x}}_{t+1|N}(\hat{\mathbf{x}}_{t+1|N} - \hat{\mathbf{x}}_{t+1|t})^{\mathrm{T}}] = 0 \tag{8.1.42}$$

as $\hat{\mathbf{x}}_{t+1|N}$ depends on $\mathbf{y}_1$ to $\mathbf{y}_N$ and $\hat{\mathbf{x}}_{t+1|t}$ on $\mathbf{y}_1$ to $\mathbf{y}_t$. Similarly,

$$E[\tilde{\mathbf{x}}_{t|N}(\hat{\mathbf{x}}_{t|N} - \hat{\mathbf{x}}_{t|t})^{\mathrm{T}}] = 0 \tag{8.1.43}$$

From (8.1.36),

$$\tilde{\mathbf{x}}_{t+1|N} - \tilde{\mathbf{x}}_{t+1|t} = \hat{\mathbf{x}}_{t+1|t} - \hat{\mathbf{x}}_{t+1|N} = P_{t+1|t}\boldsymbol{\lambda}_t \tag{8.1.44}$$

and (8.1.39) with t for $t+1$ gives

$$\tilde{\mathbf{x}}_{t|N} - \tilde{\mathbf{x}}_{t|t} = \hat{\mathbf{x}}_{t|t} - \hat{\mathbf{x}}_{t|N} = P_{t|t}\Phi_t^{\mathrm{T}}\boldsymbol{\lambda}_t \tag{8.1.45}$$

so

$$\tilde{\mathbf{x}}_{t|N} - \tilde{\mathbf{x}}_{t|t} = P_{t|t}\Phi_t^{\mathrm{T}}P_{t+1|t}^{-1}(\tilde{\mathbf{x}}_{t+1|N} - \tilde{\mathbf{x}}_{t+1|t}) \tag{8.1.46}$$

Now (8.1.42) and (8.1.43) imply respectively that

$$E[\tilde{\mathbf{x}}_{t+1|N}(\tilde{\mathbf{x}}_{t+1|t} - \tilde{\mathbf{x}}_{t+1|N})^{\mathrm{T}}] = E[\tilde{\mathbf{x}}_{t+1|N}\tilde{\mathbf{x}}_{t+1|t}^{\mathrm{T}}] - P_{t+1|N} = 0 \tag{8.1.47}$$

$$E[\tilde{\mathbf{x}}_{t|N}(\tilde{\mathbf{x}}_{t|t} - \tilde{\mathbf{x}}_{t|N})^{\mathrm{T}}] = E[\tilde{\mathbf{x}}_{t|N}\tilde{\mathbf{x}}_{t|t}^{\mathrm{T}}] - P_{t|N} = 0 \tag{8.1.48}$$

from which

$$\begin{aligned} &E[(\tilde{\mathbf{x}}_{t+1|N} - \tilde{\mathbf{x}}_{t+1|t})(\tilde{\mathbf{x}}_{t+1|N} - \tilde{\mathbf{x}}_{t+1|t})^{\mathrm{T}}] \\ &\quad = -E[\tilde{\mathbf{x}}_{t+1|t}(\tilde{\mathbf{x}}_{t+1|N} - \tilde{\mathbf{x}}_{t+1|t})^{\mathrm{T}}] \\ &\quad = -E[\tilde{\mathbf{x}}_{t+1|t}\tilde{\mathbf{x}}_{t+1|N}^{\mathrm{T}}] + P_{t+1|t} = -P_{t+1|N} + P_{t+1|t} \end{aligned} \tag{8.1.49}$$

and, similarly,

$$E[(\tilde{\mathbf{x}}_{t|N} - \tilde{\mathbf{x}}_{t|t})(\tilde{\mathbf{x}}_{t|N} - \tilde{\mathbf{x}}_{t|t}^{\mathrm{T}})] = -P_{t|N} + P_{t|t} \tag{8.1.50}$$

It follows that, multiplying each side of (8.1.46) by its transpose and taking expectations,

$$P_{t|t} - P_{t|N} = P_{t|t}\Phi_t^{\mathrm{T}}P_{t+1|t}^{-1}(P_{t+1|t} - P_{t+1|N})P_{t+1|t}^{-1}\Phi_t P_{t|t} \tag{8.1.51}$$

Summary *The backwards recursion for the covariance of the smoothed estimate* is

$$P_{t|N} = P_{t|t} + P_{t|t}\Phi_t^{\mathrm{T}}P_{t+1|t}^{-1}(P_{t+1|N} - P_{t+1|t})P_{t+1|t}^{-1}\Phi_t P_{t|t} \tag{8.1.52}$$

We have $P_{N|N}$ from the forward recursion.

Fixed-interval optimal smoothing formulae for state estimation have been derived in a wide variety of forms (references in Norton, 1975). The versions in

(8.1.36), (8.1.38), (8.1.39) and (8.1.40) entail little extra computation if the covariance is not required. For instance, the transition matrix from $\boldsymbol{\lambda}_{t+1}$ to $\boldsymbol{\lambda}_t$ is, as we saw earlier, the transpose of that from $\hat{\mathbf{x}}_{t|t-1}$ to $\hat{\mathbf{x}}_{t+1|t}$. Scalar observations and an s.r.w. or i.r.w. parameter-variation model simplify the algorithms further (Norton, 1975, 1976). The weighty covariance calculation can be organised so as to avoid matrix inversion. Economical computing arrangements for identification are discussed further by Norton (1975).

8.2 IDENTIFIABILITY

8.2.1 Factors Affecting Identifiability

Identifiability is a joint property of an identification experiment and a model. It establishes that the model parameters can be estimated adequately from the experiment. The model and experiment need not be complicated for a test of their identifiability to be non-trivial.

Example 8.2.1 Compartmental models of the type shown in Fig. 8.2.1 (Carson *et al.*, 1981; Godfrey, 1983) are often employed in biomedical studies of how various substances are metabolised. The two-compartment model in Fig. 8.2.1 represents rate equations

$$\dot{\mathbf{x}} = \begin{bmatrix} -k_{01} - k_{21} & k_{12} \\ k_{21} & -k_{02} - k_{12} \end{bmatrix} \mathbf{x} + \begin{bmatrix} u_1 \\ 0 \end{bmatrix}$$

for flow into and out of the compartments. The rate constants k_{01}, k_{02}, k_{12} and k_{21} are to be found. To make physical sense they must be non-negative. We assume that only compartment 1 can be perturbed and compartment 2 observed, both directly. The observations and estimation method allow the transfer function $X_2(s)/U_1(s)$ to be found with negligible error. Can the rate constants be determined uniquely?

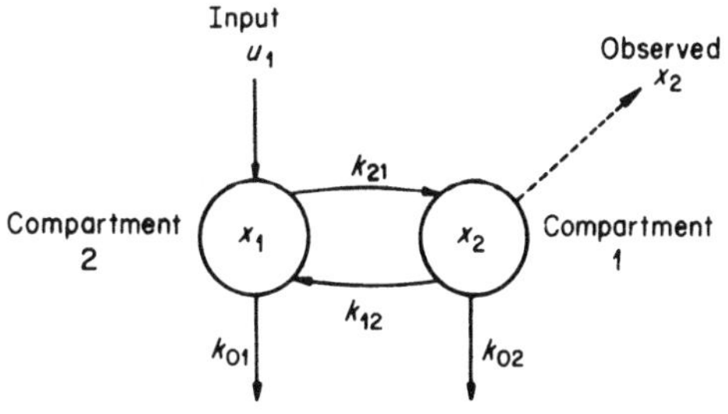

Fig. 8.2.1 Two-compartment model.

Taking Laplace transforms of the rate equations and eliminating $X_1(s)$, we find

$$X_2(s)/U_1(s) = \alpha/(s^2 + \beta s + \gamma)$$

where $\alpha = k_{21}$, $\beta = k_{01} + k_{02} + k_{12} + k_{21}$ and $\gamma = k_{01}k_{02} + k_{01}k_{12} + k_{21}k_{02}$.

The experiment finds α, β and γ. With k_{21} given by α, two equations remain in three unknowns, so we need some additional information. Let us check whether prior knowledge of one of k_{01}, k_{02} or k_{12} allows us to find the others uniquely. Consider each in turn having a known value ρ (perhaps zero).

(i) With $k_{01} = \rho$, we can find unique values $k_{02} = (\gamma - \rho(\beta - \alpha - \rho))/\alpha$ and $k_{12} = (\beta - \alpha - \rho)(1 + \rho/\alpha) - \gamma/\alpha$.

(ii) With $k_{02} = \rho$, we have $k_{01} + k_{12} = \beta - \alpha - \rho$ and $\rho k_{01} + k_{01}k_{12} = \gamma - \alpha\rho$, from which $k_{01}^2 + (\alpha - \beta)k_{01} + \gamma - \alpha\rho = 0$ and $k_{12} = \beta - \alpha - \rho - k_{01}$. In general, k_{01} and k_{12} are non-unique because of the quadratic. However, no negative value for k_{01} or k_{12} would make physical sense so, depending on the actual numbers, we may be able to pick out a unique solution for both. For example, $\alpha = 1$, $\beta = 4$ and $\gamma = 2$ give

$$k_{01} = 1.5 \pm 0.5\sqrt{1 + 4\rho} \quad \text{and} \quad k_{12} = 1.5 - \rho \mp 0.5\sqrt{1 + 4\rho}$$

To make k_{01} non-negative the ambiguous sign in k_{01} must be positive if $\rho > 2$. That would make k_{12} negative, though, so the model with $\rho > 2$ is incompatible with the observations. If $2 - \sqrt{2} < \rho \leq 2$, non-negativity of k_{12} requires the positive square root for k_{12} and hence the negative square root for k_{01}, and we obtain a unique solution. For $0 \leq \rho \leq 2 - \sqrt{2}$ both solutions for k_{01} and k_{12} are non-negative, so the solution is not unique.

(iii) With $k_{12} = \rho$ and α, β and γ as before, we have $k_{01}^2 - 2k_{01} + \rho - 1 = 0$ and $k_{02} = 3 - \rho - k_{01}$. We must take the larger solution for k_{01} to make it non-negative if $\rho < 1$, but if $\rho > 1$ the smaller solution for k_{01} is required to keep k_{02} non-negative. In both cases the other rate constant is non-negative as a result. The sign ambiguity is thereby resolved for any practicable ρ, but *only by chance*; a small change in α, β or γ would leave it unresolved, and the model not uniquely identified, for some values of ρ. △

Example 8.2.1 deals with a linear, low-order, time-invariant model whose transfer function is estimated accurately. Checking its identifiability nevertheless takes a little thought. Ideally, we should design an identification experiment with the help of analysis, running through a number of combinations of usable model and feasible experiment until we find a model we can in theory identify uniquely. Unfortunately, this may be difficult, as identifiability depends on many things: scope and quality of the observations, nature and location of the inputs, parameterisation and existing knowledge of the model, and properties of the estimation algorithm. These factors interact strongly, but we can single out and analyse some fairly restricted aspects of identifiability. Passing over the properties of estimators, which we have already enquired into, let us examine model parameterisation and input properties.

8.2.2 Deterministic Identifiability

We can start identifiability analysis by asking, as in Example 8.2.1, whether the experiment and model structure yield unique parameter values in principle, without regard to numerical accuracy or stochastic uncertainty. The topic is often called the *structural identifiability problem* (Bellman and Åström, 1970), with "structural" understood to mean "for almost all parameter values". The unique identifiability for all ρ found in part (iii) of Example 8.2.1 is not structural, as it applies only over an infinitesimal proportion of α, β and γ values. Nor is the possible uniqueness in part (ii), which applies for many parameter values but not almost all. Identification of a usable model does not always require structural identifiability, we conclude. The term "structural identifiability" is rather misleading, since identifiability may depend on prior information or on what combination of input waveforms is applied (Godfrey, 1983, Chapter 6; Problem 8.1), as well as on the model structure. We prefer to speak of *deterministic identifiability*.

The deterministic identifiability problem is quite distinct from the problem, important for multivariable models (Section 8.7), of finding an economical standard "canonical" model to represent the observed behaviour uniquely. Our choice of model is conditioned not only by a desire for uniqueness and simplicity, but also by the intended use of the model, the physical significance of its parameters and our background knowledge about it. Example 8.2.1 is a case in point, where a second-order transfer function with three parameters completely describes the relation between $u_1(t)$ and $x_2(t)$, but is less physically informative than the four-parameter compartmental model, and cannot easily take into account the non-negativity of the rate constants.

For s.i.s.o. transfer-function, differential- or difference-equation input–output models, deterministic identifiability merely requires that there is no redundancy, i.e. the model order is not too high, and that the input stimulates all the behaviour to be modelled. Section 8.2.3 considers adequacy of the input. State-space models pose a much stiffer problem, because of the great variety of possible parameterisations for any given input–output behaviour. As Example 8.2.1 has shown, Laplace-transform analysis can, with care, test deterministic identifiability of low-order state-space models, but it can be very cumbersome for model order as low as three (Norton, 1982a). The reason is that the equations relating the parameters in the preferred model to the directly identifiable transfer-function or impulse-response coefficients are of degree up to n, the model order. In continuous time the state model

$$\dot{\mathbf{x}}(t) = A\mathbf{x}(t) + B\mathbf{u}(t), \qquad \mathbf{y}(t) = C\mathbf{x}(t) + D\mathbf{u}(t) \tag{8.2.1}$$

(with $\mathbf{x}(0-)$ and $\mathbf{u}(t)$ given) is Laplace-transformed to

$$s\mathbf{X}(s) - \mathbf{x}(0-) = A\mathbf{X}(s) + B\mathbf{U}(s), \qquad \mathbf{Y}(s) = C\mathbf{X}(s) + D\mathbf{U}(s) \tag{8.2.2}$$

and the output is

$$\mathbf{Y}(s) = C(sI - A)^{-1}\mathbf{x}(0-) + \{C(sI - A)^{-1}B + D\}\mathbf{U}(s) \tag{8.2.3}$$

The elements of the transfer-function matrix $C(sI-A)^{-1}B+D$ have a common denominator $|sI-A|$ of degree n in s and the elements of A. To test if the elements of A can be found uniquely from $C(sI-A)^{-1}B+D$ requires a method of testing whether a set of simultaneous equations of degree up to n has a unique solution, whatever the numerical values. No such general method exists. We are reduced to ad hoc searching for a unique algebraic solution, as in Example 8.2.1. Similar comments apply if we look at the impulse-response matrix $Ce^{At}B$.

The difficulty can be eased, but not removed, by bridging the gap between impulse-response matrix and state-space model with a normal-mode expansion (Reid, 1983, Chapter 10; Blackman, 1977, Chapter 2). The idea is to express A and $Ce^{At}B$ in terms of the eigenvalues λ_1 to λ_n of A, the rows $\mathbf{r}_1^T$ to $\mathbf{r}_n^T$ of the modal matrix M and the columns $\mathbf{q}_1$ to $\mathbf{q}_n$ of M^{-1}. The columns of M are the eigenvectors of A, so from the defining equation of eigenvalues and eigenvectors,

$$M\Lambda \equiv M\,\mathrm{diag}(\lambda_1, \lambda_2, \ldots, \lambda_n) = AM \tag{8.2.4}$$

and element (i,j) of A is

$$a_{ij} = \text{element } (i,j) \text{ of } M\Lambda M^{-1} = \mathbf{r}_i^T\Lambda\mathbf{q}_j \tag{8.2.5}$$

For any positive integer k,

$$M\Lambda^k = AM\Lambda^{k-1} = A(AM\Lambda^{k-2}) = \cdots = A^kM \tag{8.2.6}$$

so $Me^{\Lambda t}$ is $e^{At}M$, and the impulse-response matrix $Ce^{At}B$ is $CMe^{\Lambda t}M^{-1}B$. For brevity we shall assume that individual elements of $Me^{\Lambda t}M^{-1}$ can be extracted from the impulse-response matrix. This is so in Example 8.2.1 and whenever B and C are known because of our choice of state. The observations then give one or more response components

$$h_{ij}(t) = \mathbf{r}_i^T e^{\Lambda t}\mathbf{q}_j \tag{8.2.7}$$

By definition we also have

$$\mathbf{r}_i^{\mathrm{T}}\mathbf{q}_j = \begin{cases} 1 & \text{for } i = j \\ 0 & \text{for } i \neq j \end{cases} \tag{8.2.8}$$

The eigenvalues are readily found by fitting exponentials to the observed $h_{ij}(t)$'s, so Λ in (8.2.5) and $e^{\Lambda t}$ in (8.2.7) are known. All our information about the model is now in the form of bilinear equations in the unknown rows of M and columns of M^{-1}. Prior knowledge consisting of linear equations in elements of A can be expressed in that form also, by use of (8.2.5).

If we can solve uniquely for M and M^{-1} and know Λ, we can find A uniquely through (8.2.5). We first choose the value of any one non-zero element of each eigenvector at will, since the scaling of each eigenvector is left free by its defining equation. It is usually best to make one row $\mathbf{r}$ or column $\mathbf{q}$ entirely ones, or ones and zeros if zeros are present leaving some free choices elsewhere. The positions of zeros in M and M^{-1} can be found easily from the pattern of zeros in A imposed by the model structure (Norton, 1980a). With one $\mathbf{r}$ or $\mathbf{q}$ known, some of the bilinear equations become linear, and the identifiability test has been turned into a test whether a mixed set of linear and bilinear equations has a unique solution (Norton, 1980b; Norton *et al.*, 1980). This reformulation of the original problem of testing whether equations of degree up to n have a unique solution makes a quick solution more likely. It does not, however, produce a neat general criterion for unique identifiability, and it increases the number of equations.

Example 8.2.2 The problem of Example 8.2.1 will be tackled by normal-mode analysis.

The impulse response from u_1 to x_2, with B and C known, gives

$$h_{21}(t) = \mathbf{r}_2^{\mathrm{T}} e^{\Lambda t}\mathbf{q}_1 \equiv r_{21}q_{11}e^{\lambda_1 t} + r_{22}q_{21}e^{\lambda_2 t}$$

and we choose $[1 \quad 1]$ as $\mathbf{r}_2^{\mathrm{T}}$. The amplitudes of the exponentials in $h_{21}(t)$ then give $\mathbf{q}_1$, leaving $\mathbf{r}_1$ and $\mathbf{q}_2$ unknown. From (8.2.8),

$$\mathbf{r}_1^{\mathrm{T}}\mathbf{q}_1 = 1, \qquad \mathbf{r}_1^{\mathrm{T}}\mathbf{q}_2 = 0, \qquad \mathbf{r}_2^{\mathrm{T}}\mathbf{q}_1 = 0 \quad \text{(not needed)}, \qquad \mathbf{r}_2^{\mathrm{T}}\mathbf{q}_2 = 1$$

so we have three equations in four unknowns, and the model is unidentifiable without further information.

(i) If we know $k_{01} = \rho$, then since k_{01} is $-a_{11} - a_{21}$,

$$-k_{01} = a_{11} + a_{21} = (\mathbf{r}_1 + \mathbf{r}_2)^{\mathrm{T}}\Lambda\mathbf{q}_1 = -\rho$$

This is linear in $\mathbf{r}_1$ and together with $\mathbf{r}_1^{\mathrm{T}}\mathbf{q}_2 = 0$ gives $\mathbf{r}_1$ uniquely in general. The remaining equations give $\mathbf{q}_2$ uniquely, so M and M^{-1}, and hence A and the rate constants, are found uniquely.

(ii) If we know $k_{02} = \rho$, then much as in (i), $(\mathbf{r}_1 + \mathbf{r}_2)^{\mathrm{T}}\Lambda\mathbf{q}_2 = -\rho$, and we have altogether two linear and two bilinear equations in $\mathbf{r}_1$ and $\mathbf{q}_2$. In this small example, it is easy to eliminate all but one unknown, ending up with a quadratic and two roots in general. In a larger example, unique identifiability is rapidly established when a succession of linear-equation solutions gives all the unknowns, as in (i), but not when some equations remain bilinear, as here.

(iii) If we know $k_{12} = \rho$, then $\mathbf{r}_1^{\mathrm{T}}\Lambda\mathbf{q}_2 = \rho$ and the analysis is very like (ii).

△

Surprisingly, exhaustive deterministic identifiability analysis for even a modest class of models, such as third-order linear compartmental models (Norton, 1982a), exposes quite a variety of experiment-model combinations which give determinate but non-unique solutions, as in Example 8.2.1(ii) and (iii). Sometimes the non-uniqueness has a fairly obvious cause, but sometimes not. Attempts at general deterministic identifiability analysis have been offered by many authors (Cobelli *et al.*, 1979; Delforge, 1980, 1981; Walter, 1982) with varying but incomplete success (Norton, 1982b).

8.2.3 Signal Requirements for Identifiability: Persistency of Excitation

Deterministic identifiability assures us that we are not prevented from finding the parameters by practical restrictions on what variables can act as inputs and observed outputs. We must next make sure that we are not prevented by failure of the input to excite all the dynamics. For example, a single-sinusoid input in Example 8.2.1 would allow us only to find one gain and phase change, too little to determine the three transfer-function coefficients α, β and γ. Two sinusoids would be enough.

Conditions on signals in an identification experiment to ensure adequate excitation of the dynamics are called persistency of excitation conditions. They effectively specify how many independent components must be present in the input signal, and not surprisingly the number depends on the order of the model. The conditions can be interpreted in the frequency domain, as we have just done for Example 8.2.1, or the time domain. They apply to both deterministic and stochastic signals.

We ask first what conditions must be imposed in o.l.s. The o.l.s. estimate $\hat{\boldsymbol{\theta}}$ satisfies the normal equations

$$U^{\mathrm{T}} U\hat{\boldsymbol{\theta}} = U^{\mathrm{T}}\mathbf{y} \tag{8.2.9}$$

We cannot solve for $\hat{\boldsymbol{\theta}}$ if $U^{\mathrm{T}} U$ is singular, i.e. if the regressors forming the columns of U are linearly dependent. Commonly, several regressors are lagged versions of the same signal, the input. The possibility arises that although they would not be linearly dependent for an arbitrary waveform, they are for a particular choice of waveform. The risk is significant since we prefer a simple waveform, all other things being equal.

Example 8.2.3 We contemplate o.l.s. identification of the u.p.r. $\{h\}$ in the model

$$y_t = h_1 u_{t-k-1} + h_2 u_{t-k-2} + \cdots + h_p u_{t-k-p} + e_t$$

with $\{u\}$ periodic with period P. If $p > P$, the last $p - P$ regressors repeat the first $p - P$, making $U^T U$ singular. Furthermore, if the d.c. component of $\{u\}$ (the mean over a period) is zero, P must exceed p since any P successive regressors u_{t-k-i} to $u_{t-k-i-P+1}$ would always sum to zero. Even when $P > p$ there may be trouble, e.g. if successive half-cycles of $\{u\}$ are symmetrical about zero, so that the sum of any two regressors half a cycle apart is always zero.

Further possible dependences can easily be found (Problem 8.2) but enumerating them is tedious, and it is simpler just to say that $\{u\}$ must make $U^T U$ non-singular. △

For $U^T U$ to be singular, a real, non-zero $\boldsymbol{\alpha}$ must exist such that $U^T U\boldsymbol{\alpha}$ is zero, which implies that

$$\boldsymbol{\alpha}^T U^T U \boldsymbol{\alpha} = \sum_{t=1}^{N} (\text{element } t \text{ of } U\boldsymbol{\alpha})^2 = 0 \tag{8.2.10}$$

Clearly $U\boldsymbol{\alpha}$ must be zero, i.e. the columns of U must be linearly dependent. We can write U as

$$U \equiv U' + \begin{bmatrix} 1 \\ 1 \\ \vdots \\ 1 \end{bmatrix} [\bar{u}_1 \bar{u}_2 \cdots \bar{u}_p] \tag{8.2.11}$$

where $\bar{u}_i$ is the mean of regressor i and column i of U' consists of zero-mean samples $u_{ti} - \bar{u}_i$. The condition for $U\boldsymbol{\alpha}$ to be zero is then

$$U'\boldsymbol{\alpha} = -\begin{bmatrix} 1 \\ 1 \\ \vdots \\ 1 \end{bmatrix} \bar{\mathbf{u}}^T \boldsymbol{\alpha} \tag{8.2.12}$$

where $\bar{\mathbf{u}}$ is the vector of regressor means. When each row of U' is p successive samples from a single stationary sequence $\{u - \bar{u}\}$, the elements of $U'\boldsymbol{\alpha}$ are the result of filtering $\{u - \bar{u}\}$ with a moving-average transfer function $\alpha_1 + \alpha_2 z^{-1} + \cdots + \alpha_p z^{-p+1}$. A moving average of a zero-mean sequence cannot be constant and non-zero, so if $\bar{\mathbf{u}}^T\boldsymbol{\alpha}$ is not zero, (8.2.12) cannot be satisfied if $\{u-\bar{u}\}$ is of period N or less; nor, clearly, will it be satisfied by any stochastic $\{u-\bar{u}\}$. With $\bar{\mathbf{u}}^T\boldsymbol{\alpha}$ zero, it can be satisfied only if $U'\boldsymbol{\alpha}$ is zero. We can be sure of avoiding trouble so long as $U'^T U'$ is positive-definite, since then $(U'\boldsymbol{\alpha})^T U'\boldsymbol{\alpha}$, and hence $U'\boldsymbol{\alpha}$, cannot be zero.

We now have the motivation for a definition of a usable signal $\{u\}$.

Definition A signal $\{u\}$ is *persistently exciting (p.e.) of order n* if the limits

$$\bar{u} = \lim_{N\to\infty} \frac{1}{N}\sum_{t=1}^{N} u_t, \qquad r_{uu}(i) = \lim_{N\to\infty} \frac{1}{N}\sum_{t=1}^{N} (u_t-\bar{u})(u_{t+i}-\bar{u})$$

of its sample mean and a.c.f. exist w.p. 1, and if the matrix

$$R_{uu} = [r_{uu}(i-j)]_{ij}, \qquad i=1,2,\ldots,n;\, j=1,2,\ldots,n$$

is positive-definite.

If $\{u\}$ is ergodic, expectations can replace the sample averages, and the signal is p.e. of order n if the mean and a.c.f. exist and the $n\times n$ covariance matrix is positive-definite.

The frequency-domain conditions for an ergodic $\{u\}$ to be p.e. of order n can be found from

$$\boldsymbol{\alpha}^T R_{uu}\boldsymbol{\alpha} = \boldsymbol{\alpha}^T E[(\mathbf{u}_t-\bar{\mathbf{u}})(\mathbf{u}_t-\bar{\mathbf{u}})^T]\boldsymbol{\alpha} = E[(\boldsymbol{\alpha}^T(\mathbf{u}_t-\bar{\mathbf{u}}))^2] \tag{8.2.13}$$

where $\mathbf{u}_t$ is $[u_t \quad u_{t-1} \quad \cdots \quad u_{t-n+1}]^T$ and $\bar{\mathbf{u}}$ is now $[\bar{u} \quad \bar{u} \quad \cdots \quad \bar{u}]^T$. According to (8.2.13), R_{uu} is singular only if a real, non-zero $\boldsymbol{\alpha}$ exists making $\boldsymbol{\alpha}^T(\mathbf{u}_t-\bar{\mathbf{u}})$ zero always. We can regard $\boldsymbol{\alpha}^T(\mathbf{u}_t-\bar{\mathbf{u}})$ as the output of a filter with transfer function

$$H(z^{-1}) = \alpha_1 + \alpha_2 z^{-1} + \cdots + \alpha_n z^{-n+1} \tag{8.2.14}$$

driven by $\{u-\bar{u}\}$. Its spectral density is therefore $|H(e^{-j\omega T})|^2\Phi_u(j\omega)$, where T is the sampling interval and $\Phi_u(j\omega)$ the spectral density of $\{u-\bar{u}\}$. Parseval's theorem then gives

$$\boldsymbol{\alpha}^T R_{uu}\boldsymbol{\alpha} = \frac{1}{2\pi}\int_{-\pi T}^{\pi T} |H(e^{-j\omega T})|^2\Phi_u(j\omega)\, d\omega \tag{8.2.15}$$

The spectral density $\Phi_u(j\omega)$ is non-negative at all frequencies, so the only way $\boldsymbol{\alpha}^T R_{uu}\boldsymbol{\alpha}$ can be zero is if $H(e^{-j\omega T})$ is zero at every frequency at which $\Phi_u(j\omega)$

is non-zero. Coefficients α_1 to α_n can make $H(e^{-j\omega T})$ zero at no more than $n-1$ frequencies, as they also have to fix its overall amplitude. Consequently no $\boldsymbol{\alpha}$ can make $\boldsymbol{\alpha}^T R_{uu} \boldsymbol{\alpha}$ zero provided $\Phi_u(j\omega)$ is non-zero at n or more frequencies. In other words, a scalar ergodic signal is p.e. of order n if it contains energy at n or more distinct frequencies. The proof extends to vector signals (Söderström and Stoica, 1983). The converse is also true if $\{u\}$ is scalar, since if $\Phi_u(j\omega)$ is non-zero at $n-1$ or fewer frequencies, $\boldsymbol{\alpha}$ can be chosen to make $|H(e^{-j\omega T})|^2 \Phi_u(j\omega)$ zero throughout. However, it is not generally necessary for a vector signal to contain energy at n frequencies to be p.e. of order n.

Example 8.2.4 Let us examine a possible input $\{u\}$ obtained by moving-average filtering of white noise $\{w\}$:

$$u_t = w_t + f_1 w_{t-1} + \cdots + f_p w_{t-p}$$

Any linear combination of n successive input samples, say

$$g_1 u_t + g_2 u_{t-1} + \cdots + g_n u_{t-n+1}$$

has a z-transform $G(z^{-1})(1 + F(z^{-1}))W(z^{-1})$, with obvious definitions of F and G. For an exact linear dependence to hold between every n successive input samples, $G(z^{-1})(1 + F(z^{-1}))$ must be zero. Clearly this is impossible for non-zero and causal $G(z^{-1})$, so such an input signal is persistently exciting of any required order. In the frequency domain, the p.s.d. of $\{w\}$ is flat and the filter $1 + F(z^{-1})$ can only have p spectral nulls, leaving an infinity of non-zero spectral components in $\{u\}$. △

*8.2.4 Persistency of Excitation Conditions and Convergence

Persistency of excitation conditions on regressors often play a part in proving convergence of l.s.-based estimation algorithms. Detailed convergence proofs are beyond the scope of this book (Goodwin and Sin, 1984; Ljung and Söderström, 1983), but we can afford a detailed look at one algorithm, recursive o.l.s., to get a feel for the role of persistency of excitation. In our present notation and with P interpreted as $\text{cov}\,\hat{\boldsymbol{\theta}}/\sigma^2$, the algorithm is

$$P_t^{-1} = P_{t-1}^{-1} + \mathbf{u}_t \mathbf{u}_t^T \quad \text{or} \quad P_t = P_{t-1} - \frac{P_{t-1}\mathbf{u}_t \mathbf{u}_t^T P_{t-1}}{1 + \mathbf{u}_t^T P_{t-1} \mathbf{u}_t} \tag{8.2.16}$$

$$\hat{\boldsymbol{\theta}}_t = \hat{\boldsymbol{\theta}}_{t-1} + P_t \mathbf{u}_t (y_t - \mathbf{u}_t^T \hat{\boldsymbol{\theta}}_{t-1}) \tag{8.2.17}$$

with $\hat{\boldsymbol{\theta}}_0$ and P_0 given. We might first investigate its convergence in the absence of noise and errors in model structure, i.e. with

$$y_t = \mathbf{u}_t^T \boldsymbol{\theta}, \qquad t = 1, 2, \ldots \tag{8.2.18}$$

A recursion for the parameter error $\tilde{\boldsymbol{\theta}}$ is found by subtracting each side of (8.2.17) from $\boldsymbol{\theta}$ after substituting for y_t from (8.2.18):

$$\tilde{\boldsymbol{\theta}}_t = \tilde{\boldsymbol{\theta}}_{t-1} - P_t \mathbf{u}_t \mathbf{u}_t^{\mathrm{T}} \tilde{\boldsymbol{\theta}}_{t-1} = P_t P_{t-1}^{-1} \tilde{\boldsymbol{\theta}}_{t-1} \tag{8.2.19}$$

so

$$P_t^{-1} \tilde{\boldsymbol{\theta}}_t = P_{t-1}^{-1} \tilde{\boldsymbol{\theta}}_{t-1} \tag{8.2.20}$$

With the equation for P_t^{-1} in (8.2.16), (8.2.20) is in a sense a complete statement of how the error evolves; it has very little intuitive appeal, though. A slightly better idea of how $\{\tilde{\boldsymbol{\theta}}\}$ behaves is obtainable from the scalar $\tilde{\boldsymbol{\theta}}_t^{\mathrm{T}} P_t^{-1} \tilde{\boldsymbol{\theta}}_t$. From (8.2.19) and (8.2.20)

$$\tilde{\boldsymbol{\theta}}_t^{\mathrm{T}} P_t^{-1} \tilde{\boldsymbol{\theta}}_t = \tilde{\boldsymbol{\theta}}_{t-1}^{\mathrm{T}} (I - \mathbf{u}_t \mathbf{u}_t^{\mathrm{T}} P_t) P_{t-1}^{-1} \tilde{\boldsymbol{\theta}}_{t-1} \tag{8.2.21}$$

and from (8.2.16)

$$\mathbf{u}_t^{\mathrm{T}} P_t = \left(1 - \frac{\mathbf{u}_t^{\mathrm{T}} P_{t-1} \mathbf{u}_t}{1 + \mathbf{u}_t^{\mathrm{T}} P_{t-1} \mathbf{u}_t}\right) \mathbf{u}_t^{\mathrm{T}} P_{t-1} = \frac{\mathbf{u}_t^{\mathrm{T}} P_{t-1}}{1 + \mathbf{u}_t^{\mathrm{T}} P_{t-1} \mathbf{u}_t} \tag{8.2.22}$$

so

$$\tilde{\boldsymbol{\theta}}_t^{\mathrm{T}} P_t^{-1} \tilde{\boldsymbol{\theta}}_t = \tilde{\boldsymbol{\theta}}_{t-1}^{\mathrm{T}} P_{t-1}^{-1} \tilde{\boldsymbol{\theta}}_{t-1} - \frac{\tilde{\boldsymbol{\theta}}_{t-1}^{\mathrm{T}} \mathbf{u}_t \mathbf{u}_t^{\mathrm{T}} \tilde{\boldsymbol{\theta}}_{t-1}}{1 + \mathbf{u}_t^{\mathrm{T}} P_{t-1} \mathbf{u}_t} \tag{8.2.23}$$

From (8.2.16) P_{t-1}^{-1} and hence P_{t-1} is non-negative definite, $\mathbf{u}_t^{\mathrm{T}} P_{t-1} \mathbf{u}_t$ is non-negative and the last term in (8.2.23) is positive unless the prediction error $\mathbf{u}_t^{\mathrm{T}} \tilde{\boldsymbol{\theta}}_{t-1}$ in y_t is zero. That is, $\tilde{\boldsymbol{\theta}}^{\mathrm{T}} P^{-1} \tilde{\boldsymbol{\theta}}$ decreases unless the model predicts the output exactly.

No p.e. conditions have yet been imposed on $\{u\}$. If they are, we can show that the more easily interpreted $\tilde{\boldsymbol{\theta}}^{\mathrm{T}} \tilde{\boldsymbol{\theta}}$ converges. A standard result is that the smallest (real) eigenvalue $\lambda_{\min}(A)$ of a real, symmetric matrix A satisfies

$$\mathbf{x}^{\mathrm{T}} A \mathbf{x} \geq \lambda_{\min}(A) \mathbf{x}^{\mathrm{T}} \mathbf{x} \tag{8.2.24}$$

for any real $\mathbf{x}$ (Mirsky, 1955, p. 388). Hence

$$\tilde{\boldsymbol{\theta}}_t^{\mathrm{T}} P_t^{-1} \tilde{\boldsymbol{\theta}}_t = \tilde{\boldsymbol{\theta}}_t \left(P_0^{-1} + \sum_{k=1}^{t} \mathbf{u}_k \mathbf{u}_k^{\mathrm{T}} \right) \tilde{\boldsymbol{\theta}}_t \geq \tilde{\boldsymbol{\theta}}_t \sum_{k=1}^{t} \mathbf{u}_k \mathbf{u}_k^{\mathrm{T}} \tilde{\boldsymbol{\theta}}_t$$

$$\geq \lambda_{\min} \left(\sum_{k=1}^{t} \mathbf{u}_k \mathbf{u}_k^{\mathrm{T}} \right) \tilde{\boldsymbol{\theta}}_t^{\mathrm{T}} \tilde{\boldsymbol{\theta}}_t \tag{8.2.25}$$

so

$$\tilde{\boldsymbol{\theta}}_t^{\mathrm{T}} \tilde{\boldsymbol{\theta}}_t \leq \tilde{\boldsymbol{\theta}}_t^{\mathrm{T}} P_t^{-1} \tilde{\boldsymbol{\theta}}_t \Big/ \lambda_{\min} \left(\sum_{k=1}^{t} \mathbf{u}_k \mathbf{u}_k^{\mathrm{T}} \right) \tag{8.2.26}$$

By (8.2.23), $\tilde{\boldsymbol{\theta}}_t^{\mathrm{T}} P_t^{-1} \tilde{\boldsymbol{\theta}}_t$ is a non-increasing function of t, so to ensure that $\tilde{\boldsymbol{\theta}}_t^{\mathrm{T}} \tilde{\boldsymbol{\theta}}_t$ converges to zero as t tends to infinity we need only assume that $\lambda_{\min}\left(\sum_{k=1}^{t} \mathbf{u}_k \mathbf{u}_k^{\mathrm{T}}\right)$ tends to infinity. Different though it looks, this is equivalent to a p.e. condition on $\{u\}$, for if R_{uu} is positive-definite,

$$R_{uu} = \lim_{t \to \infty} \frac{1}{t} \sum_{k=1}^{t} \mathbf{u}_k \mathbf{u}_k^{\mathrm{T}} \geq \varepsilon I > 0 \tag{8.2.27}$$

and

$$\lim_{t \to \infty} \lambda_{\min}\left(\sum_{k=1}^{t} \mathbf{u}_k \mathbf{u}_k^{\mathrm{T}}\right) = \lim_{t \to \infty} t \lambda_{\min}\left(\frac{1}{t} \sum_{k=1}^{t} \mathbf{u}_k \mathbf{u}_k^{\mathrm{T}}\right) \geq \lim_{t \to \infty} t\varepsilon = \infty \tag{8.2.28}$$

A strengthening of the p.e. condition will enable us to say something about the convergence rate. If

$$\lambda_{\min}\left(\sum_{k=1}^{M} \mathbf{u}_{t+k} \mathbf{u}_{t+k}^{\mathrm{T}}\right) \geq \varepsilon > 0, \qquad t = 0, M, 2M, \ldots \tag{8.2.29}$$

with ε independent of $\mathbf{u}_t$, that is, batches of M successive $\mathbf{u}$'s contribute positive-definite increments to P^{-1}, and if also we take P_0^{-1} as $\gamma_0 I$ with γ_0 positive,

$$\lambda_{\min}(P_{iM}^{-1}) = \lambda_{\min}\left(\gamma_0 I + \sum_{k=1}^{iM} \mathbf{u}_k \mathbf{u}_k^{\mathrm{T}}\right) \geq \gamma_0 + i\varepsilon \tag{8.2.30}$$

Here we have recognised that $\gamma_0 I$ has all its eigenvalues equal to γ_0, and that the smallest eigenvalue of the sum of two symmetric matrices is no less than the sum of their smallest eigenvalues. From (8.2.30), (8.2.24) and (8.2.23) applied from $t = 1$ to $t = iM$,

$$\begin{aligned}(\gamma_0 + i\varepsilon)\tilde{\boldsymbol{\theta}}_{iM}^{\mathrm{T}} \tilde{\boldsymbol{\theta}}_{iM} &\leq \lambda_{\min}(P_{iM}^{-1}) \tilde{\boldsymbol{\theta}}_{iM}^{\mathrm{T}} \tilde{\boldsymbol{\theta}}_{iM} \leq \tilde{\boldsymbol{\theta}}_{iM}^{\mathrm{T}} P_{iM}^{-1} \tilde{\boldsymbol{\theta}}_{iM} \\ &\leq \tilde{\boldsymbol{\theta}}_0^{\mathrm{T}} P_0^{-1} \tilde{\boldsymbol{\theta}}_0 = \gamma_0 \tilde{\boldsymbol{\theta}}_0^{\mathrm{T}} \tilde{\boldsymbol{\theta}}_0\end{aligned} \tag{8.2.31}$$

so

$$\tilde{\boldsymbol{\theta}}_{iM}^{\mathrm{T}} \tilde{\boldsymbol{\theta}}_{iM} \leq \gamma_0 \tilde{\boldsymbol{\theta}}_0^{\mathrm{T}} \tilde{\boldsymbol{\theta}}_0 / (1 + i\varepsilon) \tag{8.2.32}$$

We conclude that $\tilde{\boldsymbol{\theta}}^{\mathrm{T}} \tilde{\boldsymbol{\theta}}$ measured at intervals of M recursion steps converges to zero at a rate asymptotically inversely proportional to time.

8.3 IDENTIFICATION IN CLOSED LOOP

8.3.1 Effects of Feedback on Identifiability

Many systems have feedback which cannot be interrupted for an identification experiment. The feedback may be inherent, as in a demand–supply-price loop in economics, or externally applied but no less essential, as when an existing controller cannot safely be disconnected from an industrial process. Sometimes feedback causes no difficulty, for instance when the overall closed-loop behaviour is to be identified and the set-point can be perturbed, or when the feedback is known in advance and can be allowed for in estimating the forward-path dynamics from the closed-loop response. In other cases feedback may render the system unidentifiable.

Example 8.3.1 We want to identify by o.l.s. the forward-path dynamics of a system which has a sampled-data controller with reference input $\{r\}$. The control law, computed with negligible delay, is

$$u_t = g(r_t - y_t) - fu_{t-1}$$

We test identifiability in various cases by checking for linear dependence among the regressors.

(i) Proposed model $y_t = -ay_{t-1} + b_1u_{t-1} + b_2u_{t-2} + e_t$. The parameter vector is $[-a \quad b_1 \quad b_2]$ and the regressor matrix U is $[\mathbf{y}_{t-1} \quad \mathbf{u}_{t-1} \quad \mathbf{u}_{t-2}]$ in obvious notation. Because of the controller,

$$\mathbf{y}_{t-1} = \mathbf{r}_{t-1} - (\mathbf{u}_{t-1} + f\mathbf{u}_{t-2})/g$$

If the reference input is always zero, as it might be in a regulator, the columns of U are linearly dependent through the control law. For any α, $U\boldsymbol{\eta}$ is zero if $\boldsymbol{\eta} = \alpha[g \quad 1 \quad f]^T$, so $\alpha[g \quad 1 \quad f]^T$ could be added to the parameter estimates without affecting the model output. The model is consequently not identifiable.

If the reference input varies, the model is identifiable unless $\mathbf{r}_{t-1}$ is a linear combination of $\mathbf{u}_{t-1}$ and $\mathbf{u}_{t-2}$, and hence of $\mathbf{y}_{t-1}$ and $\mathbf{u}_{t-2}$. We need not worry about this eventuality, as the reference input will not depend on the current output.

(ii) Proposed model $y_t = -a_1y_{t-1} - a_2y_{t-2} + bu_{t-1} + e_t$. Now U is $[\mathbf{y}_{t-1} \quad \mathbf{y}_{t-2} \quad \mathbf{u}_{t-1}]$, which from the control law is

$$[\mathbf{r}_{t-1} - (\mathbf{u}_{t-1} + f\mathbf{u}_{t-2})/g \,\vdots\, \mathbf{r}_{t-2} - (\mathbf{u}_{t-2} + f\mathbf{u}_{t-3})/g \,\vdots\, \mathbf{u}_{t-1}]$$

Even with the reference input zero, the columns of U are linearly independent unless $\mathbf{u}_{t-1}$, $\mathbf{u}_{t-2}$ and $\mathbf{u}_{t-3}$ are linearly dependent. The closed-loop transfer function from $\{e\}$ to $\{u\}$ gives a relation of the form

$$-gE(z^{-1}) = (1 + \alpha_1 z^{-1} + \alpha_2 z^{-2} + \alpha_3 z^{-3})U(z^{-1})$$

Hence for $\mathbf{u}_{t-1}$, $\mathbf{u}_{t-2}$ and $\mathbf{u}_{t-3}$ to be linearly dependent, so that, say,

$$(1 + \beta_1 z^{-1} + \beta_2 z^{-2})U(z^{-1}) = 0$$

$\{e\}$ must be such that

$$(1 + \beta_1 z^{-1} + \beta_2 z^{-2})E(z^{-1}) = 0$$

Providing $\{e\}$ is p.e. of order 3, this is not so and the model is identifiable.

(iii) Model as in (i) but control law $u_t = g(r_t - y_t) - f_1 u_{t-1} - f_2 u_{t-2}$. With the reference input zero, U is

$$[-(\mathbf{u}_t + f_1 \mathbf{u}_{t-1} + f_2 \mathbf{u}_{t-2})/g \quad \mathbf{u}_{t-1} \quad \mathbf{u}_{t-2}]$$

so the model is identifiable subject to a p.e. condition on $\{e\}$ much as in (ii).

△

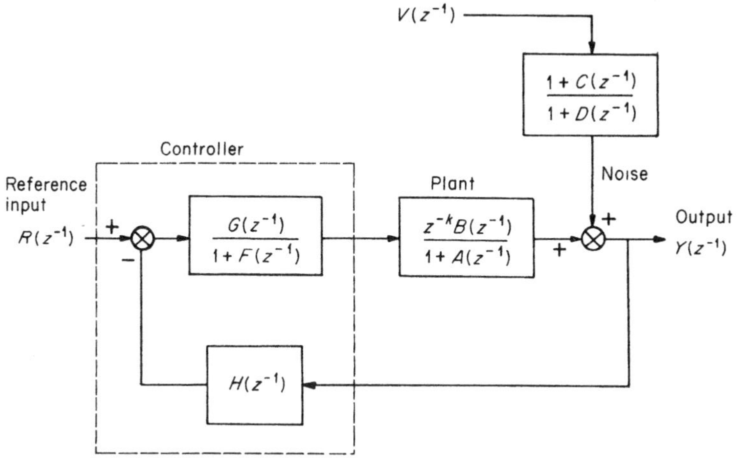

Fig. 8.3.1 Identification in closed-loop system.

The example illustrates how controller complexity and a non-zero reference input or p.e. output noise help closed-loop identifiability. A more general picture can be seen by reference to Fig. 8.3.1. The coefficients in the denominator $1 + A$ and numerator B of the plant transfer function are to be estimated. We have as usual

$$Y = -AY + z^{-k}BU + (1 + C)V \tag{8.3.1}$$

and the controller is

$$U = (G/(1+F))(R - HY) \tag{8.3.2}$$

so, eliminating U,

$$((1+A)(1+F) + z^{-k}BGH)Y = z^{-k}BGR + (1+C)(1+F)V \tag{8.3.3}$$

There are several possible approaches to identifying the system.

(i) With the reference input zero, identify the coefficients in the rational transfer function Y/V by a.r.m.a. modelling of $\{y\}$ (with no exogenous input). Pole-zero cancellation between $(1+A)(1+F) + z^{-k}BGH$ and $1+C$ would prevent calculation of A and B, but is unlikely. A more serious snag is that with R zero,

$$(1+F)U = -GHY \tag{8.3.4}$$

so we could replace model (8.3.1) by

$$Y = (-A + GHJ)Y + (z^{-k}B + (1+F)J)U + (1+C)V \tag{8.3.5}$$

with J any polynomial in z^{-1}, and still get precisely the same transfer function

$$\frac{Y}{V} = \frac{(1+C)(1+F)}{(1+A)(1+F) + z^{-k}BGH} \tag{8.3.6}$$

We infer that A and B cannot be identified from Y/V, but C can, and so can the coefficients in the characteristic equation, and hence the closed-loop poles. Switching between two or more controllers can make A and B identifiable by this method, as will be seen in Section 8.3.2.

(ii) We apply a perturbation to the reference input, identify the parameters of $(1+A)(1+F) + z^{-k}BGH$, $z^{-k}BG$ and $(1+C)(1+F)$ in (8.3.3), then solve for A, B and C knowing F, G, H and k. The first part presents no problems beyond the usual open-loop ones such as choosing a suitable perturbation and getting the model orders and dead time right. The second part involves solving two over-determined sets of linear equations, one for A and B and the other for C. The former may be ill-conditioned, especially when the controller is intended to make the closed-loop dynamics insensitive to some of the plant parameters.

(iii) With the reference input zero, we observe $\{u\}$ and $\{y\}$ and identify the parameters of (8.3.1) directly. The feedback may destroy the linear independence of the explanatory variables in (8.3.1), as in case (i) in Example 8.3.1, making the method unfeasible. The trouble, as with method (i), is indistinguishability of (8.3.5) from (8.3.1), but now $1+A$ and $z^{-k}B$ are identified separately, rather than combined in the denominator of Y/V. Thus the extra terms $-GHJ$ and $(1+F)J$ in (8.3.5) could go undetected only if they did not alter the orders of $1 + A$ and B. Depending on the orders of A, B, F, G and H, it may be that no such undetectable terms exist, as in (ii) and (iii) of Example 8.3.1, and the model is identifiable.

With this method, the identification technique is also important, for if it does not enforce causality on the model, the non-causal model

$$GHY = -(1 + F)U + \text{error} \tag{8.3.7}$$

will be obtained, impressively accurate but entirely useless if U/Y is already known. We cannot, for instance, employ the spectral or correlation methods of Chapter 3.

(iv) A method with the advantages of (ii) and (iii) is to reconstruct $\{u\}$ from $\{r\}$, $\{y\}$ and the controller equation (8.3.2). The plant parameters can then be estimated directly. Since $\{r\}$ is at our disposal we have some choice in $\{u\}$ rather than having it imposed on us by $\{v\}$ as in method (iii).

The accuracy attainable by method (ii), treating the over-determined equations as observations and finding the Markov estimates of A, B and C, is in principle the same as that achieved by a prediction-error algorithm in method (iii) (Söderström *et al.*, 1976), but round-off error may have different effects. Method (iv) avoids the potential numerical difficulties of method (ii) by estimating A, B and C directly, and $\{u\}$ does not have to be instrumented (except in the unlikely event that the controller is not known accurately enough).

*8.3.2 Conditions on Feedback and External Inputs to Ensure Identifiability

We now examine a widely applicable identifiability condition for multivariable linear feedback systems (Söderström *et al.*, 1976). The system has p_u control variables, p_y outputs and p_w external inputs apart from the output noise. The external inputs $\{\mathbf{w}\}$ may be perturbed reference inputs, signals added to the control variables, or both. The model and plant are of the form

$$\mathbf{Y}(z^{-1}) = \mathscr{B}(z^{-1})\mathbf{U}(z^{-1}) + \mathscr{C}(z^{-1})\mathbf{V}(z^{-1}) \tag{8.3.8}$$

where $\mathbf{Y}$, $\mathbf{U}$ and $\mathbf{V}$ are the z-transforms of the vector output, control and output-noise-generating variables, and $\mathscr{B}$ and $\mathscr{C}$ are matrices of rational transfer functions (with some weak assumptions ensuring good behaviour). The feedback mechanism is

$$\mathbf{U}(z^{-1}) = \mathscr{K}_i(z^{-1})\mathbf{W}(z^{-1}) + \mathscr{L}_i(z^{-1})\mathbf{Y}(z^{-1}), \qquad i = 1, 2, \ldots, s \tag{8.3.9}$$

with the controller switched between s different control laws to bring about identifiability when necessary. Each $\mathscr{K}_i$ and $\mathscr{L}_i$ is a rational-transfer-function matrix. Any correlation between $\{\mathbf{w}\}$ and $\{\mathbf{v}\}$ complicates matters slightly, without changing essentials. The identification is assumed to be indirect as in (i) and (ii) of Section 8.3.1, or direct as in (iii) and (iv) and by a prediction-error algorithm. The input $\{\mathbf{w}\}$ is p.e. of any finite order (e.g. linearly filtered white noise). The dead time in the plant-controller loop is not zero. Finally, the model is said to be *strongly system identifiable* (s.s.i.) when the parameter estimates of every parameterisation of $\mathscr{B}$ and $\mathscr{C}$ capable of the same input–output and noise-output behaviour as the actual system converges w.p. 1, as the record length tends to infinity, to the values giving that behaviour. (The reference to "every parameterisation" should become clearer in Section 8.7.)

Condition for model (8.3.8) subject to feedback (8.3.9) to be s.s.i. (Söderström *et al.*, 1976) With the stated assumptions, the model is s.s.i. if and only if the matrix

$$\mathscr{J} = \begin{bmatrix} \mathscr{K}_1 & \cdots & \mathscr{K}_s & \mathscr{L}_1 & \cdots & \mathscr{L}_s \\ 0 & \cdots & 0 & I & \cdots & I \end{bmatrix} \begin{matrix} p_u \text{ rows} \\ p_y \text{ rows} \end{matrix}$$

(with $p_w \cdots p_w, p_y \cdots p_y$ columns) is of rank $p_u + p_y$ for almost every z.

The condition tests the feedback structure but not the model structure; it tests identifiability for every model of the given dimensions. We are therefore strongly reassured when the condition is satisfied, but a model of specified order may be identifiable even when the condition is not satisfied because not *all* models with dimensions p_u and p_y are identifiable.

The need for more than one controller $\mathscr{K}$ and $\mathscr{L}$ is clear, since $\mathscr{J}$ has fewer than $p_u + p_y$ columns if p_w is less than p_u and $s = 1$, and cannot have rank $p_u + p_y$. Generally $s(p_w + p_y)$ must be at least $p_u + p_y$, and this turns out to be sufficient as well as necessary.

Example 8.3.2 (a) In Example 8.3.1, $p_u = p_y = s = 1$, so for $s(p_w + p_y) \geq p_u + p_y$, p_w must be 1 or more. That is, not every model is identifiable without an external input. Some models are, as we saw in cases (ii) and (iii). With one external input, the rank condition is satisfied and all models, even case (i), are identifiable: the model configuration (8.3.8) is s.s.i.

(b) If in general there are as many external inputs as control variables and only one control law, $p_w = p_u$ and $s = 1$, so $\mathscr{J}$ is

$$\begin{bmatrix} \mathscr{K} & \mathscr{L} \\ 0 & I \end{bmatrix}$$

with $\mathscr{K}$ square. As I contributes p_y to the rank, the rank condition requires $\mathscr{K}$ to be of full rank p_u for almost all z. $\mathscr{L}$ is immaterial. Full row rank for $\mathscr{K}$ means that the control variables must contain independent contributions from **w**, and full column rank requires the corollary that no linear combination of external inputs has zero effect on **u**.

(c) If there are no external inputs, $\mathscr{J}$ is

$$\begin{bmatrix} \mathscr{L}_1 & \cdots & \mathscr{L}_s \\ I & \cdots & I \end{bmatrix}$$

so $\mathscr{L}_1$ to $\mathscr{L}_s$ must contribute a total of p_u linearly independent rows (and columns). That is, the output feedback must provide, in time, p_u independent control signals, and some combination of outputs and feedback transfer functions, p_y in all, must be guaranteed to excite **u**. △

We next examine briefly the prime example of identification in a closed-loop system: self-tuning control.

8.3.3 Self-Tuning Control

Recursive identification allows a controller to tune itself, i.e. adjust its control-law coefficients on line by reference to a periodically updated plant model. Self-tuning, if rapid enough, makes initial manual setting-up of the controller unnecessary and enables a controller to cope with a strongly time-varying plant. Figure 8.3.2 shows a self-tuning controller. Identification and control synthesis are carried out by the same digital processor, and can sometimes be merged so that the model is only implicit, simplifying computation, as we shall see later.

Ideally, the control synthesis should optimise plant performance taking the uncertainty in the model into account as well as the performance criterion. The control signal resulting from such an overall optimisation has to compromise between conflicting requirements. It must aim at good plant behaviour, e.g. a well-regulated output, but with caution as dictated by estimation uncertainty in the model, and at the same time it must excite the plant enough for satisfactory identification. The combined optimal identification and control problem has been investigated over a long period (Feldbaum, 1960, 1965), but complete solutions have been found only for the simplest cases. An inviting if somewhat risky simplification is to treat the model as accurate during control synthesis: the *certainty equivalence* principle. Self-tuning control schemes based on this idea and implemented by microprocessor have developed rapidly since the early 1970's and are now

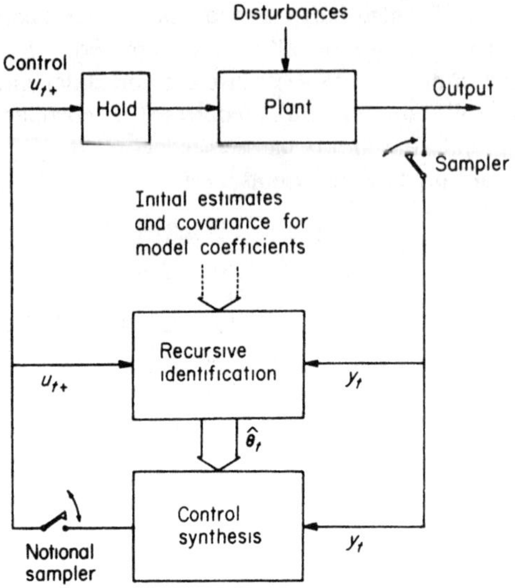

Fig. 8.3.2 Self-tuning controller.

available commercially. Åström and Wittenmark (1984) give a very readable introduction to self-tuning control, while Goodwin and Sin (1984) go into greater detail. Certainty-equivalence self-tuners can use any of a large range of control-synthesis techniques, among them pole placement specifying the closed-loop poles, linear-dynamics quadratic-cost Gaussian-disturbance optimal control, deadbeat control and minimum-variance control minimising a quadratic function of present control and expected output error at a future instant. Permuted with any of the simple recursive identification algorithms, they provide a large range of schemes.

Any adaptive control scheme, even a certainty-equivalence self-tuner applied to a linear, time-invariant plant, is a non-linear and time-varying closed-loop system. Nevertheless, stability and convergence analyses able to deal with many such schemes under reasonable assumptions are now available. Section 7.5 sketched that of Ljung (1977), and others appear in Goodwin and Sin (1984) and its references, and the references of Åström and Wittenmark (1984). An important point in implementing a certainty-equivalence self-tuning scheme is that the identification algorithm must not become over-confident and "freeze". The danger arises when an algorithm appropriate to time-invariant dynamics is applied to a plant whose dynamics drift slowly or change infrequently but abruptly. Action to prevent over-confidence, for instance by the methods of Section 8.1, must also avoid the opposite danger, of divergence. An excessive cumulative increase in the updating gain, while the control-input signal loses persistency of excitation because it derives from a well-regulated output, can cause divergence.

We conclude this brief glance at an increasingly important topic with an example. It illustrates how identification and control can be merged, and introduces the idea of recursive adaptive prediction.

Example 8.3.3 Let us develop a certainty-equivalence self-tuning controller for a system described by

$$Y(z^{-1}) = \frac{z^{-k}B(z^{-1})}{1 + A(z^{-1})}U(z^{-1}) + \frac{1 + C(z^{-1})}{1 + A(z^{-1})}V(z^{-1})$$

where $V(z^{-1})$ the transform of a zero-mean, uncorrelated noise sequence $\{v\}$, and

$$A(z^{-1}) \equiv a_1 z^{-1} + \cdots + a_n z^{-n}, \qquad B(z^{-1}) \equiv b_1 z^{-1} + \cdots + b_m z^{-m}$$
$$C(z^{-1}) \equiv c_1 z^{-1} + \cdots + c_q z^{-q}$$

The dead time k is assumed known, so b_1 is non-zero. The controller is to compute a new control value at each sampling instant, on receiving the latest output sample. At instant t, the controller computes u_t to minimise the mean-square error

$$J = E[(y_{t+k+1} - y^*_{t+k+1})^2 \,|t]$$

in y_{t+k+1}, the earliest output influenced by u_t. Here y^*_{t+k+1} is the desired value of y_{t+k+1}, and the conditioning on t indicates that J is calculated from knowledge available at instant t, including y_t.

As $\partial y_{t+k+1}/\partial u_t$ is b_1,

$$\partial J/\partial u_t = 2E[b_1(y_{t+k+1} - y^*_{t+k+1})\,|\,t] \quad \text{and} \quad \partial^2 J/\partial u_t^2 = 2b_1^2 > 0$$

so J is minimised by the u_t which makes $E[y_{t+k+1}\,|\,t]$ equal y^*_{t+k+1}. For brevity $E[y_{t+k+1}\,|\,t]$ will be called $\hat{y}_{t+k+1}$ and argument z^{-1} dropped from transforms from now on. The only uncertainty in $\hat{y}_{t+k+1}$ at instant t is due to u_t and v_{t+1} to v_{t+k+1}, since we know all inputs and outputs up to u_{t-1} and y_t, and can find all v's up to v_t via the system model. We find $\hat{y}_{t+k+1}$ by first splitting off the contribution to y_{t+k+1} due to v_{t+1} to v_{t+k+1}. This entails long division to obtain

$$(1 + C)/(1 + A) \equiv 1 + F + z^{-k}G/(1 + A)$$

with F of degree k in z^{-1} and remainder G of degree $j = \max(n, q - k)$. We then have

$$Y = z^{-k}\left(\frac{B}{1 + A}U + \frac{G}{1 + A}V\right) + (1 + F)V$$

where the term beginning z^{-k} gives that part of y_{t+k+1} known at instant t, and $(1+F)V$ the part due to later noise. The latter contributes zero to $\hat{y}_{t+k+1}$ since $\{v\}$ is zero-mean, so

$$\hat{Y} = z^{-k}\left(\frac{BU+GV}{1+A}\right)$$

$$= \frac{z^{-k}}{1+C}\left(B\left(1+F+\frac{z^{-k}G}{1+A}\right)U + G\left(Y - \frac{z^{-k}B}{1+A}U\right)\right)$$

$$= \frac{z^{-k}}{1+C}(B(1+F)U + GY)$$

We now have a recursion for the $(k+1)$-step prediction of y:

$$\hat{y}_{t+k+1} = b'_1 u_t + b'_2 u_{t-1} + \cdots + b'_{k+m} u_{t-k-m+1} + g_1 y_t + \cdots$$
$$+ g_j y_{t-j+1} - c_1 \hat{y}_{t+k} - \cdots - c_q \hat{y}_{t+k-q+1}$$

where $\hat{y}_{t+k}$ stands for $E[y_{t+k}|t-1]$ and so on, and

$$B' \equiv b'_1 z^{-1} + \cdots + b'_{k+m} z^{-k-m} \equiv B(1+F)$$

An *explicit minimum-variance self-tuner* estimates A, B and C, computes B', F and G, finds $\hat{y}_{t+k-q+1}$ to $\hat{y}_{t+k}$ recursively (which requires $1+C$ to have all its zeros inside the unit circle) and calculates u_t to make $\hat{y}_{t+k+1}$ equal y^*_{t+k+1}:

$$u_t = (y^*_{t+l+1} + \hat{c}_1 \hat{y}_{t+k} + \cdots + \hat{c}_1 \hat{y}_{t+k-q+1} - \hat{b}'_2 u_{t-1} - \cdots$$
$$- \hat{b}'_{k+m} u_{t-k-m+1} - \hat{g}_1 y_t - \cdots - \hat{g}_j y_{t-j+1})/\hat{b}'_1$$

An *implicit self-tuner* simplifies the computing by first using the $(k+1)$-step prediction equation as a regression equation

$$y_t = b'_1 u_{t-k-1} + \cdots + b'_{k+m} u_{t-2k-m} + g_1 y_{t-k-1} + \cdots$$
$$+ g_j y_{t-j-k} - c_1 \hat{y}_{t-1} - \cdots - c_q \hat{y}_{t-q}$$

to update a vector of estimates

$$\hat{\theta} = [\hat{b}'_1 \quad \cdots \quad \hat{b}'_{k+m} \quad \hat{g}_1 \quad \cdots \quad \hat{g}_l \quad \hat{c}_1 \quad \cdots \quad \hat{c}_q]^{\mathrm{T}}$$

then putting the updated estimates into the control law to find u_t as in the explicit self-tuner. In other words, the controller consists only of a recursively-updated adaptive $(k+1)$-step predictor.

In regulator problems y^* is constant, and the origin for u and y can be chosen to make y^* zero. Since the minimum-variance control law sets $\hat{y}_{t+k}$ to y^*_{t+k}, $\hat{y}_{t+k-1}$ to y^*_{t+k-1} and so on, all the terms in $\hat{y}$ in the control law are then zero. The control law simplifies to

$$u_t = -(\hat{b}'_2 u_{t-1} + \cdots + \hat{b}'_{k+m} u_{t-k-m+1} + \hat{g}_1 y_t + \cdots + \hat{g}_j y_{y-j+1})/\hat{b}'_1$$

and we need not identify c_1 to c_q. We see that it pays to match the form of the model to the job in hand. △

8.4. ADDITION OF REGRESSORS: RECURSION IN MODEL ORDER

8.4.1 Order-Incrementing Equations

We saw the ill-effects of including near-redundant regressors in l.s. in Chapter 4. Such regressors can be detected by singular-value decomposition as in Section 4.2.4 or by the model-order tests of Chapter 9. In this section we consider the reverse process of building up a model by adding regressors. It might be better to start with an over-large set of regressors and whittle it down, but sometimes the observation of extra variables is expensive or inconvenient enough to justify working from the bottom up. The results of this section are also one way of approaching lattice algorithms (Graupe *et al.*, 1980; Lee *et al.*, 1982), which are recursive in both time and model order.

We start by adding q new parameters $\boldsymbol{\phi}$ to p original ones $\boldsymbol{\theta}$, adjoining a q-column regressor matrix V_q to U_p to form the new model

$$\mathbf{y} = [U_p \quad V_q]\begin{bmatrix}\boldsymbol{\theta} \\ \boldsymbol{\phi}\end{bmatrix} + \mathbf{e} \tag{8.4.1}$$

Denoting the model order by subscripts on $\boldsymbol{\theta}$ and $\boldsymbol{\phi}$ as well as U and writing $[U_p \quad V_q]$ as U_{p+q}, we have the new normal equations

$$U_{p+q}^{\mathrm{T}} U_{p+q}\begin{bmatrix}\hat{\boldsymbol{\theta}}_{p+q} \\ \hat{\boldsymbol{\phi}}_{p+q}\end{bmatrix} = \begin{bmatrix}U_p^{\mathrm{T}}U_p & U_p^{\mathrm{T}}V_q \\ V_q^{\mathrm{T}}U_p & V_q^{\mathrm{T}}V_q\end{bmatrix}\begin{bmatrix}\hat{\boldsymbol{\theta}}_{p+q} \\ \hat{\boldsymbol{\phi}}_{p+q}\end{bmatrix} = U_{p+q}^{\mathrm{T}}\mathbf{y} = \begin{bmatrix}U_p^{\mathrm{T}}U_p\hat{\boldsymbol{\theta}}_p \\ V_q^{\mathrm{T}}\mathbf{y}\end{bmatrix} \tag{8.4.2}$$

The first row partition gives

$$\hat{\boldsymbol{\theta}}_{p+q} = \hat{\boldsymbol{\theta}}_p - [U_p^{\mathrm{T}}U_p]^{-1}U_p^{\mathrm{T}}V_q\hat{\boldsymbol{\phi}}_{p+q} \tag{8.4.3}$$

which substituted into the second gives

$$V_q^{\mathrm{T}}U_p(\hat{\boldsymbol{\theta}}_p - [U_p^{\mathrm{T}}U_p]^{-1}U_p^{\mathrm{T}}V_q\hat{\boldsymbol{\phi}}_{p+q}) + V_q^{\mathrm{T}}V_q\hat{\boldsymbol{\phi}}_{p+q} = V_q^{\mathrm{T}}\mathbf{y} \tag{8.4.4}$$

so

$$\begin{aligned}\hat{\boldsymbol{\phi}}_{p+q} &= (V_q^{\mathrm{T}}V_q - V_q^{\mathrm{T}}U_p[U_p^{\mathrm{T}}U_p]^{-1}U_p^{\mathrm{T}}V_q)^{-1}V_q^{\mathrm{T}}(\mathbf{y} - U_p\hat{\boldsymbol{\theta}}_p) \\ &\triangleq WV_q^{\mathrm{T}}\hat{\mathbf{e}}_p\end{aligned} \tag{8.4.5}$$

We now substitute $\hat{\boldsymbol{\phi}}_{p+q}$ into (8.4.3) to obtain

$$\hat{\boldsymbol{\theta}}_{p+q} = \hat{\boldsymbol{\theta}}_p - [U_p^{\mathrm{T}}U_p^{\mathrm{T}}]^{-1}U_p^{\mathrm{T}}V_qWV_q^{\mathrm{T}}\hat{\mathbf{e}}_p \tag{8.4.6}$$

The order-incrementing equation (8.4.6) resembles the time-update equation

(7.3.21) of recursive l.s. Time updating adds new rows rather than columns to the regressor matrix.

The matrix $[U^{\mathrm{T}}U]^{-1}$ must also be order-incremented, to obtain a covariance estimate for the parameters. To do so we apply a partitioned-matrix inversion lemma (Goodwin and Payne, 1977, App. E)

$$\begin{bmatrix} A & B \\ C & D \end{bmatrix}^{-1} = \begin{bmatrix} (A-BD^{-1}C)^{-1} & -(A-BD^{-1}C)^{-1}BD^{-1} \\ -(D-CA^{-1}B)^{-1}CA^{-1} & (D-CA^{-1}B)^{-1} \end{bmatrix} \tag{8.4.7}$$

and put $U_p^{\mathrm{T}}U_p$ for A, $U_p^{\mathrm{T}}V_q$ for B, $V_q^{\mathrm{T}}U_p$ for C and $V_q^{\mathrm{T}}V_q$ for D. We find that, if we denote $[U_p^{\mathrm{T}}U_p]^{-1}$ by M_p,

$$\begin{aligned} M_{p+q} &= \begin{bmatrix} U_p^{\mathrm{T}}U_p & U_p^{\mathrm{T}}V_q \\ V_q^{\mathrm{T}}U_p & V_q^{\mathrm{T}}V_q \end{bmatrix}^{-1} \\ &= \begin{bmatrix} (M_p - U_p^{\mathrm{T}}V_q(V_q^{\mathrm{T}}V_q)^{-1}V_q^{\mathrm{T}}U_p)^{-1} & -(M_p - U_p^{\mathrm{T}}V_q(V_q^{\mathrm{T}}V_q)^{-1}V_q^{\mathrm{T}}U_p)^{-1}U_p^{\mathrm{T}}V_q(V_q^{\mathrm{T}}V_q)^{-1} \\ -WV_q^{\mathrm{T}}U_pM_p & W \end{bmatrix} \end{aligned} \tag{8.4.8}$$

where W is defined in (8.4.5). The matrix inversions in partition (1, 1) can be avoided by use of the matrix-inversion lemma (7.3.26) to give

$$(M_p - U_p^{\mathrm{T}}V_q(V_q^{\mathrm{T}}V_q)^{-1}V_q^{\mathrm{T}}U_p)^{-1} = M_p + M_pU_p^{\mathrm{T}}V_qWV_q^{\mathrm{T}}U_pM_p \tag{8.4.9}$$

and partition (1, 2) is the transpose of partition (2, 1) since M_{p+q} is symmetric. Alternatively, partition (1, 1) can be found directly in the form on the right of (8.4.9) from

$$\begin{bmatrix} A & B \\ C & D \end{bmatrix}^{-1} = \begin{bmatrix} A^{-1}+A^{-1}BWCA^{-1} & -A^{-1}BW \\ -WCA^{-1} & W \end{bmatrix}$$

$$W \triangleq (D-CA^{-1}B)^{-1} \tag{8.4.10}$$

Both (8.4.7) and (8.4.10) are easy to verify.

The resemblance to covariance-updating equation (7.3.25) is quite strong if we write

$$\begin{aligned} M_{p+q} &= \begin{bmatrix} M_p & 0 \\ 0 & 0 \end{bmatrix} - \begin{bmatrix} -M_pU_p^{\mathrm{T}}V_q \\ M_pU_p^{\mathrm{T}}U_p \end{bmatrix} \\ &\quad \times (V_q^{\mathrm{T}}(U_pM_pU_p^{\mathrm{T}} - I)V_q)^{-1}[-V_q^{\mathrm{T}}U_pM_p \quad U_p^{\mathrm{T}}U_pM_p] \end{aligned} \tag{8.4.11}$$

and we would expect much of the technique for avoiding numerical difficulties in covariance updating to carry over to order incrementing.

Before interpreting the order-incrementing equations further, let us do a numerical example.

Example 8.4.1 An extra term is added to the target-position model in the radar problem of Example 4.1.1, to account for rate of change of acceleration, making the model

$$x(t) = x_0 + v_0 t + a_0 t^2/2 + bt^3/6$$

Here $\boldsymbol{\theta}$ is $[x_0 \quad v_0 \quad a_0]^{\mathrm{T}}$, already estimated, and the new parameter ϕ is b, with $q = 1$. M_p is also to hand. The new regressor vector $\mathbf{v}_1$ is

$$[0 \quad 1.3^{\cdot} \times 10^{-3} \quad 1.06 \times 10^{-2} \quad 3.6 \times 10^{-2} \quad 8.53^{\cdot} \times 10^{-2} \quad 0.16^{\cdot}]^{\mathrm{T}}$$

so to four figures

$$\begin{aligned}
\mathbf{v}_1^{\mathrm{T}}\mathbf{v}_1 &= 3.647 \times 10^{-2} \\
U_p^{\mathrm{T}}\mathbf{v}_1 &= [0.3 \quad 0.2611 \quad 0.118]^{\mathrm{T}} \\
(U_p^{\mathrm{T}}U_p)^{-1}U_p^{\mathrm{T}}\mathbf{v}_1 &= [-0.02811 \quad 0.1495 \quad 0.01837]^{\mathrm{T}} \\
W &= 269.567 \\
\hat{\phi}_{p+1} &= -4.358 = \hat{b} \\
\hat{\boldsymbol{\theta}}_{p+1} &= [4.663 \quad 235.1 \quad 55.44]^{\mathrm{T}} \\
-(U_p^{\mathrm{T}}U_p)^{-1}U_p^{\mathrm{T}}\mathbf{v}_1 W &= [7.576 \quad -40.30 \quad -4.952]^{\mathrm{T}}
\end{aligned}$$

and from (8.4.9), partition (1, 1) of M_{p+q} is

$$\begin{bmatrix} 0.7477 & -1.929 & 0.02481 \\ -1.929 & 8.067 & -0.4897 \\ 0.02481 & -0.4897 & 2.551 \end{bmatrix}$$

The extra term contributes little to $\hat{\mathbf{y}}$, and $\hat{b}$ has the large estimated standard deviation $W^{1/2} = 16.42$. Moreover, its presence increases the s.d.'s of $\hat{x}_0$ and $\hat{a}_0$ considerably but only reduces the sum of squares of output errors from 157.9 to 157.8. It is clearly not worth including.

The example shows how little computing is needed to add a single term; W is a scalar, $U_p^{\mathrm{T}}V_q$ a vector and M_{p+q} correspondingly easy to form. △

*8.4.2 Orthogonality in Order-Incrementing

We encountered orthogonality between the error $\mathbf{y} - \hat{\mathbf{y}}$ and each regressor vector in Section 4.1.3, and established its connection with the conditional-mean estimate in Section 6.3.6. The recursive l.s. algorithms of Chapter 7 can be derived elegantly by reference to orthogonality, but we opted for an algebraic derivation. To gain some idea of the power and economy of a more geometrical approach, we shall derive the order-incrementing equations by an appeal to orthogonality.

Recall that the o.l.s. estimate of $\mathbf{y}$ is

$$\hat{\mathbf{y}} = U[U^{\mathrm{T}}U]^{-1}U^{\mathrm{T}}\mathbf{y} = P(U)\mathbf{y} \tag{8.4.12}$$

where the projection matrix $P(U)$ projects $\mathbf{y}$ orthogonally on to the plane formed by all linear combinations of the columns of U. Hence the error $\mathbf{y} - P(U)\mathbf{y}$ is orthogonal to each column of U, as is easily verified:

$$U^T(\mathbf{y} - P(U)\mathbf{y}) = U^T\mathbf{y} - U^T\mathbf{y} = \mathbf{0} \tag{8.4.13}$$

By the same token $\mathbf{y} - P([U_p V_q])\mathbf{y}$ is orthogonal to each column of U_p and V_q so, with M denoting the inverse of the normal matrix as before,

$$\begin{bmatrix} U_p^T \\ V_q^T \end{bmatrix}\mathbf{y} - \begin{bmatrix} U_p^T \\ V_q^T \end{bmatrix}[U_p \quad V_q]M_{p+q}[U_p \quad V_q]\mathbf{y} = \mathbf{0} \tag{8.4.14}$$

or in terms of the partitions $M^{(11)}$, $M^{(12)}$, $M^{(21)}$ and $M^{(22)}$ of M_{p+q},

$$U_p^T\mathbf{y} - U_p^T(U_pM^{(11)} + V_qM^{(21)})U_p^T\mathbf{y} - U_p^T(U_pM^{(12)} + V_qM^{(22)})V_q^T\mathbf{y} = \mathbf{0} \tag{8.4.15}$$

$$V_q^T\mathbf{y} - V_q^T(U_pM^{(11)} + V_qM^{(21)})U_p^T\mathbf{y} - V_q^T(U_pM^{(12)} + V_qM^{(22)})V_q^T\mathbf{y} = \mathbf{0} \tag{8.4.16}$$

Now (8.4.15) must be true for any values of U_p, V_q and $\mathbf{y}$, including $\mathbf{y}$ non-zero but $U_p^T\mathbf{y}$ zero, so that

$$U_p^T(U_pM^{(12)} + V_qM^{(22)})V_q^T\mathbf{y} = \mathbf{0} \tag{8.4.17}$$

To satisfy (8.4.17) it is sufficient to make $U_p^T(U_pM^{(12)} + V_qM^{(22)})$ zero, giving

$$M^{(12)} = -M_pU_p^TV_qM^{(22)} \tag{8.4.18}$$

Similarly, (8.4.16) requires that

$$V_q^T\mathbf{y} - V_q^T(U_pM^{(12)} + V_qM^{(22)})V_q^T\mathbf{y} = \mathbf{0} \tag{8.4.19}$$

and it is enough if $I - V_q^T(U_pM^{(12)} + V_qM^{(22)})$ is zero, which on substitution of $M^{(12)}$ from (8.4.18) gives

$$M^{(22)} = (V_q^TV_q - V_q^TU_pM_pU_p^TV_q)^{-1} \equiv W \tag{8.4.20}$$

Swapping U_p and V_q all through gives $M^{(11)}$ as in (8.4.8), and $M^{(21)}$ is $M^{(12)T}$, as by definition M_{p+q} is symmetric.

The order-incrementing equations for $\hat{\theta}$ and $\hat{\phi}$ also comes from (8.4.14), but written as

$$\begin{bmatrix} U_p^T \\ V_q^T \end{bmatrix}[\mathbf{y} - U_p\hat{\theta}_{p+q} - V_q\hat{\phi}_{p+q}] = \mathbf{0} \tag{8.4.21}$$

We recognise this immediately as the normal equations (8.4.2).

8.4.3 Lattice Algorithms and Identification

A great deal of interest has been aroused in signal processing by the development of lattice algorithms (Friedlander, 1982). The algorithms employ an a.r. time-series model for l.s. signal estimation, and are implemented as a cascade of identical sections, each corresponding to an increase of one in the a.r. order. The algorithms are attractive for their computational economy and good numerical properties, and are potentially useful for identification. However, their economy depends on the model being an autoregression. For an a.r., the regressor vector at time t is that at $t-1$ shifted down one place and with one new entry at the top. The normal matrix is correspondingly updated mainly by shifting south-east. Without going into the details, we can appreciate that this simplifies a combined time-updating and order-incrementing algorithm greatly. For identification, we are rarely happy with a purely a.r. model, and almost always require an a.r.m.a. model with exogenous inputs plus a noise model, and perhaps also a constant term. The updating is much less simple, with several new samples entering at each update. The result is that computational economy is lost (Robins and Wellstead, 1981), and the lattice method has no overwhelming advantage to counterbalance its relatively complicated programming and difficulty of interpretation in identification.

8.5 MODEL REDUCTION

The fewer parameters a model has, the easier it is to understand and apply. The neatest way to ensure that a model has no more parameters than necessary is to conduct order tests during identification, as described in Chapter 9. Nevertheless, we sometimes have to reduce an existing model, perhaps to check whether order reduction alters the overall behaviour significantly. There are many approaches (Bosley and Lees, 1972) of which we shall examine a few of the most popular, applied to transfer-function rather than state-space models.

8.5.1 Moment Matching: Padé Approximation

One way to fit a reduced transfer function

$$\frac{B_m(s)}{1+A_m(s)} = \frac{b_0 + b_1 s + \cdots + b_{m-1}s^{m-1}}{1 + a_1 s + \cdots + a_m s^m} \tag{8.5.1}$$

to a larger continuous-time model is to expand the transfer function of the larger model as a power series in s:

$$H(s) = h_0 + h_1 s + h_2 s^2 + \cdots \infty \tag{8.5.2}$$

then pick the $2m$ coefficients in (8.5.1) so as to match terms of (8.5.2) up to $h_{2m-1}s^{2m-1}$. The process of rational-function approximation via a Taylor series is called *Padé approximation* (Watson, 1980). The numerator and denominator degrees can be chosen at will; we have made $B_m(s)$ of degree one less than $1 + A_m(s)$ to give a realistic finite bandwidth.

Matching h_0 matches the steady-state gain, i.e. the final value $\lim_{s\to 0} sH(s)/s$ of the step response, and we can interpret matching higher powers of s as paying attention to the response to higher derivatives of the input. The significance of the matching is best seen in terms of the impulse response. For a stable system

$$i!\,h_i = \lim_{s\to 0} s\cdot\frac{1}{s}\frac{d^i H(s)}{ds^i} = \lim_{t\to\infty} \mathscr{L}^{-1}\left\{\frac{1}{s}\frac{d^i H(s)}{ds^i}\right\} \tag{8.5.3}$$

and

$$\mathscr{L}^{-1}\left\{\frac{1}{s}\frac{d^i H(s)}{ds^i}\right\} = \int_0^t (-t)^i h(t)\,dt \tag{8.5.4}$$

(Gabel and Roberts, 1980). We see that matching h_i matches the ith time moment of the impulse response $h(t)$. That seems sensible enough.

Example 8.5.1 The model

$$Y(s) = \frac{(1+0.5s)(1+0.2s)}{(1+s)(1+0.25s)(1+0.1s)(1+0.05s)}U(s) + \text{noise}$$

is to be reduced to first or second order. To do so, we write the transfer function numerator and denominator in ascending powers of s then expand $H(s)$ by long division, quicker than repeated differentiation and (8.5.3). We obtain

$$H(s) = 1 - 0.7s + 0.6375s^2 - 0.6265s^3 + \cdots$$

An mth-order reduced model then matches the coefficients of s^0 to s^{2m-1} in

$$B_m(s) = H(s)(1 + A_m(s))$$

(i) First-order model: $b_0 = h_0$ and $b_1 = h_0 a_1 + h_1 = 0$ so $b_0 = 1$, $a_1 = 0.7$

(ii) Second-order model: $b_0 = h_0$, $b_1 = h_0 a_1 + h_1$, $b_2 = h_0 a_2 + h_1 a_1 + h_2 = 0$, $b_3 = h_1 a_2 + h_2 a_1 + h_3 = 0$ so $b_0 = 1$, $b_1 = 0.522$, $a_1 = 1.222$, $a_2 = 0.218$.

Figure 8.5.1 gives the step and impulse responses of the original and

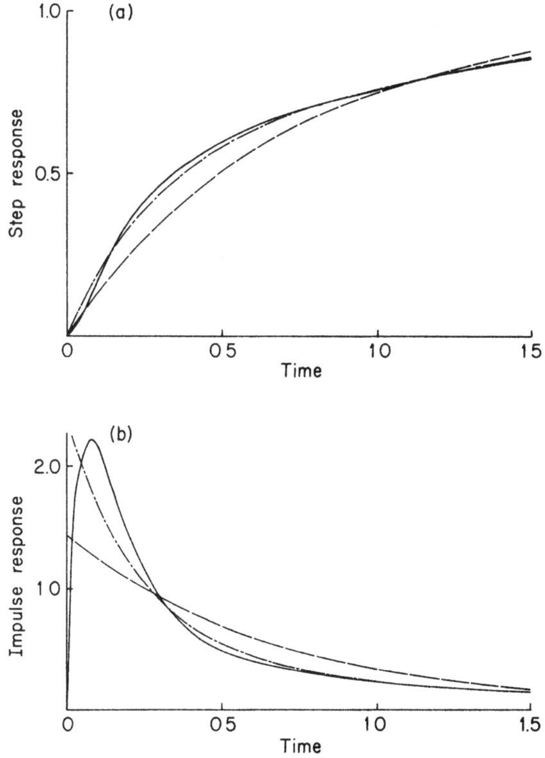

Fig. 8.5.1 (a) Step and (b) impulse responses, Example 8.5.1. ———: original model; – – –: second-order reduced model; —·—: first-order reduced model.

reduced models. Except very early on, the second-order model fits the step response well. The impulse-response fit is less impressive, with quite wrong behaviour initially. The trouble can be traced to a difference in pole-zero excess, two for the original model and one for each reduced model, invalidating the approximation

$$h(0+) = \lim_{s\to\infty} sH(s) \simeq \lim_{s\to\infty} s\frac{B_m(s)}{1 + A_m(s)} \qquad \triangle$$

8.5.2 Continued-Fraction Approximation

A model-reduction method popularised by Chen and Shieh (1968) is to expand the original rational transfer function as a continued fraction in the second

Cauer form

$$\frac{B(s)}{1+A(s)} = \cfrac{1}{c_1 + \cfrac{1}{\cfrac{c_2}{s} + \cfrac{1}{c_3 + \cfrac{1}{\cfrac{c_4}{s} + \cdots}}}} \tag{8.5.5}$$

then truncate it after $2m$ quotients and reconstruct the reduced-model rational transfer function $B_m(s)/(1 + A_m(s))$. The continued fraction can be thought of as the transfer function of the nested feedback-feedforward model in Fig. 8.5.2(a). Figures 8.5.2(b) and (c) show that truncation of the continued fraction after two and four quotients respectively gives valid first- and second-order approximations provided the innermost retained feedforward gain is much larger than the finite gain of the deleted section. At small enough s this is certainly so.

Coefficient c_1 is produced by one stage of long division on the reciprocal of the original transfer function, leaving a remainder $sD(s)/B(s)$ say:

$$\frac{B(s)}{1+A(s)} = \frac{1}{(1+A(s))/B(s)} = \frac{1}{c_1 + sD(s)/B(s)}$$
$$= \cfrac{1}{c_1 + \cfrac{1}{(1/s)(B(s)/D(s))}} \tag{8.5.6}$$

One stage of long division on $B(s)/D(s)$ then gives c_2, and so on. The coefficients turn out to be the ratios of successive elements in the first column of the Routh array

$$\left.\begin{array}{ccc} 1 & a_1 & a_2 \quad \cdots \\ b_0 & b_1 & b_2 \quad \cdots \\ a_1 - b_1/b_0 & a_2 - b_2/b_0 & \cdots \\ b_1 - b_0\dfrac{a_2 - b_2/b_0}{a_1 - b_1/b_0} & \cdots & \\ \vdots & & \end{array}\right\} \begin{array}{l} c_1 = 1/b_0 \\ c_2 = b_0/(a_1 - b_1/b_0) \\ c_3 = (a_1 - b_1/b_0)/ \\ \qquad (b_1 - b_0(a_2 - b_2/b_0)/(a_1 - b_1/b_0)) \\ \vdots \end{array} \tag{8.5.7}$$

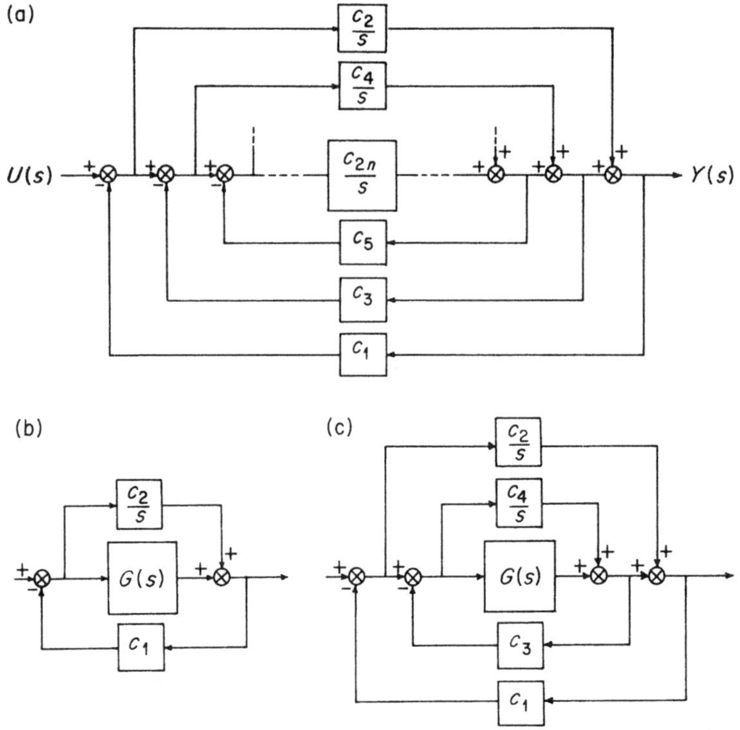

Fig. 8.5.2 Continued-fraction model reduction. (a) System with continued-fraction transfer function of $2n$ quotients, (b) system which is 1st-order when $G(s)$ is deleted.

$$\frac{Y(s)}{U(s)} \simeq \frac{c_2/s}{1 + c_1 c_2/s} \equiv \frac{1}{c_1 + 1/(c_2/s)} \qquad \text{if} \quad \frac{c_2}{s} \gg G$$

and (c) system which is 2nd-order when $G(s)$ is deleted.

$$\frac{Y(s)}{U(s)} \simeq \frac{c_2/s(1 + c_3 c_4/s) + c_4/s}{1 + c_3 c_4/s(1 + c_3 c_4/s) + c_4/s}$$

$$\equiv \cfrac{1}{c_1 + \cfrac{1}{\cfrac{c_2}{s} + \cfrac{1}{c_2 + c_4/s}}} \qquad \text{if} \quad \frac{c_4}{s} \gg G$$

Example 8.5.2 For the transfer function of Example 8.5.1, the Routh array is

$$\left.\begin{array}{llll} 1 & 1.4 & 0.4425 & 0.04375 \;\cdots \\ 1 & 0.7 & 0.1 & 0 \\ 0.7 & 0.3425 & 0.04375 \;\cdots & \\ 0.2107 & 0.0375 \;\cdots & & \\ 0.2179 & & & \\ \vdots & & & \end{array}\right\} \begin{array}{l} c_1 = 1 \\ c_2 = 1.429 \\ c_3 = 3.322 \\ c_4 = 0.9669 \\ \vdots \end{array}$$

so the first-order model, retaining only c_1 and c_2, is $1(c_1 + s/c_2) = 1/(1+0.7s)$. Retaining c_1 to c_4 gives the same second-order model as in Example 8.5.1.

△

To change the reduced-model order we simply add or delete continued-fraction coefficients; the only reworking is to turn the continued fraction back into a rational polynomial function.

8.5.3 Moment Matching in Discrete Time

Most of the literature on model reduction by moment matching concerns continuous-time systems, but we are mainly interested in discrete-time models. Expansion of a z-transform transfer function as a power series in z^{-1} just yields the unit-pulse response ordinates $g_0, g_1, \ldots$ in

$$G(z^{-1}) = g_0 + g_1 z^{-1} + g_2 z^{-2} + \cdots \infty \tag{8.5.8}$$

so the matching process of Section 8.5.1 would only match the start of the u.p.r., ignoring the rest. The time-moments of $\{g\}$ can still be related to the coefficients of the reduced transfer function

$$\frac{B_m(z^{-1})}{1 + A_m(z^{-1})} = \frac{b_0 + b_1 z^{-1} + \cdots + b_m z^{-m}}{1 + a_1 z^{-1} + \cdots + a_m z^{-m}} \tag{8.5.9}$$

but less readily than in continuous time. The ith moment is

$$M_i = \sum_{k=0}^{\infty} k^i g_k \tag{8.5.10}$$

and since

$$\left.\frac{d^r G}{dz^{-r}}\right|_{z=1} = \sum_{k=r}^{\infty} k(k-1)\cdots(k-r+1)g_k \tag{8.5.11}$$

we can express M_i as a weighted sum of G and its first i derivatives, all evaluated at $z = 1$:

$$
\begin{aligned}
M_0 &= G|_{z=1} \\
M_1 &= \left.\frac{dG}{dz^{-1}}\right|_{z=1} \\
M_2 &= \left.\frac{d^2G}{dz^{-2}}\right|_{z=1} + \left.\frac{dG}{dz^{-1}}\right|_{z=1} \\
M_3 &= \left.\frac{d^3G}{dz^{-3}}\right|_{z=1} + 3\left.\frac{d^2G}{dz^{-2}}\right|_{z=1} + \left.\frac{dG}{dz^{-1}}\right|_{z=1}
\end{aligned}
\tag{8.5.12}
$$

and so on. Hence we can match the original and reduced models through either the moments or the same number of derivatives of their u.p.r.'s. We conclude that moment-matching in discrete time is equivalent to matching the leading terms of the Taylor series expansions of the transfer functions about $z = 1$, in clear correspondence with expansion about $s = 0$ in continuous time if we interpret z as e^{sT}.

For the reduced model, $B_m(z^{-1})/(1 + A_m(z^{-1}))$ replaces $G(z^{-1})$ in the derivatives. For the original model, the derivatives or moments can be computed directly from its u.p.r. if it is available and short-lived. If not, the original model is also written in rational form and the derivatives found from that.

Example 8.5.3 The reduced-order model

$$B_m(z^{-1})/(1 + A_m(z^{-1})) = (b_0 + b_1 z^{-1})/(1 + a_1 z^{-1})$$

is to be fitted to $G(z^{-1}) = 0.5z^{-1}/(1 + 1.1z^{-1} + 0{\cdot}24z^{-2})$. Though $G(z^{-1})$ is unrealistically small, the example will bring out the main features of the procedure. The three reduced-model parameters allow us to match G, dG/dz^{-1} and d^2G/dz^{-2} at $z = 1$. Denoting the reduced transfer function by G_m,

$$
G_m|_{z=1} = \frac{b_0 + b_1}{1 + a_1}, \qquad \left.\frac{dG_m}{dz^{-1}}\right|_{z=1} = \frac{b_1 - a_1 b_0}{(1 + a_1)^2}
$$

$$
\left.\frac{d^2G_m}{dz^{-2}}\right|_{z=1} = \frac{2a_1(a_1 b_0 - b_1)}{(1 + a_1)^3}
$$

so we solve

$$
b_0 + b_1 = (1 + a_1)G|_{z=1}, \qquad b_1 - a_1 b_0 = (1 + a_1)^2 \left.\frac{dG}{dz^{-1}}\right|_{z=1}
$$

$$
2a_1(a_1 b_0 - b_1) = (1 + a_1)^3 \left.\frac{d^2G}{dz^{-2}}\right|_{z=1}
$$

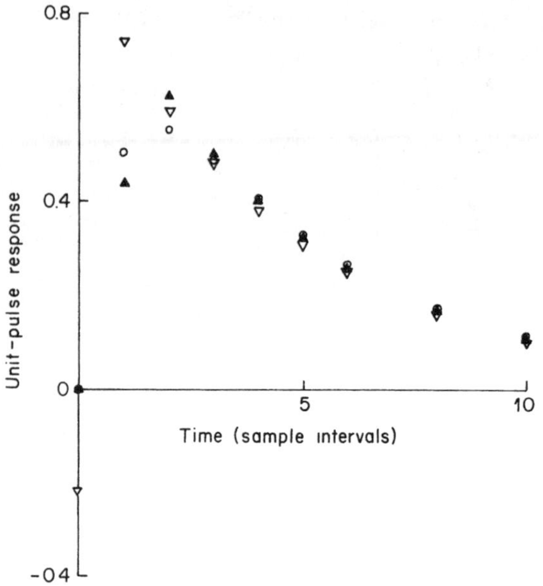

Fig. 8.5.3 Unit-impulse responses of original and reduced models. ○: original model; △: reduced model; ▲: reduced model with correct dead time.

The derivatives of G are rather tedious to find for even a second-order rational transfer function, so we might think of summing $k^i g_k$, $k=0,1,\ldots, i=0,1,2$, to get the moments directly (or computing the derivatives by (8.5.11)). In fact, $\sum_{k=0}^{15} k^2 g_k$ is 33% less than M_2, and the sum up to g_{20} still 16% less, even though g_{15} and g_{20} are only 0.0352 and 0.0115. Accuracy might be poorer still with a real u.p.r. model because of scatter and bias.

Figure 8.5.3 gives the u.p.r. of each model. Only the first few points are much in error. Matching M_0 has made the errors sum to zero, but only by driving the reduced-model u.p.r. negative initially. This implies non-minimum-phase continuous-time behaviour, contrary to the original model. A model $B_m(z^{-1})/(1+A_m(z^{-1})) = (b_1 z^{-1} + b_2 z^{-2})/(1+a_1 z^{-1})$ gives a much better fit, also shown in Fig. 8.5.3, with no negative excursion. Similar anomalous u.p.r. behaviour will appear in models estimated from real records in Chapter 10, again due to too short a model dead time. △

8.5.4 Other Reduction Methods

Padé approximation has the serious drawback that an unstable reduced model may be obtained from a stable original model (Problem 8.5). Many techniques to avoid this snag have been suggested. They usually determine $1 + A_m$ first, keeping the stable dominant poles of the original model. The simplest just throw away the fastest poles, i.e. the leftmost in the s-plane or nearest to the origin in the z-plane. The numerator coefficients can then be chosen to match m moments, achieving a poorer fit than Padé approximation does (if stable) matching $2m$ moments. These alternatives have their own drawbacks (Shamash, 1983), notably the risk of retaining a slow pole even when it is almost cancelled by a zero and so has little effect.

8.6 RECURSIVE IDENTIFICATION IN BOUNDED NOISE

The traditional noise model we have adhered to until now is white noise passed through a low-order linear filter. The white-noise sequence is characterised by its variance or covariance and its mean. When considering its p.d.f., we have usually taken it as Gaussian. Real noise is often far from Gaussian, and tends to exhibit complications such as isolated abrupt events due to unmonitored control actions, intermittent disturbances in ambient conditions or inputs such as feedstock quality, and instrument misreadings or transcription errors. More gradual changes amounting to time variation of the noise statistics also often occur, as noted in Section 8.1. Chapter 10 contains examples of real noise behaviour.

The filtered-white-Gaussian noise model can be defended on grounds of mathematical convenience, allied to a hope that an estimator with good properties in idealised noise will still perform well in real noise. With enough knowledge of the plant and its environment, a detailed noise p.d.f. might be formulated and employed by a Bayes or m.l. estimator, but shortage of prior information or excessive computing demands normally rule out that option. A more realistic aim is to match the noise representation to the extent of prior knowledge, keeping it very simple when necessary but allowing for empirical refinement of the noise model during identification. We shall now examine an alternative to probabilistic noise modelling which does not pretend to more knowledge than is actually available, but allows us to discover more about the noise as we go.

8.6.1 Bounded-Noise, Parameter-Bounding Model

Given that the noise p.d.f. and correlation structure are initially unknown, perhaps complicated, and difficult to estimate reliably by way of residuals from limited records, is there a simple non-probabilistic way to characterise the noise? The answer is yes: by bounds. We specify only the largest credible values that noise could take, so the usual model linear in the parameters becomes

$$y_t = \mathbf{u}_t^{\mathrm{T}}\boldsymbol{\theta} + v_t, \qquad |v_t| \le r_t \text{ (known)}, \qquad t = 1, 2, \ldots, N \tag{8.6.1}$$

Here the origin for y_t is chosen to make the bounds on v_t symmetrical, for convenience. The bounds are taken as constant unless we know better. Observation y_t tells us that

$$y_t - r_t \le \mathbf{u}_t^{\mathrm{T}}\boldsymbol{\theta} \le y_t + r_t \tag{8.6.2}$$

These two constraints on $\boldsymbol{\theta}$ may be interpreted as hyperplanes in $\boldsymbol{\theta}$ space, between which $\boldsymbol{\theta}$ must lie. A sequence of observations $\{y_1, y_2, \ldots, y_N\}$ gives N pairs of hyperplanes, which together confine $\boldsymbol{\theta}$ to some region D_N. After processing y_N, we know that $\boldsymbol{\theta}$ is somewhere in D_N but we assign no probabilities to different positions, and make no attempt to extract a "best" estimate.

Lack of a unique estimate of $\boldsymbol{\theta}$ is at first worrying, but we can reassure ourselves by reflecting that engineering design is largely a matter of tolerancing for adequate performance in the worst case. For this purpose, parameter bounds are just what we want.

Example 8.6.1 In the radar target problem of Example 4.1.1, we decide to bound the parameters on the (correct) assumption that the noise is between ± 10 at every sample. Hence

$$y(\tau) - 10 \le x_0 + v_0\tau + a\tau^2/2 \le y(\tau) + 10, \qquad \tau = 0, 0.2, \ldots, 1.0$$

defines the range of the parameters $\boldsymbol{\theta} = [x_0 \quad v_0 \quad a]^{\mathrm{T}}$ compatible with the observations. Figure 8.6.1 shows the cross-section of the resulting three-dimensional polyhedron D_6 at $x_0 = 5$, the correct value. The observation at $\tau = 0$ merely constrains x_0 to be between -7 and 13, i.e. puts bounding hyperplanes parallel to our cross-section, so they do not show.

The slopes $-2/\tau$ of the constraints fall in a fairly narrow range, so $[\theta_2 \quad \theta_3]$ is ill defined in one direction but well defined at right angles to it. The elongated D_6 says much the same as the estimated covariance

$$\begin{bmatrix} 958 & -1763 \\ -1763 & 3526 \end{bmatrix} \quad \text{for} \quad [\hat{\theta}_2 \quad \hat{\theta}_3]^{\mathrm{T}}$$

found by o.l.s. in Example 5.3.3. That is, θ_2 and θ_3 are individually about equally ill defined and the errors in θ_2 and θ_3 are very likely to have opposite signs.

As well as being conceptually straightforward, parameter bounding throws light on the strengths and weaknesses of the observations, and is potentially valuable in experiment design. In this example it is clear that an observation yielding constraints with $d\theta_3/d\theta_2$ positive and about 1, and $20/\tau$ apart in the θ_2

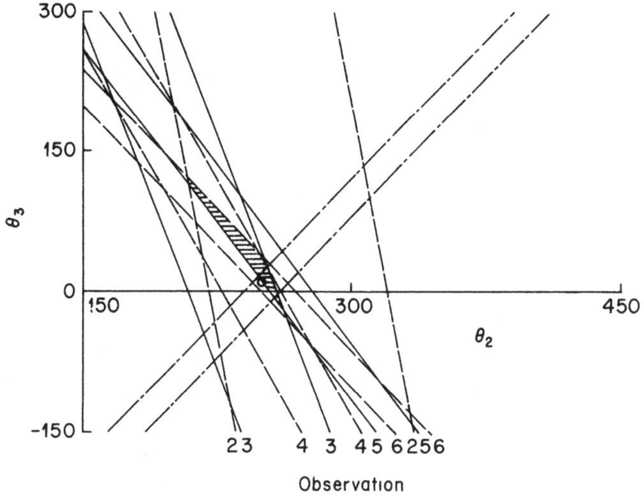

Fig. 8.6.1 Parameter bounds, Example 8.6.1. Cross-section of D_6 hatched; actual $\boldsymbol{\theta}$ circled.

direction, would have reduced D considerably. An observation at $\tau = -1$ with noise 5, for instance, would have given the chain-dotted constraints in Fig. 8.6.1 (but only if the target flew backwards). △

As we seem to be talking geometry, let us recall a few facts about vectors in Euclidean space. Those dealing with orthogonality are already familiar, at least.

Summary: Euclidean Inner-Product Space (Hadley, 1961; Luenberger, 1973; Halmos, 1958; Rockafellar, 1970) In a real inner-product space, the length (or norm) $\|\boldsymbol{\theta}\|$ of vector $\boldsymbol{\theta}$ is defined as $(\boldsymbol{\theta}^T\boldsymbol{\theta})^{1/2}$ and Euclidean geometry applies. If $\boldsymbol{\eta}$ is the orthogonal projection of $\boldsymbol{\theta}$ onto $\mathbf{u}$ (Fig. 8.6.2) and $\boldsymbol{\zeta}$ is $\boldsymbol{\theta} - \boldsymbol{\eta}$, then by Pythagoras' theorem

$$\boldsymbol{\theta}^T\boldsymbol{\theta} = (\boldsymbol{\eta} + \boldsymbol{\zeta})^T(\boldsymbol{\eta} + \boldsymbol{\zeta}) = \boldsymbol{\eta}^T\boldsymbol{\eta} + \boldsymbol{\zeta}^T\boldsymbol{\zeta} \tag{8.6.3}$$

so $\boldsymbol{\eta}$ and $\boldsymbol{\zeta}$ are orthogonal if and only if $\boldsymbol{\eta}^T\boldsymbol{\zeta}$ is zero. If $\boldsymbol{\eta}$ is $\gamma\mathbf{u}$ then

$$\mathbf{u}^T\boldsymbol{\theta} = \boldsymbol{\eta}^T(\boldsymbol{\eta} + \boldsymbol{\zeta})/\gamma = \boldsymbol{\eta}^T\boldsymbol{\eta}/\gamma = \|\boldsymbol{\eta}\|^2/\gamma = \|\boldsymbol{\eta}\|\,\|\mathbf{u}\| \tag{8.6.4}$$

and

$$\boldsymbol{\eta} = \frac{\|\boldsymbol{\eta}\|}{\|\mathbf{u}\|}\mathbf{u} = \frac{\mathbf{u}^T\boldsymbol{\theta}}{\|\mathbf{u}\|^2}\mathbf{u} = \frac{\mathbf{u}^T\boldsymbol{\theta}}{\mathbf{u}^T\mathbf{u}}\mathbf{u},$$
$$\|\boldsymbol{\eta}\| = \frac{|\mathbf{u}^T\boldsymbol{\theta}|}{\|\mathbf{u}\|} = \|\boldsymbol{\theta}\|\cos\alpha \tag{8.6.5}$$

Hence $\mathbf{u}^T\boldsymbol{\theta} = z$ says that the length of the orthogonal projection of $\boldsymbol{\theta}$ onto $\mathbf{u}$ is $z/\|\mathbf{u}\|$. This accounts for all $\boldsymbol{\theta}$ in a hyperplane onto which $z\mathbf{u}/\|\mathbf{u}\|^2$ is the perpendicular from the origin. A pair of hyperplanes $\mathbf{u}_i^T\boldsymbol{\theta} = z_{i1}$ and $\mathbf{u}_i^T\boldsymbol{\theta} = z_{i2}$ are parallel and $|z_{i1} - z_{i2}|/\|\mathbf{u}_i\|$ apart, with $\mathbf{u}_i$ normal to both. In p dimensions, any p linearly independent pairs define a polyhedron with 2^p vertices given by

$$U_p\boldsymbol{\theta} = \begin{bmatrix} \mathbf{u}_1^T \\ \vdots \\ \mathbf{u}_p^T \end{bmatrix} \boldsymbol{\theta} = \begin{bmatrix} z_{11} \text{ or } z_{12} \\ \vdots \\ z_{p1} \text{ or } z_{p2} \end{bmatrix} \tag{8.6.6}$$

with U_p non-singular.

The points in and on the polyhedron form a convex set; that is, if $z_{i1} \le z_{i2}$ for $i = 1, 2, \ldots, p$, any such points $\boldsymbol{\theta}^{(1)}$ and $\boldsymbol{\theta}^{(2)}$ satisfy

$$\mathbf{z}_1 \le U_p\boldsymbol{\theta}^{(1)} \le \mathbf{z}_2, \qquad \mathbf{z}_1 \le U_p\boldsymbol{\theta}^{(2)} \le \mathbf{z}_2 \tag{8.6.7}$$

so for any $0 \le \lambda \le 1$,

$$\mathbf{z}_1 \le \lambda\boldsymbol{\theta}^{(1)} + (1-\lambda)\boldsymbol{\theta}^{(2)} \le \mathbf{z}_2 \tag{8.6.8}$$

and the whole of the line joining $\boldsymbol{\theta}^{(1)}$ and $\boldsymbol{\theta}^{(2)}$ is in the set. The same goes for a more complicated polyhedron D formed by more than $2p$ hyperplanes, not necessarily in pairs.

Assume that distance (vector length) as defined above is appropriate to measure the uncertainty in $\boldsymbol{\theta}$. (If not, $(\boldsymbol{\theta}^T W\boldsymbol{\theta})^{1/2}$ with a specified weighting matrix W can be accommodated by transforming $\boldsymbol{\theta}$ to $\boldsymbol{\theta}^* \triangleq G\boldsymbol{\theta}$ where G^TG is W, so that $\boldsymbol{\theta}^{*T}\boldsymbol{\theta}^*$ is $\boldsymbol{\theta}^T W\boldsymbol{\theta}$.) From the discussion just before (8.6.6) we see that the hyperplanes (8.6.2) are $2r_t/\|\mathbf{u}_t\|$ apart. It follows that if we have any choice

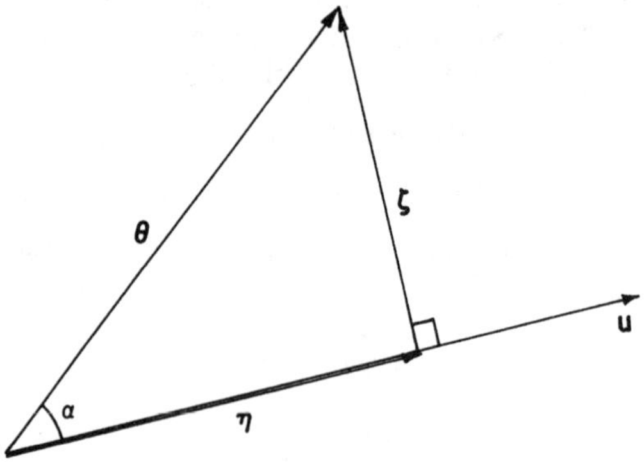

Fig. 8.6.2 Orthogonal projection.

in $\mathbf{u}_t$, the best, viewed in isolation, is that which maximises $\|\mathbf{u}_t\|$ and so brings the hyperplanes as close together as possible. The distance $2r_t/\|\mathbf{u}_t\|$ can be interpreted as a noise: signal ratio; the best choice of signal maximises signal:noise ratio, unsurprisingly.

We must return from this diversion into experiment design and see how the parameter-bounding region D can be calculated. The principle is very simple; instead of having to find $\hat{\boldsymbol{\theta}}$ to minimise some risk, log likelihood or l.s. cost, we define D by listing the constraints (8.6.2) up to date. For some applications like toleranced prediction, discussed later, this is good enough, but for others a more concise description of D is essential. The number of vertices of D_t rises far more rapidly than t, so listing them would not be practicable. It would also be difficult to update a vertex list observation by observation. We are thus led to look for some easily specified and updated approximation to D.

8.6.2 Recursive Parameter-Bounding Algorithm: Ellipsoidal Bound

An appealing idea is to use an ellipsoid

$$E_t:(\boldsymbol{\theta}-\hat{\boldsymbol{\theta}}_t)^{\mathrm{T}}P_t^{-1}(\boldsymbol{\theta}-\hat{\boldsymbol{\theta}}_t)\leq 1 \tag{8.6.9}$$

in place of D_t, with the centre $\hat{\boldsymbol{\theta}}_t$ and symmetric positive-definite matrix P_t adjusted to fit D_t as closely as possible. As $\hat{\boldsymbol{\theta}}_t$ and P_t^{-1} have p and $p(p+1)/2$ parameters, respectively, we need only update a fixed and quite small number of parameters. What is more, the updating turns out to be reasonably uncomplicated, as demonstrated by Schweppe (1968) for state estimation and Fogel and Huang (1982) for identification.

The algorithm finds an ellipsoid E_t which includes all $\boldsymbol{\theta}$ contained in both E_{t-1} and the region F_t between the hyperplanes due to the latest observation. That is, E_t contains all parameter values compatible with the latest observation and, through E_{t-1}, all earlier ones. Since the intersection of E_{t-1} and F_t is not an ellipsoid, E_t also contains some values incompatible with the observations. It is pessimistic about the uncertainty in $\boldsymbol{\theta}$, and gives an outer bound on D_t. Most of the updating process can be followed easily for the vector-observation case, with the model

$$\mathbf{y}_t = U_t\boldsymbol{\theta}+\mathbf{v}_t, \qquad \mathbf{v}_t^{\mathrm{T}}R_t^{-1}\mathbf{v}_t\leq 1, \qquad t=1,2,\ldots \tag{8.6.10}$$

Here we generalise the noise constraint to another ellipsoid, its defining matrix R_t becoming r_t^2 in (8.6.1). Figure 8.6.3 shows the updating of the bounding ellipsoid for $\boldsymbol{\theta}$. Any $\boldsymbol{\theta}$ in or on both E_{t-1} and F_t satisfies

$$(\boldsymbol{\theta}-\hat{\boldsymbol{\theta}}_{t-1})^{\mathrm{T}}P_{t-1}^{-1}(\boldsymbol{\theta}-\hat{\boldsymbol{\theta}}_{t-1})\leq 1, \qquad (\mathbf{y}_t-U_t\boldsymbol{\theta})^{\mathrm{T}}R_t^{-1}(\mathbf{y}_t-U_t\boldsymbol{\theta})\leq 1 \tag{8.6.11}$$

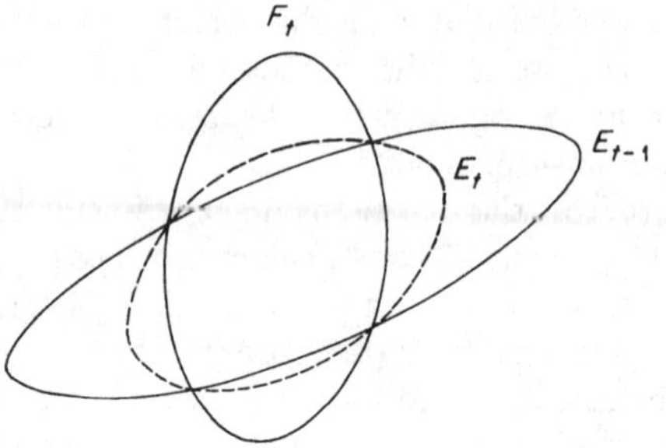

Fig. 8.6.3 Updating of ellipsoidal outer bound for $\boldsymbol{\theta}$.

so for any non-negative ρ_t,

$$(\boldsymbol{\theta} - \hat{\boldsymbol{\theta}}_{t-1})^{\mathrm{T}} P_{t-1}^{-1} (\boldsymbol{\theta} - \hat{\boldsymbol{\theta}}_{t-1}) + \rho_t (\mathbf{y}_t - U_t \boldsymbol{\theta})^{\mathrm{T}} R_t^{-1} (\mathbf{y}_t - U_t \boldsymbol{\theta}) \leq 1 + \rho_t \quad (8.6.12)$$

The left-hand side is quadratic in $\boldsymbol{\theta}$ and can be rearranged into the form (8.6.9), to give us E_t. Writing $U_t\boldsymbol{\theta}$ as $U_t\hat{\boldsymbol{\theta}}_{t-1} + U_t(\boldsymbol{\theta} - \hat{\boldsymbol{\theta}}_{t-1})$ and collecting terms, we find

$$(\boldsymbol{\theta} - \hat{\boldsymbol{\theta}}_{t-1})^{\mathrm{T}} P_{t-1}'^{-1} (\boldsymbol{\theta} - \hat{\boldsymbol{\theta}}_{t-1}) + \rho_t (\mathbf{y}_t - U_t \hat{\boldsymbol{\theta}}_{t-1})^{\mathrm{T}} R_t^{-1} (\mathbf{y}_t - U_t \hat{\boldsymbol{\theta}}_{t-1})$$
$$- 2\rho_t (\mathbf{y}_t - U_t \hat{\boldsymbol{\theta}}_{t-1})^{\mathrm{T}} R_t^{-1} U_t (\boldsymbol{\theta} - \hat{\boldsymbol{\theta}}_{t-1}) \leq 1 + \rho_t \quad (8.6.13)$$

where

$$P_{t-1}'^{-1} = P_{t-1}^{-1} + \rho_t U_t^{\mathrm{T}} R_t^{-1} U_t \quad (8.6.14)$$

Already echoes of the Kalman filter and recursive l.s. are heard. In (8.6.13), both terms in $\boldsymbol{\theta} - \hat{\boldsymbol{\theta}}_{t-1}$ can be incorporated in the new quadratic by completing the square:

$$(\boldsymbol{\theta} - \hat{\boldsymbol{\theta}}_{t-1} - \rho_t P_{t-1}' U_t^{\mathrm{T}} R_t^{-1} \mathbf{v}_t)^{\mathrm{T}} P_{t-1}'^{-1} (\boldsymbol{\theta} - \hat{\boldsymbol{\theta}}_{t-1} - \rho_t P_{t-1}' U_t^{\mathrm{T}} R_t^{-1} \mathbf{v}_t)$$
$$+ \rho_t \mathbf{v}_t^{\mathrm{T}} R_t^{-1} \mathbf{v}_t - \rho_t^2 \mathbf{v}_t^{\mathrm{T}} R_t^{-1} U_t P_{t-1}' U_t^{\mathrm{T}} R_t^{-1} \mathbf{v}_t$$
$$= (\boldsymbol{\theta} - \hat{\boldsymbol{\theta}}_t)^{\mathrm{T}} P_{t-1}'^{-1} (\boldsymbol{\theta} - \hat{\boldsymbol{\theta}}_t) + \rho_t \mathbf{v}_t^{\mathrm{T}} (R_t^{-1} - \rho_t R_t^{-1} U_t P_{t-1}' U_t^{\mathrm{T}} R_t^{-1}) \mathbf{v}_t \leq 1 + \rho_t \quad (8.6.15)$$

where

$$\mathbf{v}_t = \mathbf{y}_t - U_t \hat{\boldsymbol{\theta}}_{t-1}, \qquad \hat{\boldsymbol{\theta}}_t = \hat{\boldsymbol{\theta}}_{t-1} + \rho_t P_{t-1}' U_t^{\mathrm{T}} R_t^{-1} \mathbf{v}_t \quad (8.6.16)$$

The quadratic in $\mathbf{v}_t$ in (8.6.15) is scalar, and independent of $\boldsymbol{\theta}$. We can tidy it up by substituting P_{t-1}' from (8.6.14) then applying the matrix-inversion lemma (7.3.26) backwards, giving

$$R_t^{-1} - \rho_t R_t^{-1} U_t P_{t-1}' U_t^{\mathrm{T}} R_t^{-1}$$
$$= R_t^{-1} - \rho_t R_t^{-1} U_t P_{t-1} (I + \rho_t U_t^{\mathrm{T}} R_t^{-1} U_t P_{t-1})^{-1} U_t^{\mathrm{T}} R_t^{-1}$$
$$= (R_t + \rho_t U_t P_{t-1} U_t^{\mathrm{T}})^{-1} \quad (8.6.17)$$

By transferring the tidied-up term in $\mathbf{v}_t$ to the right-hand side of (8.6.15) then dividing through by the right-hand side, we find that the updated ellipsoid is (8.6.9) with

$$P_t = (1 + \rho_t - \rho_t \mathbf{v}_t^{\mathrm{T}}(R_t + \rho_t U_t P_{t-1} U_t^{\mathrm{T}})^{-1}\mathbf{v}_t)P'_{t-1} \tag{8.6.18}$$

We have yet to choose ρ_t to make E_t as tight a bound as possible. With D_t unknown, tightening E_t has to be interpreted as making E_t small. The volume of E_t is proportional to $|P_t|^{1/2}$, as follows. As P_t is positive-definite, all its eigenvalues λ_1 to λ_p are positive and its eigenvectors $\mathbf{m}_1$ to $\mathbf{m}_p$ are orthonormal if suitably scaled, i.e. $\mathbf{m}_i^{\mathrm{T}}\mathbf{m}_j$ is zero for $i \neq j$ and 1 for $i = j$. If

$$\Lambda \triangleq \mathrm{diag}(\lambda_1, \lambda_2, \ldots, \lambda_p), \qquad M = [\mathbf{m}_1\ \mathbf{m}_2\ \ldots\ \mathbf{m}_p] \tag{8.6.19}$$

then by definition of eigenvalues and eigenvectors

$$P_t M = M\Lambda \tag{8.6.20}$$

and M^{T} is M^{-1}. The description of E_t is simpler in terms of ξ, defined as $M^{\mathrm{T}}(\boldsymbol{\theta} - \hat{\boldsymbol{\theta}}_t)$. Since

$$\|\xi\|^2 = \xi^{\mathrm{T}}\xi = (\boldsymbol{\theta} - \hat{\boldsymbol{\theta}}_t)^{\mathrm{T}} M M^{\mathrm{T}}(\boldsymbol{\theta} - \hat{\boldsymbol{\theta}}_t) = \|\boldsymbol{\theta} - \hat{\boldsymbol{\theta}}_t\|^2 \tag{8.6.21}$$

this change of co-ordinates just shifts the origin to $\hat{\boldsymbol{\theta}}_t$ and rotates the axes, without altering distances measured from the new origin. Now

$$(\boldsymbol{\theta} - \hat{\boldsymbol{\theta}}_t)^{\mathrm{T}} P_t^{-1}(\boldsymbol{\theta} - \hat{\boldsymbol{\theta}}_t) = \xi^{\mathrm{T}} M^{-1} P_t^{-1} M \xi = \xi^{\mathrm{T}}\Lambda^{-1}\xi \leq 1 \tag{8.6.22}$$

so E_t is centred at the ξ origin, with its axes aligned with the ξ axes and of half-length $\lambda_i^{1/2}$, $i = 1, 2, \ldots, p$. Thinking of E_t as a unit hypersphere $\xi^{\mathrm{T}}\xi \leq 1$ squashed or stretched successively in each ξ_i direction by a factor $\lambda_i^{1/2}$, we see its volume is proportional to

$$V_t = \prod_{i=1}^{p} \lambda_i^{1/2} = |\Lambda|^{1/2} = |M^{-1} P_t M|^{1/2} = |M|^{-1/2}|P_t|^{1/2}|M|^{1/2} = |P_t|^{1/2} \tag{8.6.23}$$

We must express $|P_t|$ as a function of ρ_t. First we tackle $|P'_{t-1}|$. The matrix-inversion lemma turns (8.6.14) into

$$P'_{t-1} = P_{t-1} - \rho_t P_{t-1} U_t^{\mathrm{T}}(R_t + \rho_t U_t P_{t-1} U_t^{\mathrm{T}})^{-1} U_t P_{t-1} \tag{8.6.24}$$

For scalar observations, we can write this as

$$P'_{t-1} = (I - \rho_t P_{t-1}\mathbf{u}_t\mathbf{u}_t^{\mathrm{T}}/(r_t^2 + \rho_t \mathbf{u}_t^{\mathrm{T}} P_{t-1}\mathbf{u}_t))P_{t-1} \tag{8.6.25}$$

then find $|P'_{t-1}|$ via the lemma (Goodwin and Payne, 1977, Appendix E)

$$|I + \mathbf{b}\mathbf{c}^{\mathrm{T}}| = |1 + \mathbf{c}^{\mathrm{T}}\mathbf{b}| \tag{8.6.26}$$

with $\rho P_{t-1}\mathbf{u}_t$ for $\mathbf{b}$ and $-\mathbf{u}_t/(r_t^2+\rho_t\mathbf{u}_t^T P_{t-1}\mathbf{u}_t)$ for $\mathbf{c}$. Denoting $\mathbf{u}_t^T P_{t-1}\mathbf{u}_t$ by g_t,

$$|P'_{t-1}| \equiv (1-\rho_t g_t/(r_t^2+\rho_t g_t))|P_{t-1}| \tag{8.6.27}$$

Taking determinants of P'_{t-1} and P_t in (8.6.18), we then have $|P_t|$ in terms of ρ_t. The ρ_t which minimises V_t follows by routine algebra, setting $\partial V_t/\partial\rho_t$ or $\partial|P_t|/\partial\rho_t$ to zero. It is the positive root of

$$(p-1)g_t^2\rho_t^2 + ((2p-1)r_t^2 - g_t + v_t^2)g_t\rho_t + r_t^2(p(r_t^2-v_t^2)-g_t) = 0 \tag{8.6.28}$$

If both roots are negative or complex, we infer that E_{t-1} and F_t do not intersect, in other words y_t and $\mathbf{u}_t$ are jointly incompatible with E_{t-1} and the model at the assumed r_t. This explicit warning allows us to revise the model or r_t, or set ρ_t to zero and ignore y_t as an outlier.

Recursive Ellipsoidal-Outer-Bounding Algorithm for Parameter-Bounding Identification (Fogel and Huang, 1982) With model (8.6.1), the algorithm updates the ellipsoid

$$E{:}(\boldsymbol{\theta}-\hat{\boldsymbol{\theta}})^T P^{-1}(\boldsymbol{\theta}-\hat{\boldsymbol{\theta}}) \le 1$$

as the outer bound for the feasible-parameter region.

With $\hat{\boldsymbol{\theta}}_0, P_0$ specified (e.g. $\hat{\boldsymbol{\theta}}_0=\mathbf{0}$, $P_0=\alpha I$, $\alpha \gg 1$),

(i) Calculate $g_t=\mathbf{u}_t^T P_{t-1}\mathbf{u}_t$ and $v_t = y_t - \mathbf{u}_t^T\hat{\boldsymbol{\theta}}_{t-1}$.

(ii) Find ρ_t as positive root of (8.6.28); if no positive real root, set ρ_t to zero or stop and review model.

(iii) Calculate P'_{t-1} from (8.6.25).

(iv) Set $\hat{\boldsymbol{\theta}}_t=\hat{\boldsymbol{\theta}}_{t-1}+\rho_t P'_{t-1}\mathbf{u}_t v_t/r_t^2$ (scalar-y version of (8.6.16)).

(v) Set $P_t=(1+\rho_t-\rho_t v_t^2/(r_t^2+\rho_t g_t))P'_{t-1}$ (scalar-y version of (8.6.18)).

Belforte and Bona (1985) have recently pointed out that the bound E_t can be tightened, at observations for which only one of the two hyperplanes forming F_t intersects E_{t-1}, by replacing the one which does not intersect E_{t-1} by a parallel hyperplane tangent to E_{t-1} before computing ρ_t.

Example 8.6.2 We apply the recursive algorithm to the problem of Example 8.6.1 with $\hat{\boldsymbol{\theta}}_0$ zero and $P_0=10^6 I$. The first three steps give, rounded to three figures,

t	g_t	v_t	ρ_t	$\dfrac{P_t}{P'_{t-1}}$	P_t	$\hat{\theta}_t$
1	10^6	3	0.500	1.50	$\begin{bmatrix} 300 & 0 & 0 \\ 0 & 1.50\times 10^6 & 0 \\ 0 & 0 & 1.50\times 10^6 \end{bmatrix}$	$\begin{bmatrix} 3.00 \\ 0 \\ 0 \end{bmatrix}$
2	6.09×10^4	56.0	0.472	1.42	$\begin{bmatrix} 424 & -2.09\times 10^3 & -209 \\ -2.09\times 10^3 & 3.88\times 10^4 & -2.09\times 10^5 \\ -209 & -2.09\times 10^5 & 2.11\times 10^6 \end{bmatrix}$	$\begin{bmatrix} 3.27 \\ 275 \\ 27.5 \end{bmatrix}$
3	5.03×10^3	-17.4	0.445	1.39	$\begin{bmatrix} 540 & -3.28\times 10^3 & 9.33\times 10^3 \\ -3.28\times 10^3 & 5.09\times 10^4 & -2.16\times 10^5 \\ 9.33\times 10^3 & -2.16\times 10^5 & 1.03\times 10^6 \end{bmatrix}$	$\begin{bmatrix} 4.70 \\ 286 \\ -254 \end{bmatrix}$

As in recursive l.s., updating P takes the bulk of the computing. The extra work to find ρ is insignificant.

An interesting thing happens if we take $10^4 I$ as P_0. The first step is little affected and in the second we get a reasonable-looking ρ_2, 0.333, but according to (8.6.18) P_2 is $-1.29P'_1$. Negative principal-diagonal elements appear in P_2, and the algorithm breaks down (E_2 is not an ellipsoid). The explanation is that F_2 does not intersect E_1, which is too small because E_0 was too small. We must make E_0 large enough to be sure it contains all $\boldsymbol{\theta}$ compatible with the observations. △

An alternative to bounding the parameters by an ellipsoid is to compute a "box" of bounds on the individual parameters (Milanese and Belforte, 1982). The computation comprises a number of linear programming problems, one per bound.

Our look at parameter bounding ends with an example of an application for which the feasible-parameter region need not be explicitly calculated.

*8.6.3 Toleranced Prediction

The output y_s due to a specified $\mathbf{u}_s$ in a bounded-noise model, with $\boldsymbol{\theta}$ known to be within a region D, is predicted by stating bounds between which it will fall. For a purpose like alarm scanning or checking whether the output will be within specification, such a prediction is attractive, since it states definitely whether, according to the model, an alarm condition could arise or whether the output specification will certainly be met. Similar comments apply whenever the object of the prediction is to facilitate a yes/no decision.

With

$$\min_{\boldsymbol{\theta}\in D}(\mathbf{u}_s^{\mathrm{T}}\boldsymbol{\theta}) - r_s \le y_s \le \max_{\boldsymbol{\theta}\in D}(\mathbf{u}_s^{\mathrm{T}}\boldsymbol{\theta}) + r_s \qquad (8.6.29)$$

y_s is predicted by finding the extrema of a linear function of $\boldsymbol{\theta}$, subject to linear inequality constraints if D is a polyhedron. This is the standard linear programming problem (Hadley, 1961; Luenberger, 1973) for which efficient methods able to handle hundreds of constraints exist.

Example 8.6.3 We want to predict the target position at $\tau = 2$ from the results of Example 8.6.1. To stay on Fig. 8.6.1 we assume $x_0 = 5$. In practice we would keep x_0 free, of course. At $\tau = 2$

$$\min_{\theta \in D}(5 + 2\theta_2 + 2\theta_3) - r_s \leq y_s \leq \max_{\theta \in D}(5 + 2\theta_2 + 2\theta_3) + r_s$$

so we need only find the extrema of $2\theta_2 + 2\theta_3$ in the feasible region. They are where lines of slope -1 touch the ends of D, at $(\theta_2, \theta_3) = (261.25, -18.75)$ and $(207.75, 122.5)$, the southernmost and northernmost vertices in Fig. 8.6.1. The corresponding range for y_s, with $r_s = 10$, is from 480 to 675.5. △

8.7 IDENTIFICATION OF MULTIVARIABLE SYSTEMS

Our limited aim in this section is to introduce two new aspects of identification opened up by m.i.m.o. systems: choice of parameterisation and choice of cost function. One of the most striking things about the literature on identification is its concentration on s.i.s.o. systems (or occasionally m.i.s.o. or s.i.m.o., which raise few new issues). Only a small proportion deals with m.i.m.o. identification, in spite of its importance, and this book is no exception. The reason is not mere faintheartedness. As well as generally having more parameters than s.i.s.o. systems, m.i.m.o. systems raise substantial new problems. One, considered later, is how to base a cost function or risk on a vector of output variables. Other difficulties arise when we start to choose a model structure. First, our fundamental need in any identification exercise is to understand what goes on in the system well enough to judge the validity of the results and their practical implications. A m.i.m.o. system need not have many inputs and outputs for its overall behaviour to be too complex to grasp all at once. We are then forced to investigate one s.i.s.o. (or conceivably m.i.s.o.) relation at a time. Scientific method itself owes much of its analytical nature to this fact. By breaking a m.i.m.o. problem down in this way, we also sidestep another new difficulty, which is to choose an acceptable parameterisation.

8.7.1 Parameterisation of Multi-Input–Multi-Output Models

It is a non-trivial matter to decide how to parameterise a m.i.m.o. model even when its general type has been selected. The difficulty is best appreciated through examples.

Example 8.7.1 A multivariable system with input **u** and output **y** could be described by the z-transform model

$$\mathbf{Y}(z^{-1}) = G(z^{-1})\mathbf{U}(z^{-1}) + \text{noise}$$

where G is a matrix of rational transfer functions. There is more than one way to write this model.

(i) If we start with a difference-equation model and take z-transforms, getting an a.r.m.a.x. model

$$A_1(z^{-1})\mathbf{Y}(z^{-1}) = B_1(z^{-1})\mathbf{U}(z^{-1}) + \text{noise}$$

with A_1 and B_1 matrices of polynomials in z^{-1} and A_1 non-singular except at isolated values of z^{-1}, we obtain the *left matrix-fraction description* (m.f.d.) (Kailath, 1980; Goodwin and Payne, 1977)

$$\mathbf{Y}(z^{-1}) = A_1^{-1}(z^{-1})B_1(z^{-1})\mathbf{U}(z^{-1}) + \text{noise}$$

For example, we might have a two-input, two-output model with

$$A_1(z^{-1}) = \begin{bmatrix} 1 - 0.32z^{-1} & 1 - 0.8z^{-1} \\ 1 - 0.88z^{-1} & 1 - 0.4z^{-1} \end{bmatrix}$$

$$B_1(z^{-1}) = \begin{bmatrix} 1 + 1.6z^{-1} & 1 - 1.8z^{-1} - 1.6z^{-2} \\ 1 + 2z^{-1} & 1 - 1.4z^{-1} - 2z^{-2} \end{bmatrix}$$

The corresponding rational-transfer-function matrix is

$$\begin{aligned} G(z^{-1}) &\equiv A_1^{-1}(z^{-1})B_1(z^{-1}) \\ &= \begin{bmatrix} z^{-1} & -z^{-2} \\ 1 + 0.8z^{-1} & 1 - 1.6z^{-1} - 0.8z^{-2} \end{bmatrix} \Big/ (1 - 0.6z^{-1}) \end{aligned}$$

with many fewer parameters than we might have expected from the degrees of elements in A_1 and B_1. Nothing similar can happen in s.i.s.o. models where, in the absence of pole–zero cancellation (which incidentally does not occur here), all coefficients in A_1 and B_1 appear in G.

(ii) An equally valid model is the *right m.f.d.*

$$\mathbf{Y}(z^{-1}) = B_2(z^{-1})A_2^{-1}(z^{-1})\mathbf{U}(z^{-1}) + \text{noise}$$

If

$$A_2(z^{-1}) = \begin{bmatrix} 1 - 0.6z^{-1} & z^{-1} \\ 0 & 1 \end{bmatrix}, \qquad B_2(z^{-1}) = \begin{bmatrix} z^{-1} & 0 \\ 1 + 0.8z^{-1} & 1 \end{bmatrix}$$

the model has exactly the same transfer-function matrix G as in (i), but A_2 and B_2 have fewer parameters than A_1 and B_1, and, depending on how far the degrees of the elements are known in advance, arguably fewer than G. Single-input–single-output systems have no such choice between two m.f.d.'s.

(iii) The right m.f.d. with

$$A_3(z^{-1}) = \begin{bmatrix} 1-0.9z^{-1} & 0.45z^{-2} \\ -0.3 & 0.5+0.15z^{-1} \end{bmatrix}$$

$$B_3(z^{-1}) = \begin{bmatrix} z^{-1} & -0.5z^{-2} \\ 0.7+0.8z^{-1} & 0.5-0.35z^{-1}-0.4z^{-2} \end{bmatrix}$$

also gives the same G is in (i) and (ii). The differences from A_2 and B_2 are due to the existence of a polynomial matrix $M(z^{-1})$, non-singular for all but isolated values of z^{-1}, such that

$$A_3 \triangleq A_2 M^{-1}, \qquad B_3 \triangleq B_2 M^{-1}$$

are still polynomial matrices. Such an M is called a *right divisor* of A_2 and B_2. It does not alter G because

$$B_3 A_3^{-1} = B_2 M^{-1} M A_2^{-1} = B_2 A_2^{-1}$$

We define the degree of the m.f.d. as the degree of $|A|$, since the common denominator polynomial of the elements of G is $|A|$, arising from A^{-1}. In model (iii),

$$|A_3| = |A_2|/|M| \qquad \text{so} \quad \deg|A_3| = \deg|A_2| - \deg|M|$$

and going from A_2 and B_2 to A_3 and B_3 reduces the m.f.d. degree unless $|M|$ is of degree zero, i.e. a constant. That is why M is called a divisor. In the present case

$$M(z^{-1}) = \begin{bmatrix} 1+0.3z^{-1} & z^{-1} \\ 0.6 & 2 \end{bmatrix}$$

so $|M|$ is 2, of degree zero, and the degree of the m.f.d. is not reduced from (ii) to (iii). A polynomial matrix with constant determinant is called *unimodular*.

(iv) A left m.f.d. of degree one less than in (i) and the same G can be obtained using the left divisor

$$M(z^{-1}) = \begin{bmatrix} 1 & 0 \\ 1-z^{-1} & z^{-1} \end{bmatrix}$$

which is not unimodular since $|M|$ is z^{-1}. The details are left to Problem 8.7

△

The example shows that redundancy in an m.f.d. multivariable model, introducing unnecessary parameters and consequent ill-conditioning, may be non-trivial to detect even if the degree of the matrix-fraction description is as small as possible, as in (ii), (iii) and (iv). An infinity of apparently different models has the same input–output behaviour. We must restrict the model structure enough to make it uniquely identifiable if we are interested in the parameters themselves, or minimal (in number of parameters) if we are only interested in the model output but do not want to risk ill-conditioned estimation. The choice of structure is further complicated by doubt over the minimal number of parameters to fit the actual system adequately, and by the desire for a structure which suits the application, e.g. control design. Selection of a parameterisation is discussed by Glover and Willems (1974), Denham (1974) and Gevers and Wertz (1984). Analogous comments apply to the other main family of multivariable models, state-space descriptions.

These examples and references also indicate that a large body of multivariable theory underlies m.i.m.o. identification. Its unfamiliarity is another reason why we cannot pursue m.i.m.o. systems further.

8.7.2 Cost Functions for Multivariable Models

Many s.i.s.o. identification algorithms minimise a scalar l.s. cost function of the errors between observed and model outputs. The vector-output generalisation of the sum of squares of output errors is the matrix

$$S_N = \sum_{t=1}^{N} (\mathbf{y}_t - \hat{\mathbf{y}}_t)(\mathbf{y}_t - \hat{\mathbf{y}}_t)^{\mathrm{T}} \tag{8.7.1}$$

so we are presented with an extra choice, how to derive from S_N a scalar V_N to minimise. We might decide to weigh all elements of $\mathbf{y}_t - \hat{\mathbf{y}}_t$ equally and use

$$V_N = \sum_{t=1}^{N} \sum_{i=1}^{p} \{(y_{ti} - \hat{y}_{ti})^2\} = \operatorname{tr} S_N \tag{8.7.2}$$

Weighted l.s. with a weighting matrix W can also be implemented through

$$\begin{aligned} V_N &= \sum_{t=1}^{N} (\mathbf{y}_t - \hat{\mathbf{y}}_t)^{\mathrm{T}} W (\mathbf{y}_t - \hat{\mathbf{y}}_t) = \sum_{t=1}^{N} \operatorname{tr} W (\mathbf{y}_t - \hat{\mathbf{y}}_t)(\mathbf{y}_t - \hat{\mathbf{y}}_t)^{\mathrm{T}} \\ &= \operatorname{tr} W S_N \end{aligned} \tag{8.7.3}$$

Finally, the cost function

$$V_N = |S_N/N| \tag{8.7.4}$$

turns out to have considerable appeal. If we assume that the error sequence $\{\mathbf{y} - \hat{\mathbf{y}}\}$ is composed of independent, Gaussian, zero-mean random variables with constant but unknown covariance, the parameter estimates which minimise $|S_N/N|$ are the m.l. estimates. They are therefore asymptotically efficient. Furthermore, S_N/N is the m.l. estimate of $\text{cov}(\mathbf{y} - \hat{\mathbf{y}})$ (Ljung, 1976).

8.8 IDENTIFICATION OF NON-LINEAR SYSTEMS

Techniques for analysing linear dynamical systems are effective and relatively easy to use. We can swap readily from one to another as convenience dictates, from Laplace transforms to state equations say, or from difference equations to z-transform transfer functions. We rapidly acquire an intuitive understanding of linear systems and are happy to think in terms of poles and zeros, power spectra and correlation functions, step and impulse responses. The contrast with analysis of non-linear dynamics (Vidyasagar, 1978) is sharp. Methods for non-linear systems tend to apply only to restricted classes, to give only partial or approximate information and to be cumbersome. The reason is that non-linear behaviour is vastly diverse and complex. Linearity and time-invariance impose tight constraints on possible behaviour. If we say a linear system is stable, we don't have to add ifs and buts about initial conditions or size of disturbance. We need not worry about limit cycles, bifurcation and chaos (Mees and Sparrow, 1981). Superposition makes it routine, almost trivial, to relate linear-system response to excitation and initial conditions. An impulse response or transfer function says all there is to say about the input–output behaviour of a linear, time-invariant system. Jump resonance, subharmonic oscillation and generation of harmonics do not occur.

We might expect that non-linear systems are also generally much harder to identify than linear systems, and this is so. A coherent body of economical, well tried and widely applicable identification technique for non-linear systems does not exist. The rest of this section reviews briefly some of the difficulties in identification posed by non-linearity, and some situations where progress can be made.

8.8.1 Volterra-Series Model

The first difficulty, once non-linearity has been detected, is to find a versatile but economical form of model. We know that for linear s.i.s.o. systems versatility conflicts with economy. For example, an impulse response (estimated as a discrete-time u.p.r.) will cope with any such system without the need to determine the model order and dead time explicitly, but requires many more coefficients than an equivalent rational transfer function. The conflict is worse for non-linear systems. The non-linear generalisation of the input–output convolution

$$y(t) = \int_0^\infty h(\tau)u(t-\tau)\,d\tau \qquad (8.8.1)$$

is the *Volterra series* (Schetzen, 1980)

$$y(t) = \int_0^\infty h_1(\tau)u(t-\tau)\,d\tau + \int_0^\infty \int_0^\infty h_2(\tau_1,\tau_2)u(t-\tau_1)u(t-\tau_2)\,d\tau_1\,d\tau_2 + \cdots$$
$$+ \int_0^\infty \cdots \int_0^\infty h_r(\tau_1,\tau_2,\ldots,\tau_r)\prod_{i=1}^{r} u(t-\tau_i)\,d\tau_i + \cdots \infty \qquad (8.8.2)$$

Much effort has gone into ways to identify the *Volterra kernels* h_1, h_2, etc., mostly based on generalising the correlation methods of Chapter 3 (Billings, 1980). For example, for a Gaussian white-noise input (but not *any* white noise)

$$E[u(t-\tau)y(t)] = h_1(\tau)$$
$$E[u(t-\tau_1)u(t-\tau_2)y(t)] = \delta(\tau_1-\tau_2)Ey(t) + 2h_2(\tau_1,\tau_2) \qquad (8.8.3)$$

and with ergodic signals time averages can be computed to approximate the left-hand sides.

A representation as general as (8.8.2) would only be contemplated if the system were too poorly understood to suggest a more specific model. However, in those circumstances only an extreme optimist would expect much from a model like (8.8.2). Unconstrained estimation of the nth Volterra kernel is impracticable for $n > 2$ and uninviting for $n = 2$, simply because of the number of points (or component functions) required to specify a function of n independent variables. If a typical impulse response takes about 15–30 points to describe it in the absence of good prior information on its shape, h_2 might take several hundred and h_3 many thousands. Large quantities of observations and computing are required to estimate the points adequately. Even then, the prospects of extracting any meaning from the results are not good. It is far from trivial to pick out the important features from a linear-system u.p.r. estimate in the presence of noise. Interpretation of a detailed and noise-affected $h_2(\tau_1,\tau_2)$ is much harder, and the crucial step of inferring the form of a parametric model from the unparameterised results is daunting. Simple non-linearities such as hard saturation give rise to non-simple kernel behaviour, to add to this difficulty. Marmarelis and Marmarelis (1978) present case studies which illustrate these points well. A number of ways of alleviating the difficulties of Volterra-series identification have been investigated, including use of a deterministic pseudo-noise input (Barker and Davy, 1978) or a mixture of sinusoids together with a rational-transfer-function version of the series (Lawrence, 1981), but nothing can nullify the essential wastefulness of the representation.

Faced with these difficulties, we shall pass on to more restricted models.

8.8.2 Block-Oriented Models

Many systems display significant non-linearity only in one or two memoryless relations, and can be represented as cascades of linear dynamical and non-linear instantaneous sub-systems. Figure 8.8.1 shows one *block-oriented* model of this sort. With a suitable input, the contents of each block in this model can be identified from cross-correlation functions by exploiting the property of *separability* (Billings and Fakhouri, 1978 and 1982). If two processes $x_1(t)$ and $x_2(t)$ have a joint p.d.f. $p(x_1(t_1), x_2(t_2))$ such that

$$\begin{aligned}\int_{-\infty}^{\infty} & x_1(t_1)p(x_1(t_1), x_2(t_2))\,dx_1(t_1)\\ &= \int_{-\infty}^{\infty} x_1(t_1)p(x_1(t_1) \mid x_2(t_2))p(x_2(t_2))\,dx_1(t_1)\\ &= p(x_2(t_2))E[x_1(t_1) \mid x_2(t_2)] = g_1(x_2(t_2))g_2(t_2 - t_1) \qquad (8.8.4)\end{aligned}$$

then $x_1(t_1)$ is said to be separable with respect to $x_2(t_2)$. Among others, jointly Gaussian processes are separable. We can find expressions for g_1 and g_2 in the case where x_1 and x_2 are the same process x. Putting t_1 equal to t_2, (8.8.4) gives

$$g_1(x(t_2))g_2(0) = p(x(t_2))x(t_2) \qquad (8.8.5)$$

and so the a.c.f. of x at lag $t_2 - t_1$ is

$$\begin{aligned}r_{xx}(t_2 - t_1) &= \int_{-\infty}^{\infty} x(t_2) \int_{-\infty}^{\infty} x(t_1)p(x(t_1), x(t_2))\,dx(t_1)\,dx(t_2)\\ &= \int_{-\infty}^{\infty} x(t_2)g_1(x(t_2))g_2(t_2 - t_1)\,dx(t_2)\\ &= \int_{-\infty}^{\infty} x^2(t_2)p(x(t_2))(g_2(t_2 - t_1)/g_2(0))\,dx(t_2)\\ &= (g_2(t_2 - t_1)r_{xx}(0))/g_2(0) \qquad (8.8.6)\end{aligned}$$

Separability allows us to replace the instantaneous non-linearity by an equivalent gain, in equations analogous to the Wiener–Hopf equation

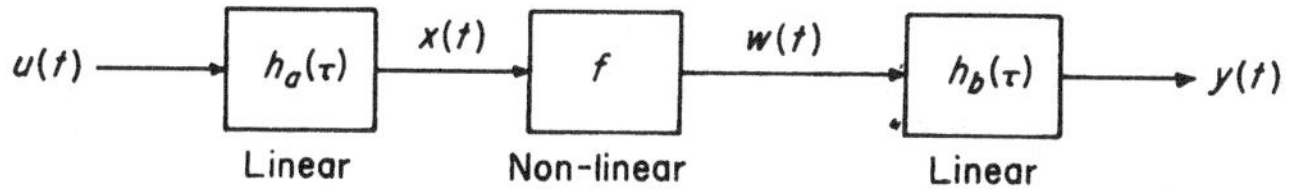

Fig. 8.8.1 Block-oriented cascade model.

(Section 3.1). Assuming $x(t)$ in Fig. 8.8.1 to be separable with respect to itself, the cross-correlation across the non-linearity is

$$r_{xw}(t_2-t_1)=\int_{-\infty}^{\infty}\int_{-\infty}^{\infty} x(t_1)f(x(t_2))p(x(t_1),x(t_2))\,dx(t_1)\,dx(t_2)$$

$$=\int_{-\infty}^{\infty} f(x(t_2))g_1(x(t_2))g_2(t_2-t_1)\,dx(t_2) \qquad (8.8.7)$$

With $g_1(x(t_2))$ from (8.8.5) and $g_2(t_2-t_1)$ from (8.8.6),

$$r_{xw}(t_2-t_1)=\int_{-\infty}^{\infty} f(x(t_2))\frac{p(x(t_2))x(t_2)}{g_2(0)}\frac{g_2(0)r_{xx}(t_2-t_1)}{r_{xx}(0)}\,dx(t_2)$$

$$=\frac{r_{xx}(t_2-t_1)}{r_{xx}(0)}E[x(t_2)f(x(t_2))]=cr_{xx}(t_2-t_1) \qquad (8.8.8)$$

where c depends on $f(\cdot)$ and the p.d.f. of $x(t)$ but not on the time-structure of $x(t)$.

According to (8.8.8) the non-linearity has the same effect on r_{xw} as a gain c. This enables us eventually to write r_{uy} in terms of h_a and h_b, the linear-block impulse responses in Fig. 8.8.1, as follows. Providing linear operators do not destroy separability, so that $u(t)$ is separable with respect to $x(t)$, the Fourier transform of (8.8.8) implies that

$$R_{uw}(j\omega)=R_{xw}(j\omega)/H_a(j\omega)=cR_{xx}(j\omega)/H_a(j\omega)=cR_{ux}(j\omega) \qquad (8.8.9)$$

Back in the time domain, (8.8.9) says we can replace r_{uw} by cr_{ux}, so

$$r_{uy}(\tau)=\int_0^{\infty} h_b(\tau_1)r_{uw}(\tau-\tau_1)\,d\tau_1$$

$$=c\int_0^{\infty} h_b(\tau_1)\int_0^{\infty} h_a(\tau_2)r_{uu}(\tau-\tau_1-\tau_2)\,d\tau_2\,d\tau_1 \qquad (8.8.10)$$

where the last step follows from the Wiener–Hopf equation. If we employ an input which is white as well as separable, $r_{uu}(\tau-\tau_1-\tau_2)$ is zero except at $\tau_2=\tau-\tau_1$, where it is σ_u^2, and

$$r_{uy}(\tau)=c\sigma_u^2\int_0^{\infty} h_b(\tau_1)h_a(\tau-\tau_1)\,d\tau_1 \qquad (8.8.11)$$

The integral is the convolution of h_a and h_b, that is, the overall impulse response of the linear sections cascaded. The non-linearity shows up only as gain c. The overall linear dynamics can therefore be identified, to within a scale factor, by cross-correlation just as in Chapter 3, but with a more restricted choice of input, separable as well as white.

Billings and Fakhouri (1978) show that with suitable assumptions cross-correlating $u^2(t)$ with $y(t)$ gives

$$r_{u^2y}(\tau) = \text{const} \times \int_0^\infty h_b(\tau_1)h_a^2(\tau - \tau_1)\,d\tau_1 \tag{8.8.12}$$

and that the z-transform transfer functions of the linear sections can then be estimated straightforwardly from the transfer-function products corresponding to the convolutions in (8.8.11) and (8.8.12). An important byproduct of this approach is a model-structure test. If the first linear block is absent, i.e. $h_a(t)$ is $\delta(t)$, (8.8.11) and (8.8.12) give the same result to within a scale factor, and if the second linear block is absent, the result of (8.8.12) is a constant times the square of that of (8.8.11).

8.8.3 Regression Models for Non-Linear Systems

As observed in Chapter 4, a regression model for o.l.s. or its variants may be non-linear in the variables, although it must be linear in the parameters. Non-linearity may, however, cause a new difficulty in noise modelling (Billings,

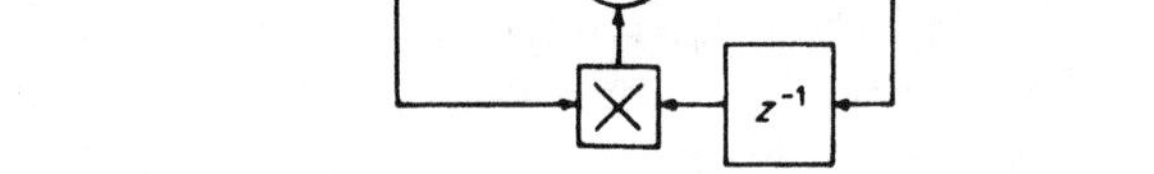

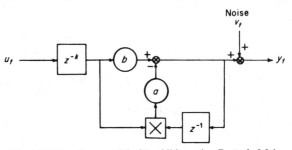

Fig. 8.8.2 Non-linear model with additive noise, Example 8.8.1.

1984). Noise which is physically additive at the output will appear non-additively in the regression equation whenever earlier values of the *noise-free* output enter non-additively into the regressors.

Example 8.8.1 The system modelled by Fig. 8.8.2 is described by

$$y_t = bu_{t-k} - au_{t-k}(y_{t-1} - v_{t-1}) + v_t$$

and has a regressor vector $[u_{t-k} \quad u_{t-k}(y_{t-1} - v_{t-1})]^T$ at instant t. The noise is partly additive, partly multiplicative in the regression equation although physically additive.

The regressor containing the unmeasured v_{t-1} might be replaced by a recursive estimate

$$u_{t-k}(y_{t-1} - \hat{v}_{t-1}) = u_{t-k}(\hat{b}u_{t-k-1} - \hat{a}u_{t-k-1}(y_{t-2} - \hat{v}_{t-2}))$$

where $\hat{a}$, $\hat{b}$ and $y_{t-2} - \hat{v}_{t-2}$ come from the previous step. An interesting alternative would be to treat $b + av_{t-1}$ as a time-varying first parameter, with the regressor vector $[u_{t-k} \quad u_{t-k}y_{t-1}]^T$. The temptation to extemporise is considerable. △

8.9 SIMULTANEOUS ESTIMATION OF PARAMETERS AND STATE

The standard state-estimation problem (Kalman, 1960; Jazwinski, 1970; Maybeck, 1979) is to estimate $\mathbf{x}_t$ recursively from known inputs $\{\mathbf{u}\}$ up to $\mathbf{u}_{t-1}$ and observations $\{\mathbf{y}\}$ up to $\mathbf{y}_t$, with given initial conditions ($\hat{\mathbf{x}}_0$ and its covariance) and the model

$$\begin{aligned} \mathbf{y}_t &= H_t\mathbf{x}_t + \mathbf{v}_t, & \operatorname{cov} \mathbf{v}_t &= R_t \\ \mathbf{x}_t &= \Phi_{t-1}\mathbf{x}_{t-1} + B_{t-1}\mathbf{u}_{t-1} + \Gamma_{t-1}\mathbf{w}_{t-1}, & \operatorname{cov} \mathbf{w}_{t-1} &= Q_{t-1} \end{aligned} \tag{8.9.1}$$

Matrices $H_t, \Phi_{t-1}, B_{t-1}, \Gamma_{t-1}, Q_{t-1}$ and R_t are taken as completely known. In the aerospace applications where state estimation was so successful in the 1960's, the state model often described Newtonian dynamics and most parameters were indeed well known. Even in those circumstances, filter performance can sometimes be improved by refining uncertain parameter values. An example is when approximations have been made to simplify the model, so that some parameters hide non-linearity or high-order dynamics, and vary as a result. For other applications such as process control, state estimation may well not be feasible without on-line estimation of uncertain parameters.

State estimation in general, and combined parameter and state estimation in particular, are heavily technological subjects (Bierman, 1977; Maybeck, 1982). We can afford, however, a brief look at the main ways of tackling the combined estimation problem. Estimation of Q_{t-1} and R_t on-line will not be considered, as specialised techniques tend to be employed (Mehra, 1974; Maybeck, 1982).

8.9.1 State Augmentation and Extended Kalman Filtering

The unknown parameters $\boldsymbol{\theta}$ can be regarded, as in Section 8.1.3, as state variables distinguished only by our preference to regard them as such and by their simple dynamics. They can be adjoined to the other state variables to give a state equation

$$\mathbf{x}'_t \triangleq \begin{bmatrix} \mathbf{x}_t \\ \boldsymbol{\theta}_t \end{bmatrix} = \begin{bmatrix} \Phi_{t-1} & 0 \\ 0 & I \end{bmatrix} \mathbf{x}'_{t-1} + \begin{bmatrix} B_{t-1} \\ 0 \end{bmatrix} \mathbf{u}_{t-1} + \begin{bmatrix} \Gamma_{t-1} & 0 \\ 0 & I \end{bmatrix} \mathbf{w}'_{t-1} \tag{8.9.2}$$

where $\mathbf{w}'_{t-1}$ is $\mathbf{w}_{t-1}$ augmented by non-zero elements for any elements of $\boldsymbol{\theta}$ modelled as time-varying. The augmented-state observation equation is

$$\mathbf{y}_t = [H_t \quad 0]\mathbf{x}'_t + \mathbf{v}_t \tag{8.9.3}$$

Simultaneous estimation of $\mathbf{x}_t$ and $\boldsymbol{\theta}_t$ can now go ahead, once a suitable covariance has been specified for the elements of $\mathbf{w}'_{t-1}$ which represent the changes in $\boldsymbol{\theta}$. Trial and error may be necessary in finding this covariance, but a more serious difficulty is that any unknown parameter in Φ_{t-1}, Γ_{t-1} or H_t gives rise to a non-linear term in the augmented state $\mathbf{x}'_t$ or augmented state and noise $\mathbf{w}_{t-1}$. As the standard recursive state-estimation algorithm, the Kalman filter, has covariance-updating equations which rely on linearity of the state and observation equations and additivity of the noise, we must remedy the situation by local linearisation.

Example 8.9.1 The position of the accelerating target of Example 4.1.1 is observed every 0.1 s by two instruments. At sample instant t one instrument gives $y_{1,t}$ subject to zero-mean error, uncorrelated with earlier errors. The other gives $y_{2,t}$, noise-free but affected by unknown gain and constant bias. The acceleration of the target varies unpredictably from one sample instant to the next, but the changes are assumed to have zero mean and known variance. The target position x and velocity $\dot{x}$ are to be estimated after each observation instant.

We define the state as $[x \quad \dot{x} \quad \ddot{x}]^{\mathrm{T}}$ and derive discrete-time state equations by trapezoidal integration of $\ddot{x}$ then $\dot{x}$, yielding

$$\mathbf{x}_t = \begin{bmatrix} 1 & 0.1 & 0.005 \\ 0 & 1 & 0.1 \\ 0 & 0 & 1 \end{bmatrix} \mathbf{x}_{t-1} + \begin{bmatrix} 1 & 0.05 & 0.0025 \\ 0 & 1 & 0.05 \\ 0 & 0 & 1 \end{bmatrix} \mathbf{w}_{t-1}$$

where elements 1 and 2 of $\mathbf{w}_{t-1}$ account for the errors in integrating velocity and acceleration. The gain and bias of the second instrument are treated as extra state variables x_4 and x_5, so

$$x_{4,t} = x_{4,t-1}, \qquad x_{5,t} = x_{5,t-1}$$

$$y_{1,t} = x_{1,t} + v_{1,t}, \qquad y_{2,t} = x_{4,t}x_{1,t} + x_{5,t}$$

The augmented state equations are still linear, but the second observation equation must be linearised to

$$\delta y_{2,t} = \bar{x}_{4,t}\,\delta x_{1,t} + \bar{x}_{1,t}\,\delta x_{4,t} + x_{5,t} + v_{2,t}$$

where $\delta x_{1,t}$ and $\delta x_{4,t}$ are deviations from nominal values $\bar{x}_{1,t}$ and $\bar{x}_{4,t}$, and $\delta y_{2,t}$ is the deviation of $y_{2,t}$ from $\bar{x}_{1,t}\bar{x}_{4,t}$. Suitable values for $\bar{x}_{1,t}$ and $\bar{x}_{4,t}$ are the estimates based on observations up to sample $t-1$. Noise $v_{2,t}$ is optional, and accounts for the linearisation error $\delta x_{1,t}\,\delta x_{4,t}$ (not uncorrelated with the state as the filter assumes, in fact).

The presence of x_4 and x_5 increases the computing substantially and introduces some doubt over the effects of the linearisation error. △

If the state and observation equations before linearisation are

$$\begin{aligned} \mathbf{x}'_t &= \mathbf{f}(\mathbf{x}'_{t-1}, \mathbf{u}_{t-1}, \mathbf{w}_{t-1}) \\ \mathbf{y}_t &= \mathbf{g}(\mathbf{x}'_t, \mathbf{v}_t) \end{aligned} \tag{8.9.4}$$

where $\mathbf{f}$ and $\mathbf{g}$ are differentiable with respect to $\mathbf{x}'_{t-1}$ and $\mathbf{x}'_t$ respectively, the linearised equations are

$$\begin{aligned} \delta\mathbf{x}'_t &= F_{t-1}\,\delta\mathbf{x}'_{t-1} + \mathbf{f}(\bar{\mathbf{x}}'_{t-1}, \mathbf{u}_{t-1}, \mathbf{w}_{t-1}) - \mathbf{f}(\bar{\mathbf{x}}'_{t-1}, \mathbf{u}_{t-1}, \mathbf{0}) \\ \delta\mathbf{y}_t &= G_t\,\delta\mathbf{x}'_t + \mathbf{g}(\bar{\mathbf{x}}'_t, \mathbf{v}_t) - \mathbf{g}(\bar{\mathbf{x}}'_t, \mathbf{0}) \end{aligned} \tag{8.9.5}$$

where elements (i,j) of F_{t-1} and G_t are

$$[F_{t-1}]_{ij} = \left.\frac{\partial f_i}{\partial x'_{j,t-1}}\right|_{\bar{\mathbf{x}}'_{t-1}}, \qquad [G_t]_{ij} = \left.\frac{\partial g_i}{\partial x'_{j,t}}\right|_{\bar{\mathbf{x}}'_t} \tag{8.9.6}$$

and the deviations are

$$\delta\mathbf{x}'_{t-1} \triangleq \mathbf{x}'_{t-1} - \bar{\mathbf{x}}'_{t-1}, \qquad \delta\mathbf{x}'_t \triangleq \mathbf{x}'_t - \bar{\mathbf{x}}'_t, \qquad \delta\mathbf{y}_t \triangleq \mathbf{y}_t - \bar{\mathbf{y}}_t \tag{8.9.7}$$

with the nominal values given by

$$\bar{\mathbf{y}}_t = \mathbf{g}(\bar{\mathbf{x}}'_t, \mathbf{0}), \qquad \bar{\mathbf{x}}'_t = \mathbf{f}(\bar{\mathbf{x}}'_{t-1}, \mathbf{u}_{t-1}, \mathbf{0}) \tag{8.9.8}$$

Notice that the noise terms are allocated to the linearised state and observation equations rather than to the non-linear time-update (8.9.8).

The obvious choice for $\bar{\mathbf{x}}'_{t-1}$ is the up-to-date estimate $\hat{\mathbf{x}}'_{t-1|t-1}$. The time-updated nominal state $\bar{\mathbf{x}}'_t$ is then used in place of $\hat{\mathbf{x}}'_{t|t-1}$, and $\bar{\mathbf{y}}_t$ in place of the one-step-ahead prediction $H_t\hat{\mathbf{x}}'_{t|t-1}$, in calculating the innovation and updating the state estimate when observation y_t is received. The remainder of the Kalman filter, with F_{t-1} for Φ_{t-1} and G_t for H_t, is (8.1.11) and, with $P_{t|t-1}$ for P_{t-1}, (7.3.12). This state estimator based on equations linearised about $\hat{\mathbf{x}}_{t-1|t-1}$, is called the *extended Kalman filter*.

8.9.2 Alternated Parameter and State Estimation

The need for local linearisation can be avoided by separating parameter updating from state updating. Figure 8.9.1 shows a scheme suggested by Goodwin and Sin (1984). A similar arrangement has been used in adaptive filtering for equalisation of data-communication channels (Nicholson and Norton, 1979). A Kalman-filter step, with the current estimate $\hat{\boldsymbol{\theta}}_{t-1}$ as $\boldsymbol{\theta}$, computes $\hat{\mathbf{x}}_{t|t-1}$ and passes the prediction error (innovation) $\mathbf{y}_t - \hat{\mathbf{y}}_t(\hat{\mathbf{x}}_{t|t-1}, \hat{\boldsymbol{\theta}}_{t-1})$ to a prediction-error recursive identification algorithm of the type discussed in Section 7.4.6. A step of the identification algorithm then produces $\hat{\boldsymbol{\theta}}_t$ for the next state-estimation step.

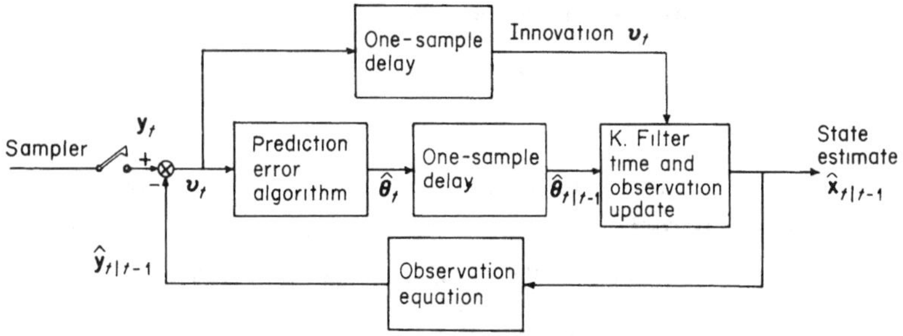

Fig. 8.9.1 Alternated parameter and state estimation.

The arrangement incidentally requires less computing than extended Kalman filtering with an augmented state. The computing demands of the parameter estimation algorithm are comparable with those of the state estimator for any given number of unknowns, whereas the demand of the state estimator rises more than linearly with the number of state variables.

8.9.3 Maximum-Likelihood Estimator

Both techniques discussed so far involve substituting nominal parameter values for best values when estimating the state. (We say "best" rather than "actual" values because the model is not an exact representation of the underlying dynamical process.) Whatever their asymptotic properties, the schemes are approximate in finite samples and rather heuristic. A more formal procedure (Maybeck, 1982) would be to minimise at each recursion step the log-likelihood function of the parameters. The likelihood function embraces the entire history of the state estimates and covariances, since they are all affected by any change in $\boldsymbol{\theta}$ applying throughout the recursion.

At sample time t, the likelihood of $\boldsymbol{\theta}$ is, by Bayes' rule applied repeatedly,

$$\begin{aligned} p(Y \mid \boldsymbol{\theta}) &\equiv p(\mathbf{x}_t, \mathbf{y}_t, \mathbf{y}_{t-1}, \ldots, \mathbf{y}_1 \mid \boldsymbol{\theta}) \\ &= p(\mathbf{x}_t \mid \mathbf{y}_t, \ldots, \mathbf{y}_1, \boldsymbol{\theta}) p(\mathbf{y}_t, \ldots, \mathbf{y}_1 \mid \boldsymbol{\theta}) \\ &= p(\mathbf{x}_t \mid \mathbf{y}_t, \ldots, \mathbf{y}_1, \boldsymbol{\theta}) \prod_{i=0}^{t-1} p(\mathbf{y}_{t-i} \mid \mathbf{y}_{t-i-1}, \ldots, \mathbf{y}_1, \boldsymbol{\theta}) \end{aligned} \tag{8.9.9}$$

Further progress depends on assuming a form for the p.d.f.'s. If they are assumed Gaussian, each is fully specified by its mean and covariance. The mean of $p(\mathbf{x}_t \mid \mathbf{y}_t, \ldots, \mathbf{y}_1, \boldsymbol{\theta})$ is simply the conditional-mean estimate $\hat{\mathbf{x}}_{t|t}$ of $\mathbf{x}_t$ given $\boldsymbol{\theta}$ and the observations up to $\mathbf{y}_t$. Both $\hat{\mathbf{x}}_{t|t}$ and the covariance $P_{t|t}$ would be given by a Kalman filter with parameter $\boldsymbol{\theta}$. The mean of $p(\mathbf{y}_{t-i} | \mathbf{y}_{t-i-1}, \ldots, \mathbf{y}_1, \boldsymbol{\theta})$ is the conditional-mean estimate $H_{t-i}\hat{\mathbf{x}}_{t-i|t-i-1}$ of $\mathbf{y}_{t-i}$, and its covariance, with contributions from the mutually uncorrelated error in $\hat{\mathbf{x}}_{t-i|t-i-1}$ and noise $\mathbf{v}_{t-i}$, is $H_{t-i}P_{t-i|t-i-1}H_{t-i}^{\mathrm{T}} + R_{t-i}$, which would also be given by a Kalman filter. Hence the log likelihood is

$$\begin{aligned} L(\boldsymbol{\theta}, \mathbf{x}_t) = &-\tfrac{1}{2}\ln((2\pi)^n |P_{t|t}|) - \tfrac{1}{2}(\mathbf{x}_t - \hat{\mathbf{x}}_{t|t})^{\mathrm{T}} P_{t|t}^{-1} (\mathbf{x}_t - \hat{\mathbf{x}}_{t|t}) \\ &- \frac{1}{2} \sum_{i=0}^{t-1} \{\ln((2\pi)^q |H_{t-i}P_{t-i|t-i-1}H_{t-i}^{\mathrm{T}} + R_{t-i}|) \\ &+ (\mathbf{y}_{t-i} - H_{t-i}\hat{\mathbf{x}}_{t-i|t-i-1})^{\mathrm{T}} (H_{t-i}P_{t-i|t-i-1}H_{t-i}^{\mathrm{T}} + R_{t-i})^{-1} \\ &\times (\mathbf{y}_{t-i} - H_{t-i}\hat{\mathbf{x}}_{t-i|t-i-1})\} \end{aligned} \tag{8.9.10}$$

Setting $\partial L(\boldsymbol{\theta}, \mathbf{x}_t)/\partial \mathbf{x}_t$ to zero we obtain

$$-P_{t|t}^{-1}(\mathbf{x}_t - \hat{\mathbf{x}}_{t|t}) = \mathbf{0} \tag{8.9.11}$$

which merely says that $\hat{\mathbf{x}}_{t|t}$ is the m.l. estimate of $\mathbf{x}_t$, under the Gaussian assumption, provided the m.l. estimate of $\boldsymbol{\theta}$ is used in computing $\hat{\mathbf{x}}_{t|t}$. It is clearly impracticable to run a Kalman filter over the whole observation set for each conceivable value of $\boldsymbol{\theta}$ and find what value maximises $L(\boldsymbol{\theta}, \mathbf{x}_t)$, so the maximisation must be performed algebraically, setting $\partial L(\boldsymbol{\theta}, \mathbf{x}_t)/\partial \boldsymbol{\theta}$ to zero. The resulting equations are complicated, mainly because each $P_{t-i|t-i-1}$ depends on $\boldsymbol{\theta}$. They do not allow an explicit solution for $\boldsymbol{\theta}$, but have to be solved iteratively. Maybeck (1982) gives a full account of how they can be solved.

The drawback of this procedure is its very high computing demand, which necessitates a succession of approximations detailed by Maybeck (1982). These include approximating the Hessian matrix $\partial^2 L/\partial \boldsymbol{\theta}^2$ by a matrix using the gradients but not second derivatives of L, using the matrix

$(\partial \mathbf{x}_{t-i|t-i-1}/\partial \boldsymbol{\theta})(\partial \hat{\mathbf{x}}_{t-i|t-i-1}/\partial \boldsymbol{\theta})^{\mathrm{T}}$ in place of its expected value, updating $\hat{\boldsymbol{\theta}}$ less frequently than $\hat{\mathbf{x}}$ and, more drastically, neglecting the dependence of P on $\boldsymbol{\theta}$ and updating $\partial^2 L/\partial \boldsymbol{\theta}^2$ infrequently.

8.9.4 Bayes Estimator

Recall from Chapter 6 that Bayes estimators are derived from the posterior p.d.f. of the unknowns. If we denote the observation history $\mathbf{y}_t, \mathbf{y}_{t-1}, \ldots, \mathbf{y}_1$ by Y_t, the posterior p.d.f. we are interested in is $p(\mathbf{x}_t, \boldsymbol{\theta} \mid Y_t)$. The most attractive Bayes estimator is the conditional mean (Section 6.3.1), which is optimal for a wide range of loss functions. If we supply $\boldsymbol{\theta}$ and assume that $p(\mathbf{x}_t \mid \boldsymbol{\theta}, Y_t)$ is Gaussian, the Kalman filter computes the conditional mean as $\hat{\mathbf{x}}_{t|t}$. The question is, can we exploit this when $\boldsymbol{\theta}$ is not given, but is also unknown?

The joint posterior p.d.f. of $\mathbf{x}_t$ and $\boldsymbol{\theta}$ is

$$p(\mathbf{x}_t, \boldsymbol{\theta} \mid Y_t) = p(\mathbf{x}_t \mid \boldsymbol{\theta}, Y_t) p(\boldsymbol{\theta} \mid Y_t) \tag{8.9.12}$$

and by Bayes' rule,

$$p(\boldsymbol{\theta} \mid Y_t) = \frac{p(\mathbf{y}_t \mid \boldsymbol{\theta}, Y_{t-1}) p(\boldsymbol{\theta} \mid Y_{t-1})}{p(\mathbf{y}_t \mid Y_{t-1})} \propto p(\mathbf{y}_t \mid \boldsymbol{\theta}, Y_{t-1}) p(\boldsymbol{\theta} \mid Y_{t-1}) \tag{8.9.13}$$

As $p(\mathbf{y}_t | \boldsymbol{\theta}, Y_{t-1})$ has mean $H\hat{\mathbf{x}}_{t|t-1}$ and covariance $H_t P_{t|t-1} H_t^{\mathrm{T}} + R_t$ calculable, given $\boldsymbol{\theta}$, by a Kalman filter, (8.9.13) looks like a usable recursion for $p(\boldsymbol{\theta} \mid Y)$, assuming that $p(\mathbf{y}_t \mid \boldsymbol{\theta}, Y_{t-1})$ is Gaussian and fully specified by its mean and covariance. With both p.d.f.'s on the right-hand side of (8.9.12) accounted for, we seem to be home and dry. However, each Kalman filter gives the conditional mean and covariance for only one specific value of $\boldsymbol{\theta}$. In both (8.9.13) and (8.9.12) we need them for the entire range of possible values of $\boldsymbol{\theta}$. The idea is therefore only computationally feasible if $\boldsymbol{\theta}$ is is restricted to a sufficiently small number of discrete values for one Kalman filter to be assigned to each possible value. It might be possible to reduce the number of values of $\boldsymbol{\theta}$ as the recursion goes on, as $p(\boldsymbol{\theta} \mid Y_t)$ is likely to become more sharply peaked as t increases. One virtue of the Bayes approach, the ability to cope with a changing shape of $p(\boldsymbol{\theta} \mid Y)$, would be lost by doing so.

REFERENCES

Adby, P. R., and Dempster, M. A. H. (1974). "Introduction to Optimization Methods". Chapman & Hall, London.

Åström, K. J., and Wittenmark, B. (1984). "Computer Controlled Systems". Prentice-Hall, Englewood Cliffs, New Jersey.

Barker, H. A., and Davy, R. W. (1978). Measurement of second-order Volterra kernels using pseudorandom ternary signals. *Int. J. Control* **27**, 277–291.

Belforte, G., and Bona, B. (1985). An improved parameter identification algorithm for signals with unknown-but-bounded errors, *IFAC Symp. Identification System Parameter Estimation, York, England*, 1507–1512.

Bellman, R. A., and Åström, K. J. (1970). On structural identifiability. *Math. Biosci.* **7**, 329–339.

Bierman, G. J. (1977). "Factorization Methods for Discrete Sequential Estimation". Academic Press, New York and London.

Billings, S. A. (1980). Identification of nonlinear systems—a survey. *Proc. IEE, Pt. D* **127**, 272–285.

Billings, S. A. (1984). Identification of nonlinear systems. *In* "Non-linear System Design" (S. A. Billings, J. O. Gray and D. H. Owens, eds.). Peter Peregrinus, London.

Billings, S. A., and Fakhouri, S. Y. (1978). Identification of nonlinear systems using correlation analysis. *Proc. IEE* **125**, 691–697.

Billings, S. A., and Fakhouri, S. Y. (1982). Identification of systems containing linear dynamic and static nonlinear elements. *Automatica* **18**, 15–26.

Blackman, P. F. (1977). "Introduction to State-variable Analysis". Macmillan, London.

Bosley, M. J., and Lees, F. P. (1972). A survey of simple transfer function derivations from high-order state-variable models. *Automatica* **8**, 765–775.

Carson, E. R., Cobelli, C., and Finkelstein, L. (1981). The identification of metabolic systems—a review. *Am. J. Physiol.* **240**, R120–R129.

Chen, C. F., and Shieh, L. S. (1968). A novel approach to linear model simplification. *Int. J. Control* **8**, 561–570.

Cobelli, C., Lepschy, A., and Romanin-Jacur, G. (1979). Identifiability of compartmental systems and related structural properties. *Math. Biosci.* **44**, 1–18.

D'Azzo, J. J., and Houpis, C. H. (1981). "Linear Control System Analysis and Design", 2nd ed. McGraw-Hill Kogakusha, Tokyo.

Delforge, J. (1980). New results on the problem of identifiability of a linear system. *Math. Biosci.* **52**, 73–96.

Delforge, J. (1981). Necessary and sufficient structural conditions for local identifiability of a system with linear compartments. *Math. Biosci.* **54**, 159–180.

Denham, M. J. (1974). Canonical forms for the identification of multivariable linear systems. *IEEE Trans. Autom. Control* **AC-19**, 646–656.

Feldbaum, A. A. (1960). Dual control theory. *Automation Remote Control* **21**, 874–880, 1033–1039; **22**, 1–12, 109–121.

Feldbaum, A. A. (1965). "Optimal Control Systems". Academic Press, New York and London.

Fogel, E., and Huang, Y. F. (1982). On the value of information in system identification—bounded noise case. *Automatica* **18**, 229–238.

Friedlander, B. (1982). Lattice filters for adaptive processing. *Proc. IEEE* **70**, 829–867.

Gabel, R. A., and Roberts, R. A. (1980). "Signals and Linear Systems". Wiley, New York.

Gevers, M., and Wertz, V. (1984). Uniquely identifiable state-space and ARMA parametrizations for multivariable linear systems. *Automatica* **20**, 333–347.

Glover, K., and Willems, J. C. (1974). Parametrizations of linear dynamical systems: Canonical forms and identifiability. *IEEE Trans. Autom. Control* **AC-19**, 640–646.

Godfrey, K. R. (1983). "Compartmental Models and Their Application". Academic Press, New York and London.

Goodwin, G. C., and Payne, R. L. (1977). "Dynamic System Identification Experiment Design and Data Analysis". Academic Press, New York and London.

Goodwin, G. C., and Sin, K. S. (1984). "Adaptive Filtering Prediction and Control". Prentice-Hall, Englewood Cliffs, New Jersey.

Graupe, D., Jain, V. K., and |Salahi, J. (1980). A comparative analysis of various least-squares identification algorithms. *Automatica* **16**, 663–681.

Hadley, G. (1961). "Linear Algebra". Addison-Wesley, Reading, Massachusetts.

Halmos, P. R. (1958). "Finite-dimensional Vector Spaces". Van Nostrand Reinhold, Princeton, New Jersey.

Jazwinski, A. H. (1970). "Stochastic Processes and Filtering Theory". Academic Press, New York and London.

Kailath, T. (1980). "Linear Systems". Prentice-Hall, Englewood Cliffs, New Jersey.

Kalman, R. E. (1960). A new approach to linear filtering and prediction problems. *J. Basic Eng. Trans. ASME Ser. D* **82**, 35–45.

Lawrence, P. J. (1981). Estimation of the Volterra functional series of a nonlinear system using frequency-response data. *Proc. IEE Pt. D* **128**, 206–210.

Lee, D. T. L., Friedlander, B., and Morf, M. (1982). Recursive ladder algorithms for ARMA modelling. *IEEE Trans. Autom. Control* **AC-27**, 753–763.

Ljung, L. (1976). On the consistency of prediction error identification methods. *In* "System Identification: Advances and Case Studies" (R. K. Mehra and D. G. Lainiotis, eds.). Academic Press, New York and London.

Ljung, L. (1977). Analysis of recursive stochastic algorithms. *IEEE Trans. Autom. Control* **AC-22**, 551–575.

Ljung, L., and Söderström, T. (1983). "Theory and Practice of Recursive Identification". MIT Press, Cambridge, Massachusetts.

Luenberger, D. G. (1973). "Introduction to Linear and Nonlinear Programming". Addison-Wesley, Reading, Massachusetts.

Marmarelis, P. Z., and Marmarelis, V. Z. (1978). "Analysis of Physiological Systems". Plenum, New York and London.

Maybeck, P. S. (1979). "Stochastic Models, Estimation, and Control", Vol. 1. Academic Press, New York and London.

Maybeck, P. S. (1982). "Stochastic Models, Estimation, and Control", Vol. 2. Academic Press, New York and London.

McGarty, T. P. (1974). "Stochastic Systems and State Estimation". Wiley, New York.

Mees, A., and Sparrow, C. T. (1981). Chaos *Proc. IEE Pt. D* **128**, 201–205.

Mehra, R. K. (1972). Approaches to adaptive filtering. *IEEE Trans. Autom. Control* **AC-17**, 693–698.

Milanese, M., and Belforte, G. (1982). Estimation theory and uncertainty intervals evaluation in presence of unknown but bounded errors: linear families of models and estimators. *IEEE Trans. Autom. Control* **AC-27**, 408–414.

Mirsky, L. (1955). "An Introduction to Linear Algebra". Oxford University Press, London.

Nicholson, G., and Norton, J. P. (1979). Kalman filter equalization for a time-varying communication channel. *Aust. Telecomm. Res.* **13**, 3–12.

Norton, J. P. (1975). Optimal smoothing in the identification of linear time-varying systems. *Proc. IEE* **122**, 663–668.

Norton, J. P. (1976). Identification by optimal smoothing using integrated random walks. *Proc. IEE* **123**, 451–452.

Norton, J. P. (1980a). Structural zeros in the modal matrix and its inverse. *IEEE Trans. Autom. Control* **AC-25**, 980–981.

Norton, J. P. (1980b). Normal-mode identifiability analysis of linear compartmental systems in linear stages. *Math. Biosci.* **50**, 95–115.

Norton, J. P. (1982a). An investigation of the sources of non-uniqueness in deterministic identifiability. *Math. Biosci.* **60**, 89–108.

Norton, J. P. (1982b). Letter to the Editor. *Math. Biosci.* **61**, 295–298.

Norton, J. P., Brown, R. F., and Godfrey, K. R. (1980). Modal analysis of identifiability of linear compartmental models. *Proc. IEE Pt. D* **127**, 83–92.
Reid, J. G. (1983). "Linear System Fundamentals". McGraw-Hill, New York.
Robins, A. J., and Wellstead, P. E. (1981). Recursive system identification using fast algorithms. *Int. J. Control* **33**, 455–480.
Rockafellar, R. T. (1970). "Convex Analysis". Princeton Univ. Press, Princeton, New Jersey.
Schetzen, M. (1980). "The Volterra and Wiener Theories of Nonlinear Systems". Wiley, New York.
Schweppe, F. C. (1968). Recursive state estimation: unknown but bounded errors and system inputs. *IEEE Trans. Autom. Control* **AC-13**, 22–28.
Shamash, Y. (1983). Critical review of methods for deriving stable reduced-order models. *In* "Identification and System Parameter Estimation 1982" (G. A. Bekey and G. N. Saridis, eds.). Pergamon, Oxford.
Söderström, T., Gustavsson, I., and Ljung, L. (1976). Identifiability conditions for linear multivariable systems operating under feedback. *IEEE Trans. Autom. Control* **AC-21**, 837–840.
Söderström, T., and Stoica, P. G. (1983). "Instrumental Variable Methods for System Identification". Springer-Verlag, Berlin and New York.
Vidyasagar, M. (1978). "Nonlinear Systems Analysis". Prentice-Hall, Englewood Cliffs, New Jersey.
Walter, E. (1982). "Identifiability of State Space Models". Springer-Verlag, Berlin and New York.
Watson, G. A. (1980). "Approximation Theory and Numerical Methods". Wiley, Chichester and New York.

PROBLEMS

8.1 Consider the two-compartment system of Example 8.2.1 with only compartment 1 observed, the observation gain being c_1. Verify that the model

$$\dot{\mathbf{x}} = \begin{bmatrix} -k_{01} - k_{21} & k_{12} \\ k_{21} & -k_{02} - k_{12} \end{bmatrix} \mathbf{x} + \begin{bmatrix} u_1 \\ u_2 \end{bmatrix}, \qquad y = c_1 x_1$$

gives

$$Y(s) = \frac{c_1((s + k_{02} + k_{12})U_1(s) + k_{12}U_2(s))}{s^2 + (k_{01} + k_{21} + k_{02} + k_{12})s + (k_{01}k_{02} + k_{01}k_{12} + k_{02}k_{21})}$$

Show that, if $U_1(s)$ and $U_2(s)$ are exactly known, the deterministic identifiability of the model from $Y(s)$ depends on the choice of input waveforms as follows:

(i) If inputs u_1 and u_2 have steps or impulses applied separately and the impulse responses from u_1 to y and u_2 to y are found, the model is completely identifiable, including c_1.

(ii) If simultaneous steps or impulses are applied as u_1 and u_2, the rate constants in the model cannot be identified.

(iii) If a step is applied as u_1 and an impulse as u_2 (treated as a δ function), then the model is completely identifiable only if c_1 is known in advance.

(iv) If an impulse is applied as u_1 and a step as u_2, the model is completely identifiable even if c_1 is not initially known.

8.2 A proposed input signal for o.l.s. identification of a u.p.r. model as in Example 8.2.3 has period P samples. Is it persistently exciting of order P if (i) it has non-zero mean, and successive half-cycles are symmetrical about the mean; (ii) successive half-cycles are not symmetrical about the mean but each cycle is symmetrical about its half-way point in time?

8.3 Consider updating of the covariance of the recursive w.l.s. estimate discussed in Section 8.1.3. The batch w.l.s. estimate has covariance $E[(U^T WU)^{-1} U^T WRW^T U(U^T WU)^{-1}]$ according to (5.3.14), where U is the regressor matrix, R the covariance of the regression-equation error and W the weighting matrix, which in Section 8.1.3 is diagonal. First verify that (8.1.5) updates $(U^T WU)^{-1}$ by writing $U_t^T W_t U_t$ in terms of $U_{t-1}^T W_{t-1} U_{t-1}$ and an increment at time t and using the matrix-inversion lemma, much as in Section 7.3.3. Then produce an updating scheme for the covariance of the w.l.s. estimate in the case of zero-mean regression-equation error of constant variance σ^2.

8.4 With g and f known in advance, is model (i) of Example 8.3.1 identifiable from observations of the reference input and the output by fitting a closed-loop transfer function to the response to a deterministic reference input and then solving for a, b_1 and b_2 from the coefficients in the transfer function? Is the model still identifiable if either g or f is unknown? [This is not so easy to answer as it seems at first sight.]

Alternatively, could the model be identified by exciting the system with a suitable reference input sequence and rewriting the model as a regression equation relating y_t to earlier output samples and to reference-input samples (eliminating $\{u\}$)? Would o.l.s. do?

8.5 Show that a Padé approximation $b/(1+as)$ to the transfer function $H(s) = (1+\beta s)/((1+s)(1+\alpha s))$ is unstable if $\beta > 1+\alpha$. Can this happen even if $H(s)$ is both stable and minimum-phase (i.e. if all the zeros and poles of $H(s)$ are in the left-hand half plane)? For such an $H(s)$, can the alternative Padé approximation $(b_0 + b_1 s)/(1+as)$ be unstable?

8.6 Construct the cross-section at $\theta_1 = 5$ of the polyhedron formed by bounding $\boldsymbol{\theta}$ as in Example 8.6.1, with the same observations as in that example but measuring time from midway between the third and fourth samples. Notice whether the feasible-parameter region ends up larger or smaller than in Example 8.6.1.

8.7 Find $A(z^{-1})$ and $B(z^{-1})$ of the reduced-degree left-matrix-fraction description of the system of Example 8.7.1 using the left divisor given in part (iv) of the example. Verify that the degree is reduced and explain the reduction in terms of removing a common factor from the numerator and denominator of G as given by $A_1^{-1}B_1$ in part (i).

8.8 Consider alternated state and parameter estimation as described in Section 8.9.2, with recursive l.s. as the identification algorithm, and recall that this algorithm is identical in form, although not in interpretation, to a Kalman filter. The identification algorithm is applied to (8.9.1) with $\boldsymbol{\theta}$ appearing linearly in Φ_{t-1} and H_t, and not appearing in B_{t-1} or Γ_t; assume $\boldsymbol{\theta}$ is known to be constant. Write out the updating equations for $\hat{\boldsymbol{\theta}}$ and its covariance, and for $\hat{\mathbf{x}}$ and its covariance.

Compare the combined set of equations with the extended Kalman filter for the same problem, as described in Section 8.9.1. Would the extended Kalman filter with the covariance replaced by a block-diagonal matrix

$$\begin{bmatrix} P_x & 0 \\ 0 & P_\theta \end{bmatrix}$$

give the same set of equations?

8.9* Investigate the conditions for the model (8.3.8) of a linear feedback system to be strongly system identifiable if the system has feedback of the form (8.3.9), two control variables, two outputs and one external input apart from output noise, and the controller is switched between two different control laws. Specifically, find which elements of the transfer-function matrix $\mathscr{L}_1$ in (8.3.9) must differ from the corresponding elements of $\mathscr{L}_2$ if the matrices $\mathscr{K}_1$ and $\mathscr{K}_2$ are both of the form

$$\begin{bmatrix} k_i \\ 0 \end{bmatrix}$$

and the system is to be s.s.i.

Chapter 9

Experimental Design and Choice of Model Structure

9.1 INTRODUCTION

As already emphasised, identification is not a matter of applying standard techniques in a specified way and getting guaranteed results. Rather we look at the intended use of the model, the observations obtainable, the possibilities for experimentation and the time and effort available, then, if not dissuaded from going any further, put together a model by an untidy combination of experiment, computation, analysis and revision. Every stage is uncertain, and common experience is that each identification exercise raises some new problem. This is a reflection not primarily of immaturity in identification techniques, but rather of the immense variety of dynamical behaviour, experimental constraints and purposes for modelling. For these reasons, designing a software package for identification is extremely difficult, and the most successful packages presuppose a great deal of intervention by the user (Box and Jenkins, 1976; Young, 1984).

There is little point in trying to prescribe a comprehensive list of steps in identifying a model and techniques for each step. Instead, the next two chapters consider a number of aspects of experiment design, model structure selection and model validation, with no pretence that all eventualities are covered.

9.2 EXPERIMENT DESIGN

9.2.1 Adequacy of Input and Output

We encountered persistency of excitation conditions in Section 8.2.3, with regard to possible redundancy among the lagged input samples forming the regressors in least-squares estimation of a unit-pulse response. The input sequence $\{u\}$ was persistently exciting (p.e.) of order p if no selection from every p successive samples was exactly linearly dependent. The frequency-domain counterpart is that the input contains power at p or more frequencies. An input which is p.e. of order p allows us to estimate p u.p.r. ordinates by least squares.

Much recent effort has gone into deriving persistency of excitation conditions to ensure convergence of recursive algorithms for both identification and adaptive control (Yuan and Wonham, 1977; Moore, 1983; Goodwin *et al.*, 1985; Solo, 1983; Goodwin and Sin, 1984). The main significance of such conditions in open-loop identification is to warn against over-simple choices of input, such as a deterministic signal with a short period. It is not normally difficult to ensure adequate excitation by employing, for instance, a linearly filtered white sequence as input (Example 8.2.4), so long as we can choose the input. The situation is different in adaptive control, where the input depends on the output, but we shall not pursue that problem further than it was taken in Section 8.3. Persistency of excitation also plays a part in proving consistency of identification algorithms, as we saw for recursive o.l.s. in Section 8.2.4. An early invocation of p.e. conditions (Åström and Bohlin, 1966) was to prove that the "maximum-likelihood" algorithm of Section 7.2.3 is consistent if the system is stable and completely state controllable from the input $\{u\}$ or noise $\{e\}$ (D'Azzo and Houpis, 1981) allowing response components due to every pole of the system to be excited, and if also $\{u\}$ is p.e. of order 2n and the model is

$$y_t = -a_1 y_{t-1} - \cdots - a_n y_{t-n} + b_1 u_{t-k-1} + \cdots + b_n u_{t-k-n}$$
$$+ v_t + c_1 v_{t-1} + \cdots + c_q v_{t-q} \qquad (9.2.1)$$

with $\{e\}$ zero-mean, uncorrelated and Gaussian.

Given that an input is p.e. of the required order, the question remains whether it has enough power in the pass band of the system to yield acceptable model-coefficient estimates from a record of realistic length. The modest p.e. requirements, non-zero power at some minimum number of frequencies (Section 8.2.3), may ensure asymptotic convergence but do not guarantee satisfactory finite-sample performance. Theoretically the input bandwidth can be much smaller than that of the system and still allow a model to be identified, since the model structure relates behaviour beyond the input bandwidth to behaviour within it. Practically, the bandwidths must be comparable. The spectral distribution of the noise power may also have to be taken into account.

Example 9.2.1 The influence of input bandwidth can be seen in an example (Robins, 1984) of the identification of aircraft dynamics. Lateral acceleration l and yaw rate r are measured as the aircraft rudder deflection ζ is perturbed. The aircraft velocity along its roll axis is u, constant during the test, and along its pitch axis is v. The problem is to identify the aerodynamic derivatives y_v, y_ζ, n_v, n_r and n_ζ in the model (for negligible roll motion)

$$l \triangleq \dot{v} + ru = y_v v + y_\zeta \zeta$$

$$\dot{r} = n_v v + n_r r + n_\zeta \zeta$$

from noisy measurements. Figure 9.2.1a shows the estimates of y_ζ obtained from a simulated test with a 0.5 Hz bandwidth rudder-perturbation signal. The estimates converge, but rather slowly. The reason becomes clear on inspecting the transfer functions from ζ to l and r:

$$\frac{L(j\omega)}{Z(j\omega)} = \frac{-y_\zeta \omega^2 - y_\zeta n_r j\omega + (y_\zeta n_v - y_v n_\zeta)u}{-\omega^2 - (y_v + n_r)j\omega + n_v u + y_v n_r}$$

$$\frac{R(j\omega)}{Z(j\omega)} = \frac{n_\zeta j\omega + y_\zeta n_v - y_v n_\zeta}{-\omega^2 - (y_v + n_r)j\omega + n_v u + y_v n_r}$$

with $y_\zeta n_v$ much smaller than $y_v n_\zeta$ in practice. At low frequencies the numerators are dominated by $-y_v n_\zeta u$ and $-y_v n_\zeta$, respectively, and y_ζ does not appear in the denominator, so the measurements contain little information about y_ζ. However, at sufficiently high frequencies, $L(j\omega)/Z(j\omega) \cong y_\zeta$. A higher-bandwidth input signal should therefore give better estimates of y_ζ. Figure 9.2.1b confirms that it does; a 5 Hz bandwidth rudder-deflection signal results in much faster convergence of the y_ζ estimate. △

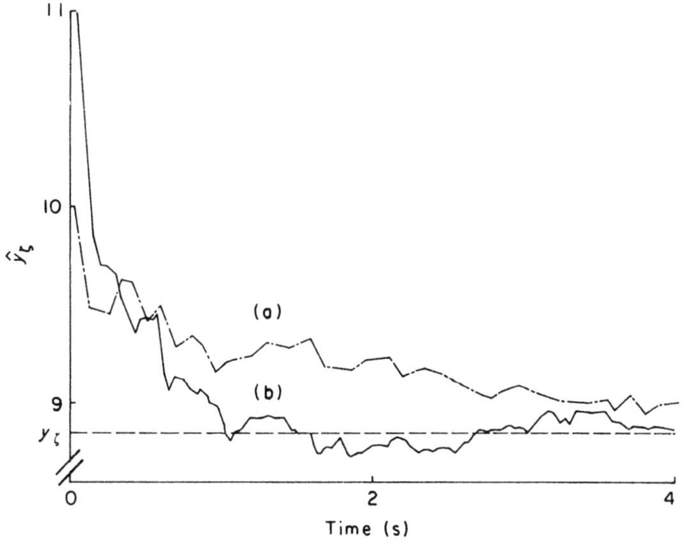

Fig. 9.2.1 Recursive estimates of y_ζ, Example 9.2.1.

The need for a high enough sampling rate for input and output, long enough records to determine the slower components of the response and allow the initial-condition response to subside, and a long enough period if the input is periodic, have been discussed in Chapters 2 and 3. Checking these points requires rough prior knowledge of the dynamics, but in engineering applications this is very often at hand, and is in any case highly desirable for assessing the credibility of the final model before it is put into use. Some prior experimentation may be required, to get information such as the spread of time constants, approximate gains, the nature and frequency of disturbances and the incidence of drift.

When serious data-logging starts, instrument and data-logger unreliability may be troublesome. For systems with slow dynamics, breaks in records are very likely, often as simultaneous short breaks in several records. When two simultaneously interrupted records are shifted relative to one another as required by their relative timing within the model, each break results in two equal intervals or one longer interval during which one or other record is unavailable. The time shifts in computing correlation functions have a similar effect. Reliability is a worse problem in multi-input or multi-output models, of course, and this is one reason for the concentration on methods for s.i.s.o. identification.

*9.2.2 Optimisation of Input

If a specially designed input perturbation is allowed or the output-sampling schedule can be chosen, it is worth considering whether they can be designed to optimise model accuracy. Consider input optimisation first (Goodwin and Payne, 1977). The basic idea is to maximise some scalar measure of estimation accuracy derived from the information matrix P^{-1} or covariance P of an unbiased estimate $\hat{\boldsymbol{\theta}}$ of the parameter vector $\boldsymbol{\theta}$. Recall that for an unbiased estimator, $E\hat{\boldsymbol{\theta}}$ is $\boldsymbol{\theta}$ and

$$\operatorname{cov}\hat{\boldsymbol{\theta}} \equiv P = E[(\hat{\boldsymbol{\theta}} - \boldsymbol{\theta})(\hat{\boldsymbol{\theta}} - \boldsymbol{\theta})^{\mathrm{T}}] \tag{9.2.2}$$

The model structure is assumed to have been selected already. The optimisation is subject to constraints on, for instance, input or output power or amplitude, number of output samples or experiment duration. To avoid specialising the results to a particular estimator, $\hat{\boldsymbol{\theta}}$ is assumed to achieve the Cramér–Rao bound on accuracy (Section 5.4.1), so

$$P^{-1} = F = \mathop{E}_{Y|\boldsymbol{\theta}}\left[\frac{\partial}{\partial\boldsymbol{\theta}}(\ln p(Y|\boldsymbol{\theta})\frac{\partial^{\mathrm{T}}}{\partial\boldsymbol{\theta}}(\ln p(Y|\boldsymbol{\theta}))\right] \tag{9.2.3}$$

where Y denotes the entire set of observations. One reasonable scalar measure of estimation accuracy is

$$E\left[\sum_{i=1}^{p} w_i(\hat{\theta}_i - \theta_i)^2\bigg|_{\theta=\hat{\theta}_0}\right] \equiv \operatorname{tr} WP\bigg|_{\theta=\hat{\theta}_0} \tag{9.2.4}$$

where the weights w_i making up the diagonal matrix W are chosen to suit the model application, and $\hat{\boldsymbol{\theta}}_0$ is a prior estimate of $\boldsymbol{\theta}$. An alternative is $-\log|P^{-1}|$, which is not affected by the scaling of individual $\hat{\theta}_i$'s.

Any numerical optimisation of an identification experiment has the drawback that it depends on prior knowledge of $\boldsymbol{\theta}$, so a good design is guaranteed only when it is least needed. However, a variety of helpful qualitative results concerning input design have been obtained, particularly in the frequency domain (Goodwin and Payne, 1977).

*9.2.3 Optimisation of Output-Sampling Schedule

Formal optimisation has also been applied successfully to the design of output-sampling schedules for experiments in which only very limited observations can be made (Di Stefano, 1980). With these applications in mind, we shall look into the process of optimising the Fisher information matrix F.

First we must make some assumptions about the model form and noise probability distribution, to enable us to write down $p(Y\,|\,\boldsymbol{\theta})$ in (9.2.3). An important case leading to tractable algebra is the linear state-variable model (time-invariant for simplicity, although this is not essential):

$$\dot{\mathbf{x}}(t) = A(\boldsymbol{\theta})\mathbf{x}(t) + B(\boldsymbol{\theta})\mathbf{u}(t) \tag{9.2.5}$$

with specified forcing inputs $\mathbf{u}(t)$ and initial state $\mathbf{x}(0)$, e.g. zero with the system quiescent before being perturbed. We consider noisy scalar observations

$$y(t) = y^c(t) + e(t) = \mathbf{h}(\boldsymbol{\theta})^{\mathrm{T}}\mathbf{x}(t) + e(t) \tag{9.2.6}$$

sampled at times

$$t = t_k, \qquad k = 1, 2, \ldots, N \tag{9.2.7}$$

For conciseness $y(t_k)$ will be written y_k, and similarly y_k^c and e_k. The noise samples will be assumed Gaussian, with mean zero and variance σ_k^2 at time t_k, and independent of one another. Since y^c depends only on $\mathbf{x}$ and $\mathbf{u}$, it is deterministic, so y_k is Gaussian with mean $\mathbf{h}^T\mathbf{x}_k$ and variance σ_k^2, and the y samples are independent. Consequently,

$$p(Y\,|\,\boldsymbol{\theta}) = p(y_1, y_2, \ldots, y_N\,|\,\boldsymbol{\theta}) \prod_{k=1}^{N} p(y_k\,|\,\boldsymbol{\theta})$$

$$= \prod_{k=1}^{N} \left\{ \frac{1}{\sqrt{2\pi}\,\sigma_k} \exp\left(-\frac{(y_k - \mathbf{h}^T\mathbf{x}_k)^2}{2\sigma_k^2} \right) \right\} \tag{9.2.8}$$

and so

$$\ln p(Y\,|\,\boldsymbol{\theta}) = \sum_{k=1}^{N} \left\{ -\tfrac{1}{2}\ln(2\pi\sigma_k^2) - \frac{(y_k - \mathbf{h}^T\mathbf{x}_k)^2}{2\sigma_k^2} \right\} \tag{9.2.9}$$

Hence

$$\frac{\partial}{\partial \boldsymbol{\theta}}(\ln p(Y\,|\,\boldsymbol{\theta})) = \sum_{k=1}^{N} \left\{ \frac{y_k - \mathbf{h}^T\mathbf{x}_k}{\sigma_k^2} \frac{\partial}{\partial \boldsymbol{\theta}}(\mathbf{h}^T\mathbf{x}_k) \right\} \tag{9.2.10}$$

As the noise samples $y_k - \mathbf{h}^T\mathbf{x}_k$ are independent and zero-mean, the expected value of the product of any two of them is zero, so on substituting (9.2.10) into the expression for the Cramér–Rao bound, we are left with

$$F = E\left[\sum_{k=1}^{N} \left\{ \frac{(y_k - \mathbf{h}^T\mathbf{x}_k)^2}{\sigma_k^4} \frac{\partial}{\partial \boldsymbol{\theta}}(\mathbf{h}^T\mathbf{x}_k) \frac{\partial^T}{\partial \boldsymbol{\theta}}(\mathbf{h}^T\mathbf{x}_k) \right\} \right]$$

$$= \sum_{k=1}^{N} \left\{ \frac{1}{\sigma_k^2} \frac{\partial}{\partial \boldsymbol{\theta}}(\mathbf{h}^T\mathbf{x}_k) \frac{\partial^T}{\partial \boldsymbol{\theta}}(\mathbf{h}^T\mathbf{x}_k) \right\} \tag{9.2.11}$$

The only remaining problem in relating F to design parameters such as the sampling times or input waveforms is to calculate $\partial(\mathbf{h}^T\mathbf{x}_k)/\partial\boldsymbol{\theta}$ for each sample instant. It is

$$\frac{\partial}{\partial \boldsymbol{\theta}}(\mathbf{h}^T\mathbf{x}_k) = [J_\theta x_k]^T\mathbf{h} + [J_\theta h]^T\mathbf{x}_k \tag{9.2.12}$$

and $[J_\theta x_k]$, the Jacobian matrix of $\mathbf{x}_k$ with respect to $\boldsymbol{\theta}$, can be found column by column by differentiating the state equation (9.2.5) with respect to each unknown θ_j, yielding

$$\frac{\partial \dot{\mathbf{x}}}{\partial \theta_j} = \frac{\partial A}{\partial \theta_j}\mathbf{x} + A\frac{\partial \mathbf{x}}{\partial \theta_j} + \frac{\partial B}{\partial \theta_j}\mathbf{u} + B\frac{\partial \mathbf{u}}{\partial \theta_j}, \qquad j = 1, 2, \ldots, p \tag{9.2.13}$$

The whole procedure for evaluating the information matrix F at a given $\boldsymbol{\theta}$ comprises integrating first the state equation (9.2.5), then, with $\mathbf{x}(t)$ known, the sensitivity equations (9.2.13), and finally substituting the sample-time values of the derivatives into (9.2.12) and thence (9.2.11). Trial-and-error optimisation of F by repetition of this procedure is plainly a heavy computational task. On the other hand, as Di Stefano points out, useful guidance can be obtained by a few trial evaluations of F without finding the optimal solution. The effects of any change in the experiment can be predicted by comparing two evaluations. Di Stefano comments that in his medical applications trials of this sort can be performed "prior to drawing a single drop of biological fluid".

Example 9.2.2 We consider once more a compartmental model as employed in biomedical studies. The model shown in Fig. 9.2.2 represents the flow of material into and out of the compartments by rate equations

$$\dot{\mathbf{x}} = \begin{bmatrix} -k_{21} & k_{12} \\ k_{21} & -k_{12} - k_{02} \end{bmatrix} \mathbf{x} + \begin{bmatrix} 1 \\ 0 \end{bmatrix} u$$

in the compartment contents $x_1(t)$ and $x_2(t)$. Rate constant k_{02} for loss from compartment 2 to the environment is known in advance, and we have to design an experiment to estimate k_{12} and k_{21}, the rate constants for flow between the compartments. A bolus dose (impulse) is introduced into compartment 1 of the previously empty system, and the ensuing variation of the amount in that compartment sampled. We wish to compare two schedules, taking samples at times 1, 2, 3 and 2, 3, 4. Our measure of estimation accuracy will be tr F, which will be evaluated at nominal (guessed) values $k_{12} = 0.05$, $k_{21} = 0.4$. The

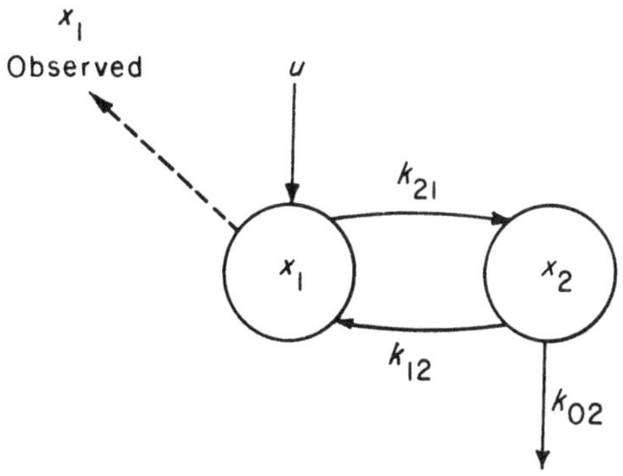

Fig. 9.2.2 Compartmental model, Example 9.2.2.

known k_{02} is 0.25, the observation gain is 1 and the input dose is 1, so the observations are

$$y_k = [1 \quad 0]\mathbf{x}_k + e_k, \qquad \operatorname{var} e_k = \sigma_k^2$$

and standard linear-system theory gives

$$\mathbf{x}(t) = \int_0^t e^{A(t-\tau)} B\mathbf{u}(\tau)\,d\tau = e^{At}\begin{bmatrix}1\\0\end{bmatrix}$$

here, with

$$A = \begin{bmatrix} -k_{21} & k_{12} \\ k_{21} & -k_{12} - k_{02} \end{bmatrix}$$

The transition matrix e^{At} is found, by inverse Laplace transformation of $[sI - A]^{-1}$ or otherwise, to be

$$e^{At} = \begin{bmatrix} \frac{1}{3}\exp(-0.2t) + \frac{2}{3}\exp(-0.5t) & \frac{1}{6}\exp(-0.2t) - \frac{1}{6}\exp(-0.5t) \\ \frac{4}{3}\exp(-0.2t) - \frac{4}{3}\exp(-0.5t) & \frac{2}{3}\exp(-0.2t) + \frac{1}{3}\exp(-0.5t) \end{bmatrix}$$

The first column gives $\mathbf{x}(t)$. Gradient $\partial(\mathbf{h}^{\mathrm{T}}\mathbf{x}_k)/\partial\boldsymbol{\theta}$ in the expression (9.2.11) for F is $\partial x_1(t_k)/\partial\boldsymbol{\theta}$, so only the derivatives of $x_1(t)$ need be computed from the sensitivity equations. With $\theta_1 \equiv k_{12}$ and $\theta_2 \equiv k_{21}$,

$$\frac{\partial A}{\partial \theta_1} = \begin{bmatrix} 0 & 1 \\ 0 & -1 \end{bmatrix}, \qquad \frac{\partial A}{\partial \theta_2} = \begin{bmatrix} -1 & 0 \\ 1 & 0 \end{bmatrix}$$

and with B, C and $u(t)$ all independent of $\boldsymbol{\theta}$, the sensitivity equations are

$$\frac{\partial \dot{\mathbf{x}}}{\partial \theta_1} = A\frac{\partial \mathbf{x}}{\partial \theta_1} + \begin{bmatrix} x_2 \\ -x_2 \end{bmatrix}$$

$$= \begin{bmatrix} -0.4 & 0.05 \\ 0.4 & -0.3 \end{bmatrix}\frac{\partial \mathbf{x}}{\partial \theta_1} + \frac{4}{3}\begin{bmatrix} \exp(-0.2t) - \exp(-0.5t) \\ -\exp(-0.2t) + \exp(-0.5t) \end{bmatrix}$$

$$\frac{\partial \dot{\mathbf{x}}}{\partial \theta_2} = A\frac{\partial \mathbf{x}}{\partial \theta_2} + \begin{bmatrix} -x_1 \\ x_1 \end{bmatrix}$$

$$= \begin{bmatrix} -0.4 & 0.05 \\ 0.4 & -0.3 \end{bmatrix}\frac{\partial \mathbf{x}}{\partial \theta_2} + \frac{1}{3}\begin{bmatrix} -\exp(-0.2t) - 2\exp(-0.5t) \\ \exp(-0.2t) + 2\exp(-0.5t) \end{bmatrix}$$

Their initial conditions are zero, because $\mathbf{x}(0)$ is independent of $\boldsymbol{\theta}$. They are solved easily, if tediously, obtaining from the convolution integral

$$\frac{\partial x_1}{\partial \theta_1} = \tfrac{2}{9}\{(t + \tfrac{40}{3})\exp(-0.2t) - (5t + \tfrac{40}{3})\exp(-0.5t)\}$$

$$\frac{\partial x_1}{\partial \theta_2} = \tfrac{1}{9}\{(-2t + \tfrac{10}{3})\exp(-0.2t) + (10t - \tfrac{10}{3})\exp(-0.5t)\}$$

Inserting the chosen sample times t_k, we can now evaluate F for each trial schedule:

$$F = \sum_k \left\{ \frac{1}{\sigma_k^2} \begin{bmatrix} (\partial x_1/\partial\theta_1)^2 & (\partial x_1/\partial\theta_1)(\partial x_1/\partial\theta_2) \\ (\partial x_1/\partial\theta_1)(\partial x_1/\partial\theta_2) & (\partial x_1/\partial\theta_2)^2 \end{bmatrix}_{t=t_k} \right\}$$

t_k	1	2	3	4
$\sigma_k^2 \times$ (increase in F)	$\begin{bmatrix} 0.0187 & 0.0780 \\ 0.0780 & 0.3256 \end{bmatrix}$	$\begin{bmatrix} 0.1418 & 0.2378 \\ 0.2378 & 0.3989 \end{bmatrix}$	$\begin{bmatrix} 0.3447 & 0.2927 \\ 0.2927 & 0.2485 \end{bmatrix}$	$\begin{bmatrix} 0.5304 & 0.2319 \\ 0.2319 & 0.1014 \end{bmatrix}$

The contributions to F from samples at times 1, 2, 3 and 4 are given in the accompanying tabulation, but the influence of an individual sample on the accuracy of $\hat{\boldsymbol{\theta}}$ cannot be seen from these figures. The most informative indicator is the covariance F^{-1} of $\hat{\boldsymbol{\theta}}$. If the noise variance is σ^2 for all samples, F^{-1} is

$$\sigma^2 \begin{bmatrix} 8.027 & -5.020 \\ -5.020 & 4.167 \end{bmatrix}$$

for sampling at 1, 2 and 3, compared with

$$\sigma^2 \begin{bmatrix} 4.155 & -4.230 \\ -4.230 & 5.642 \end{bmatrix}$$

for sampling at 2, 3 and 4. The reason why the former schedule allows more accurate estimation of k_{21}, as shown by element (2, 2) of F^{-1}, is that the faster decaying of the two exponential components in $x_1(t)$ is much more sensitive in both amplitude and time constant to k_{21} than is the slower component, and it is better defined by the earlier sampling schedule. The better definition of k_{12} by the later schedule, indicated by element (1, 1) being smaller, is less readily explained. The slower time constant is actually less sensitive to k_{12} than the faster one, and the absolute sensitivities of the exponential amplitudes to k_{12} are equal. However, the slower component has only about half the amplitude of the faster one, so is relatively more sensitive. It is evidently not very easy to predict which schedule will be better for k_{12} without the full analysis to find F^{-1}.

For sampling at 1, 2 and 3, tr F^{-1} is $12.19\sigma^2$, compared with $9.80\sigma^2$ for the other schedule. The other measure $-\log|F|$ is $0.9165 + 4\log\sigma$ and $0.7442 + 4\log\sigma$, respectively. A scalar measure of accuracy is rather unsatisfactory when, as here and as usual, some parameters improve and others get worse when schedules are changed. Weighting individual parameter variances as in (9.2.4) supposes that we know in advance how seriously we take individual errors. Unfortunately we do not know until we see what variances are obtained, so trial-and-error adjustment of the weights is necessary. △

9.3 SELECTION OF MODEL STRUCTURE

The selection of a model structure starts before design of the identification experiment, and continues during and after it. Decisions on the scope and form of the model must be made. The scope determines what variables should be included, what time scale the model is to operate on, what range of operating conditions should be covered, what observations should be used and what information the model has to provide. Model form was discussed in detail in Chapter 1.

Both form and scope are greatly limited by what turns out to be practicable in collecting observations. Anyone concerned with identification in industry has had the deflating experience of having what seems a perfectly reasonable suggestion dismissed on unanswerable grounds such as unserviceable instrumentation, unwillingness to interrupt normal operation, inability to wait for the results, lack of manpower, missing records, or a conviction that the results are already known or impossible to obtain. As rational plans are so often frustrated, there is little point in generalising further about these informal aspects of structure selection.

Before going on to the topic of model order determination, we should note that very often one starts with a strong preference for one particular model structure. The preference may stem from familiarity, proven effectiveness or mathematical tractability of the structure. The belief "better the devil you know than the devil you don't know" at times explains retention of a structure with admitted weaknesses. In other cases one hesitates to drop a "classical" model on which much effort has been spent, even when it has obvious defects.

9.3.1 Model Order Determination

The determination of model order is an important problem, for which techniques are well developed. Model order will be taken, rather loosely, to mean either the total number of parameters or the number of the input, output or noise terms, as appropriate. Sometimes it coincides with the order of the dynamics, but more often not. Folklore has something to say about model order determination. All other things being equal, the simpler of two models each encompassing the actual system behaviour is felt to be better. This feeling has been dignified with the title of the parsimony principle. Although it has some basis in the numerical ill-conditioning and statistical inefficiency associated with indiscriminate addition of terms to a model, it is an oversimplification. Whether a simpler model is on average better depends on the intended use of the model, the family of alternative models contemplated and the estimator employed.

Stoica and Söderström (1982) discuss under exactly what conditions the parsimony principle applies. They compare structures by calculating for each of them the mean value $\bar{V}$, over all possible values of the model estimates $\hat{\boldsymbol{\theta}}$, of any scalar measure $V(\hat{\boldsymbol{\theta}})$ of model performance (smaller for better performance) which is differentiable twice with respect to $\hat{\boldsymbol{\theta}}$. By Taylor series expansion they show that as the record length N tends to infinity,

$$\bar{V} = \underset{\hat{\boldsymbol{\theta}}}{E}[V(\hat{\boldsymbol{\theta}}^*)] \simeq V(\hat{\boldsymbol{\theta}}) + \frac{1}{2N}\operatorname{tr}(V''(\hat{\boldsymbol{\theta}}^*)P) \tag{9.3.1}$$

where $\hat{\boldsymbol{\theta}}^*$ minimises V, V'' is the second-derivative matrix of V and P the asymptotic covariance of the normalised estimation error $\sqrt{N}(\hat{\boldsymbol{\theta}} - \hat{\boldsymbol{\theta}}^*)$. If two model structures $\mathcal{M}_1$ and $\mathcal{M}_2$ being compared are hierarchical in the sense that $\mathcal{M}_1$ is contained in $\mathcal{M}_2$, and the system generating the observations has the structure $\mathcal{M}_1$, $\bar{V}(\hat{\boldsymbol{\theta}}_1)$ will be no larger than $\bar{V}(\hat{\boldsymbol{\theta}}_2)$ if the estimator achieves either the Cramér–Rao bound on P^{-1} or

$$P^{-1} = E[[J_{\theta}\hat{\mathbf{e}}]^{\mathrm{T}} R^{-1} [J_{\theta}\hat{\mathbf{e}}]] \tag{9.3.2}$$

where R is the covariance of the white noise $\mathbf{e}$ making up the unpredictable part of the observed system output, and $\hat{\mathbf{e}}$ is the model-output error.

Hierarchical structures include important cases like autoregressions or moving averages of different orders, but not, for example, a two-term moving average as $\mathcal{M}_1$ and a third-order autoregression as $\mathcal{M}_2$. The assumption that the structures are hierarchical can be dropped if the performance measure is

$$V(\hat{\boldsymbol{\theta}}) = |E[\hat{\mathbf{e}}\hat{\mathbf{e}}^{\mathrm{T}}]| \tag{9.3.3}$$

and the covariance satisfies (9.3.2). In that case $V''(\hat{\boldsymbol{\theta}}^*)$ is $2|R|P^{-1}$ and $\bar{V}$ is $|R| \dim \hat{\boldsymbol{\theta}}$, so a model structure with fewer unknowns to estimate is better, hierarchical or not. Counterexamples are presented to show that the parsimony principle does not apply in general unless the conditions above are met.

These results are very general; they restrict the model structure and proposed application very little, and apply to large classes of estimators and noise and input distributions. However, the assumption that $\mathcal{M}_1$ includes the process generating the records is doubtful. The most common practical situation is for a model to be estimated in the full knowledge that it is only a simplified and partial representation of the system behaviour, adequate for a stated purpose. For instance, a low-order model may be required so that a simple controller can be designed from it, or non-linear behaviour may be treated as time variation of a simple linear model rather than modelled explicitly. The consequence is a blurring of questions of model goodness. A larger model might well be a better fit to the observations, and a better predictor, but unacceptable because of its complexity. We should not lose sight of the fact that tests of model structure are to help us compromise between complexity and performance, exclude grossly deficient models and avoid ill-conditioned computation, rather than to determine the "correct" structure.

We shall review three popular ways of testing model structure: F tests, the Akaike information criterion and comparison of product-moment matrices.

9.3.2 *F* Test

The F test operates on the sample mean-square model-output errors

$$V_i(\hat{\boldsymbol{\theta}}_i) = \frac{1}{N} \sum_{t=1}^{N} \hat{e}_t^2(\boldsymbol{\theta}_i), \qquad i = 1, 2 \tag{9.3.4}$$

from two alternative model structures $\mathcal{M}_1$ and $\mathcal{M}_2$, which usually differ only in their numbers of terms. If $\mathcal{M}_1$ has p_1 parameters and $\mathcal{M}_2$ a larger number p_2, and if the output errors from $\mathcal{M}_1$ and $\mathcal{M}_2$ form sequences of independent, Gaussian, zero-mean, constant-variance random variables, then V_2 and $V_1 - V_2$ are independent χ^2 variates with $N - p_2$ and $p_2 - p_1$ degrees of freedom (Wadsworth and Bryan, 1974). It follows first that $\mathcal{M}_1$ is an adequate model, in that its output errors have no time structure, and second that the statistic

$$f = \frac{V_1 - V_2}{V_2} \frac{N - p_2}{p_2 - p_1} \tag{9.3.5}$$

has an $F(p_2 - p_1, N - p_2)$ distribution. The hypothesis that $\mathcal{M}_1$ is adequate can be tested at any chosen significance level by comparing f with the value exceeded with the corresponding probability.

Example 9.3.1 The rainfall–river flow model

$$y_t = -a_1 y_{t-1} - \cdots - a_n y_{t-n} + b_0 u_t + \cdots + b_{m-1} u_{t-m} + c_1 e_{t-1} + \cdots$$
$$+ c_q e_{t-q} + d + e_t$$

was fitted by the extended least-squares algorithm to 1455 hourly samples $\{y\}$ of flow in the Afon Hirnant in Wales. The input samples $\{u\}$ are means of readings from six rain gauges. The gauges register total rainfall over an hour, and the effective pure delay of the flow response to the areal mean rainfall is under an hour, so the model has b_0 non-zero. For all runs the noise order q was 3. Three models gave the values in the accompanying tabulation, where p is the

Model	n	m	p	V
1	2	3	9	0.074286
2	2	4	10	0.073542
3	3	4	11	0.073248

total number of coefficients estimated and V the mean-square output error. For models 1 and 2, f is

$$\frac{7.44 \times 10^{-4}}{0.073542} \frac{1445}{1} = 14.6$$

If this is greater than the value exceeded by a $F(1, 1445)$ variate with probability α, the hypothesis that the extra coefficient in model 2 is redundant is rejected at level α. For $\alpha = 0.05$ the value is 3.84, and for $\alpha = 0.01$ it is 6.63, so the hypothesis is rejected at either level. Comparing models 2 and 3, f is

$$\frac{2.94 \times 10^{-4}}{0.073248} \frac{1444}{1} = 5.80$$

so model 3 is taken as significantly better at level 0.05 but not at level 0.01.

△

9.3.3 The Akaike Information Criterion

The log-likelihood function discussed in Section 6.4 is the basis of an alternative test of model structure (Akaike, 1974). The Kullback–Liebler mean information is defined as

$$I(\hat{\boldsymbol{\theta}}_i, \boldsymbol{\theta}_0) \triangleq \underset{Y}{E}[L(\boldsymbol{\theta}_0) - L(\hat{\boldsymbol{\theta}}_i)] \tag{9.3.6}$$

where $L(\boldsymbol{\theta})$ is the log-likelihood function of $\boldsymbol{\theta}$ given the set of observations Y and $\boldsymbol{\theta}_0$ is the "true" value of $\boldsymbol{\theta}$. If we suppress any doubts about the meaning of the "true" $\boldsymbol{\theta}$, it turns out that $I(\hat{\boldsymbol{\theta}}_i, \boldsymbol{\theta}_0)$ can be approximated by the sum of a term in $\boldsymbol{\theta}_0$, the same for any candidate model structure, and a term proportional to the *information criterion*

$$C_i \triangleq -2L(\hat{\boldsymbol{\theta}}_i) + 2p_i \tag{9.3.7}$$

The test then consists of comparing C_2 with C_1 and accepting $\mathcal{M}_1$ as adequate if C_1 is smaller. (The information criterion is abbreviated to AIC. According to Akaike, the A stands for A, as distinct from B, C, etc.; according to everyone else, it stands for Akaike.)

From Section 6.4.5, we know that scalar observations affected by Gaussian errors of unknown constant variance give rise to a maximum log-likelihood

$$L(\hat{\boldsymbol{\theta}}_i) = -\frac{N}{2}\left(1 + \ln 2\pi + \ln\left(\frac{1}{N}\sum_{t=1}^{N} \hat{e}_t^2(\hat{\boldsymbol{\theta}}_i)\right)\right) \tag{9.3.8}$$

If we substitute this into (9.3.7) and drop the part independent of $\hat{\boldsymbol{\theta}}_i$ we obtain the statistic

$$C_i' = N \ln V_i + 2p_i \tag{9.3.9}$$

to test. Notice the assumption that $\hat{\boldsymbol{\theta}}_i$ is the m.l. estimate for $\mathcal{M}_i$.

Example 9.3.2 The models quoted in Example 9.3.1 were obtained by the e.l.s. algorithm, which is approximately m.l. if the noise is assumed Gaussian. The models give the values in the accompanying tabulation, so model 3 is preferred. A reduction of 0.26% in V_2 would be enough to make $C_2' < C_3'$.

Model	V	p	C'
1	0.07429	9	−3765
2	0.07354	10	−3777
3	0.07325	11	−3781

Both the F test and the AIC test seem to favour the larger model unless the m.s. output errors are very close. This point is followed up in problems 9.4 and 9.5.

△

The attraction of the AIC test is that it does not require a significance level to be chosen subjectively. As Söderström (1977) has pointed out, the AIC test can be viewed as an F test, for if by the AIC test

$$N \ln V_1 + 2p_1 < N \ln V_2 + 2p_2 \tag{9.3.10}$$

then

$$\frac{V_1}{V_2} < \exp\left(\frac{2}{N}(p_2 - p_1)\right) \tag{9.3.11}$$

and so the AIC test is equivalent to an F test giving

$$f = \left(\frac{V_1}{V_2} - 1\right)\frac{N - p_2}{p_2 - p_1} < \frac{N - p_2}{p_2 - p_1}\left(\exp\left(\frac{2}{N}(p_2 - p_1)\right) - 1\right) \tag{9.3.12}$$

Moreover, with p_2 and $p_2 - p_1$ much smaller than N,

$$\frac{N-p_2}{p_2-p_1}\left(\exp\left(\frac{2}{N}(p_2-p_1)\right)-1\right) \simeq \frac{N-p_2}{p_2-p_1}\frac{2}{N}(p_2-p_1) \simeq 2 \qquad (9.3.13)$$

This value corresponds to a not very stringent significance level of roughly 16% for $p_2 - p_1 = 1$ (addition of one parameter) and little lower for $p_2 - p_1 = 2$, both for large N.

The main doubt about the usefulness of the F and AIC tests arises from their assumption of Gaussian $\{\hat{e}\}$. In many identification problems the noise and residuals are distinctly non-Gaussian. An approach to model-order testing which relies less on an assumption about the probability distribution of the noise would be of interest. For a linear model, the product-moment matrix provides just such an approach.

9.3.4 Product-Moment Matrix Test

The idea behind model-structure testing using product-moment matrices (Lee, 1964) is that the noise-free output x_t from a system with input–output dynamics

$$x_t = -a_1 x_{t-1} - \cdots - a_n x_{t-n} + b_1 u_{t-1} + \cdots + b_n u_{t-n}, \qquad t = 1, 2, \ldots, N \qquad (9.3.14)$$

is exactly linearly dependent on the set of samples x_{t-1} to x_{t-n} and u_{t-1} to u_{t-n}. For any trial model order $\hat{n}$ greater than n, the dependence reduces the rank of

$$U(u, x, \hat{n}) = \left[\begin{array}{cccc|cccc} x_{\hat{n}} & x_{\hat{n}-1} & \cdots & x_1 & u_{\hat{n}} & u_{\hat{n}-1} & \cdots & u_1 \\ x_{\hat{n}+1} & \cdots & & & u_{\hat{n}+1} & \cdots & & \\ \vdots & & & & \vdots & & & \\ x_N & x_{N-1} & \cdots & & u_N & u_{N-1} & \cdots & \end{array}\right] \qquad (9.3.15)$$

by the number $\hat{n} - n$ of x columns linearly dependent on the other columns. If the 'clean' output x were accessible, a simple test of system order would be to check the rank of U or, more conveniently, the small ($2\hat{n}$ square) matrix $U^T U/N$, henceforth called A. Assuming that the input is p.e. of order $n + 1$ or more, the u columns are still linearly independent when $\hat{n}$ is $n + 1$, so the onset of singularity of A at that $\hat{n}$ indicates unequivocally that the system order is n, the largest $\hat{n}$ for which A remains non-singular. Unfortunately, $\{x\}$ is not accessible, and the rank deficiency of A is obscured by noise $\{e\}$ in the observed output samples $\{y\}$:

$$y_t = x_t + e_t, \qquad t = 1, 2, \ldots, N \qquad (9.3.16)$$

For high output signal-to-noise ratios, it may be possible to use $\{y\}$ in place of $\{x\}$ in U and still detect the onset of ill-conditioning of A as $\hat{n}$ is raised, but for realistic amounts of noise some modification is normally required. An early suggestion (Woodside, 1971) was to estimate and remove explicitly the effect of noise on A. If we denote $U(u, y, \hat{n})$ by

$$U(u, y, \hat{n}) = [\mathbf{y}_{\hat{n}} \quad \mathbf{y}_{\hat{n}-1} \quad \mathbf{y}_1 \quad \mathbf{u}_{\hat{n}} \quad \mathbf{u}_{\hat{n}-1} \quad \cdots \quad \mathbf{u}_1] \tag{9.3.17}$$

where

$$\mathbf{y}_{\hat{n}-i}^{\mathrm{T}} = [y_{\hat{n}-i} \quad y_{\hat{n}-i+1} \quad \cdots \quad y_{N-i}], \qquad i = 0, 1, \ldots, \hat{n}-1 \tag{9.3.18}$$

and similarly for $\mathbf{u}_{\hat{n}-i}$, then

$$A(u, y, \hat{n}) = \frac{1}{N}\left[\begin{array}{cccc|cccc} \mathbf{y}_{\hat{n}}^{\mathrm{T}}\mathbf{y}_{\hat{n}} & \mathbf{y}_{\hat{n}}^{\mathrm{T}}\mathbf{y}_{\hat{n}-1} & \cdots & \mathbf{y}_{\hat{n}}^{\mathrm{T}}\mathbf{y}_1 & \mathbf{y}_{\hat{n}}^{\mathrm{T}}\mathbf{u}_{\hat{n}} & \mathbf{y}_{\hat{n}}^{\mathrm{T}}\mathbf{u}_{\hat{n}-1} & \cdots & \mathbf{y}_{\hat{n}}^{\mathrm{T}}\mathbf{u}_1 \\ \vdots & & & & & & & \\ \mathbf{y}_1^{\mathrm{T}}\mathbf{y}_{\hat{n}} & \cdots & & \mathbf{y}_1^{\mathrm{T}}\mathbf{y}_1 & \mathbf{y}_1^{\mathrm{T}}\mathbf{u}_{\hat{n}} & \cdots & & \mathbf{y}_1^{\mathrm{T}}\mathbf{u}_1 \\ \hline \mathbf{u}_{\hat{n}}^{\mathrm{T}}\mathbf{y}_{\hat{n}} & \cdots & & \mathbf{u}_{\hat{n}}^{\mathrm{T}}\mathbf{y}_1 & \mathbf{u}_{\hat{n}}^{\mathrm{T}}\mathbf{u}_{\hat{n}} & \cdots & & \mathbf{u}_{\hat{n}}^{\mathrm{T}}\mathbf{u}_1 \\ \vdots & & & & & & & \\ \mathbf{u}_1^{\mathrm{T}}\mathbf{y}_{\hat{n}} & \cdots & & \mathbf{u}_1^{\mathrm{T}}\mathbf{y}_1 & \mathbf{u}_1^{\mathrm{T}}\mathbf{u}_{\hat{n}} & \cdots & & \mathbf{u}_1^{\mathrm{T}}\mathbf{u}_1 \end{array}\right] \tag{9.3.19}$$

We assume that $\{e\}$ is uncorrelated with $\{u\}$ and therefore with $\{x\}$, which depends only on $\{u\}$, and also assume that $\{e\}$ and $\{u\}$ are ergodic and the system time-invariant. As the number of observations N rises, these assumptions give

$$A(u, y, \hat{n}) \rightarrow A(u, x, \hat{n}) + \begin{bmatrix} R_{ee}(\hat{n}) & 0 \\ 0 & 0 \end{bmatrix} \tag{9.3.20}$$

where element (i, j) of $R_{ee}(\hat{n})$ is the autocorrelation of $\{e\}$ at lag $i-j$. Noise on the observed input $\{u\}$ can be treated by allowing an input-noise autocorrelation matrix as the bottom right partition. Estimates of the noise autocorrelation ordinates have to be supplied if we are to reconstruct the noise-free A from

$$\hat{A}(u, x, \hat{n}) = A(u, y, \hat{n}) - \begin{bmatrix} R_{ee}(\hat{n}) & 0 \\ 0 & 0 \end{bmatrix} \tag{9.3.21}$$

It may be possible to measure the noise a.c.f. by holding the input steady, so that all output variation is noise. If not, informed guesswork may yield a usable estimate of the normalised a.c.f. (normalised by $r_{ee}(0)$, the noise m.s. value). A possible way of providing $r_{ee}(0)$, suggested by Woodside, is to take the smallest solution of

$$|A(u, x, \hat{n})| = \left| A(u, y, \hat{n}) - r_{ee}(0) \begin{bmatrix} R'_{ee}(\hat{n}) & 0 \\ 0 & \end{bmatrix} \right| = 0 \tag{9.3.22}$$

where $R'_{ee}(\hat{n})$ is the normalised autocorrelation matrix and $\hat{n}$ is large enough to justify (9.3.22). Equation (9.3.22) says $r_{ee}(0)$ is a generalised eigenvalue of $A(u, y, \hat{n})$.

However $\hat{A}(u, x, \hat{n})$ is obtained, it will be ill-conditioned rather than singular, because of the approximation error. Some way has to be found of deciding when the matrix is ill enough conditioned to indicate that $\hat{n}$ is greater than n. Woodside suggests forming either the determinant ratio

$$\rho(\hat{n}) = |\hat{A}(u, x, \hat{n})| / |\hat{A}(u, x, \hat{n}+1)| \tag{9.3.23}$$

as a normalised scalar statistic, or another determinant ratio $\rho_s(\hat{n})$ which approximates the sum of the squared model-output errors obtainable in the absence of noise. As $\hat{n}$ is increased, $\rho(\hat{n})$ should jump upwards at $\hat{n} = n$, and $\rho_s(\hat{n})$ should drop suddenly at $\hat{n} = n$, as output modelling error due to too low a model order vanishes. The derivation of $\rho_s(\hat{n})$ is a bit tedious, and you may prefer just to note the results, (9.3.29) and (9.3.31).

First we write down the sum S_{n-1} of squared regression-equation errors for the o.l.s. model of order $n-1$. Next we express S_{n-1} as the quotient of two determinants which relate closely to $A(u, y, n-1)$ and $A(u, y, n)$. Finally, we replace $A(u, y, n-1)$ and $A(u, y, n)$ by $\hat{A}(u, x, n-1)$ and $\hat{A}(u, x, n)$ so as to reduce the effects of the noise in $\{y\}$, and compute an "enhanced" sum of squares of errors. The details are as follows.

The regressor matrix for the model given by (9.3.14) and (9.3.16), but with order $n-1$ rather than n, is

$$H_{n-1} \triangleq \begin{bmatrix} y_{n-1} & \cdots & y_1 & u_{n-1} & \cdots & u_1 \\ y_{N-1} & \cdots & y_{N-n+1} & u_{N-1} & \cdots & u_{N-n+1} \end{bmatrix} \equiv [Y_{n-1} \quad U_{n-1}] \tag{9.3.24}$$

The o.l.s. model outputs $\hat{y}_n$ to $\hat{y}_N$ form the vector

$$\hat{\mathbf{y}} = H_{n-1}[H_{n-1}^{\mathrm{T}} H_{n-1}]^{-1} H_{n-1}^{\mathrm{T}} \mathbf{y}_n \tag{9.3.25}$$

and the sum of squares of errors is

$$S_{n-1} = (\mathbf{y}_n - \hat{\mathbf{y}})^{\mathrm{T}}(\mathbf{y}_n - \hat{\mathbf{y}}) = \mathbf{y}_n^{\mathrm{T}}(\mathbf{y}_n - \hat{\mathbf{y}}) \tag{9.3.26}$$

since $\hat{\mathbf{y}}$ is orthogonal to $\mathbf{y}_n - \hat{\mathbf{y}}$. All the quantities in (9.3.25) and (9.3.26) may be computed from the partitioned matrix

$$B(u, y, n) \triangleq \begin{bmatrix} \mathbf{y}_n^{\mathrm{T}} \mathbf{y}_n & \mathbf{y}_n^{\mathrm{T}} H_{n-1} \\ H_{n-1}^{\mathrm{T}} \mathbf{y}_n & H_{n-1}^{\mathrm{T}} H_{n-1} \end{bmatrix} \tag{9.3.27}$$

Expansion of $|B(u, y, n)|$ by its first row gives

$$|B(u, y, n)| = \mathbf{y}_n^{\mathrm{T}} \mathbf{y}_n |H_{n-1}^{\mathrm{T}} H_{n-1}| - \mathbf{y}_n^{\mathrm{T}} H_{n-1} \operatorname{adj}(H_{n-1}^{\mathrm{T}} H_{n-1}) H_{n-1}^{\mathrm{T}} \mathbf{y}_n \tag{9.3.28}$$

so

$$S_{n-1} = \mathbf{y}_n^{\mathrm{T}}\mathbf{y}_n - \mathbf{y}_n^{\mathrm{T}}H_{n-1}[H_{n-1}^{\mathrm{T}}H_{n-1}]^{-1}H_{n-1}^{\mathrm{T}}\mathbf{y}_n = \frac{|B(u,y,n)|}{|H_{n-1}^{\mathrm{T}}H_{n-1}|} \qquad (9.3.29)$$

We can relate $H_{n-1}^{\mathrm{T}}H_{n-1}$ to $A(u, y, n-1)$ by noticing that

$$U(u,y,n-1) = \begin{bmatrix} & Y_{n-1} & & & U_{n-1} & \\ y_N & \cdots & y_{N-n+2} & u_N & \cdots & u_{N-n+2} \end{bmatrix} \qquad (9.3.30)$$

so that

$$\begin{aligned} NA(u,y,n-1) &= U(u,y,n-1)^{\mathrm{T}}U(u,y,n-1) \\ &= H_{n-1}^{\mathrm{T}}H_{n-1} + [y_N \quad \cdots \quad y_{N-n+2}u_N \quad \cdots \quad u_{N-n+2}]^{\mathrm{T}} \\ &\quad \times [y_N \quad \cdots \quad y_{N-n+2}u_N \quad \cdots \quad u_{N-n+2}] \end{aligned} \qquad (9.3.31)$$

For N large, the right-hand side of (9.3.31) is close to $H_{n-1}^{\mathrm{T}}H_{n-1}$. Also,

$$U(u,y,n) = [\mathbf{y}_n \quad Y_{n-1} \quad \mathbf{u}_n \quad U_{n-1}] \qquad (9.3.32)$$

so $B(u, y, n)$ could be obtained by deleting the $\mathbf{u}_n$ column and $\mathbf{u}_n^{\mathrm{T}}$ row from $NA(u, y, n)$. If we reconstruct the approximate noise-free $\hat{A}(u, x, n-1)$ and $\hat{A}(u, x, n)$, and pick out $\hat{B}(u, x, n)$ from the latter, we can compute in place of (9.3.29)

$$\rho_s(n-1) = |\hat{B}(u,x,n)|/|N\hat{A}(u,x,n-1)| \qquad (9.3.33)$$

Woodside found that this "enhanced" statistic indicates the correct order of simulated records with mean-square signal:noise ratios down to about 10.

Instead of finding $\hat{A}(u, x, n)$ by estimating $R_{ee}(n)$, Wellstead (1978) suggests calculating

$$A'(u,y,n) = (U^{\mathrm{T}}(u,z,n)U(u,y,n))/N \qquad (9.3.34)$$

where $\{z\}$ is an instrumental-variable sequence which, as in section 7.4.5, is strongly correlated with $\{u\}$ and $\{x\}$ but uncorrelated with the noise sequence $\{e\}$. Provided that z_t and u_t are uncorrelated with e_{t-n+1} up to e_{t+n-1},

$$E[A'(u,y,n)] = E[A'(u,x,n)] \qquad (9.3.35)$$

Now $A'(u, x, n)$ suffers rank deficiency in exactly the same way as does $A(u, x, n)$ since $U(u, x, n)$ is present in both, so the determinant-ratio tests described above can be performed equally well using $A'(u, x, n)$ in place of $A(u, x, n)$. At the small price of having to treat the input as noise-free, this provides a computationally cheap alternative to Woodside's enhancement method. Caution is necessary if $\{z\}$ is generated by passing $\{u\}$ through a linear filter; if the order of the filter is less than the trial order $\hat{n}$, the determinant ratios will detect the filter order rather than the system order.

Generalisations of the instrumental product-moment matrix method are presented by Wellstead and Rojas (1982). They point out that minor amendments to U allow the test to cover different orders n and m for the autoregressive and moving-average parts of the model (9.3.14) and to find the dead time k. The testing proceeds in stages. First $\hat{n}$ and $\hat{m} + \hat{k}$ are increased together to identify the larger of n and $m + k$, then $\hat{n}$ and $\hat{m} + \hat{k}$ are reduced alternately to find which is the smaller of n and $m + k$, and establish its value. Finally $\hat{m}$ and $\hat{k}$ are varied to find m and k.

Taking a more statistical view, the product-moment matrix tests are methods of detecting ill-conditioning of the covariance matrix of the model-coefficient estimates. The normal matrix for o.l.s. with the model (9.3.14) and (9.3.16) is $NA(u, y, n)$, so on the assumption of uncorrelated regression-equation error ("white noise"), the covariance of the coefficient estimates is $r_{ee}(0)A^{-1}(u, y, n)/N$. It is estimated as an intrinsic part of the e.l.s. algorithm, and can also be obtained easily from the product-moment matrix inverse updated by the recursive instrumental-variable algorithm (Young *et al.*, 1980). Scalar measures of covariance ill-conditioning are discussed in this reference also.

We should recall at this point that Chapter 4 has already provided ways of testing the structure of linear models. The utility of each term can be tested by singular-value decomposition of U, as described in Section 4.2.4. Alternatively, the Golub–Householder method described in Section 4.2.2 makes trial deletion of terms from the model easy; one need only place those terms last, so that deleting them merely removes corresponding columns from the extreme right of the triangular matrix V in (4.2.10), leaving the rest of the computation unchanged.

9.4 SUMMARY

In this chapter we have reviewed several results and techniques which may be some help in choosing a model structure and designing an experiment to estimate its coefficients. The reason for such diffident wording is that success in identifying a useful model depends much more on accurate appreciation of what the model must do and recognition of what is going on in the results of identification experiments than on virtuosity in applying analytical techniques. The next chapter will illustrate some of the problems that arise when we start experimenting in earnest.

FURTHER READING

Optimal experiment design is covered by Silvey (1980), Beck and Arnold (1977), Zarrop (1979), Goodwin and Payne (1977) and Kalaba and Springarn

(1982). Mehra and Lainiotis (1976) have two substantial sections on the topic, along with interesting identification case studies. We have not considered multivariable systems, although the basic ideas concerning experiment design and parsimony apply to them too. Model structure and identification accuracy for multivariable systems are the subjects of a chapter in Kashyap and Rao (1976). Guidorzi *et al.* (1982) offer a test for multivariable model structure which does not require prior fitting of a selection of models.

Comparative studies of a number of structure selection methods have been carried out by van den Boom and van den Enden (1974).

REFERENCES

Akaike, H. (1974). A new look at the statistical model identification. *IEEE Trans. Autom. Control* **AC-19**, 716–723.

Åström, K. J., and Bohlin, T. (1966). Numerical identification of linear dynamic systems from normal operating records. *In* "Theory of Self-Adaptive Control Systems" (P. H. Hammond, ed.). Plenum, New York.

Beck, J. V., and Arnold, K. J. (1977). "Parameter Estimation in Engineering and Science". Wiley, New York.

Box, G. E. P., and Jenkins, G. M. (1976). "Time Series Analysis Forecasting and Control". Holden-Day, San Francisco, California.

D'Azzo, J. J., and Houpis, C. H. (1981). "Linear Control System Analysis and Design", 2nd ed. McGraw-Hill, New York.

Di Stefano, J. J. (1980). Design and optimisation of tracer experiments in physiology and medicine. *Fed. Proc.* **39**, 84–90.

Goodwin, G. C., and Payne, R. L. (1977). "Dynamic System Identification: Experiment Design and Data Analysis". Academic Press, New York and London.

Goodwin, G. C., and Sin, K. S. (1984). "Adaptive Filtering Prediction and Control". Prentice-Hall, Englewood Cliffs, New Jersey.

Goodwin, G. C., Norton, J. P., and Viswanathan, M. N. (1985). Persistency of excitation for nonminimal models of systems having purely deterministic disturbances. *IEEE Trans. Autom. Control* **AC-30**, 589–592.

Guidorzi, R. P., Losito, M. P., and Muratori, T. (1982). The range error test in the structual identification of linear multivariable systems. *IEEE Trans. Autom. Control* **AC-27**, 1044–1053.

Kalaba, R., and Springarn, K. (1982). "Control, Identification and Input Optimization". Plenum, New York.

Kashyap, R. L., and Rao, A. R. (1976). "Dynamic Stochastic Models from Empirical Data". Academic Press, New York and London.

Lee, R. C. K. (1964). "Optimal Estimation, Identification and Control". MIT Press, Cambridge, Massachusetts.

Mehra, R. K., and Lainiotis, D. G. (eds.) (1976). "System Identification". Academic Press, New York and London.

Moore, J. B. (1983). Persistence of excitation in extended least squares. *IEEE Trans. Autom. Control* **AC-28**, 60–68.

Robins, A. J. (1984). Identification of aerodynamic derivatives using an extended Kalman filter. Lecture notes, graduate course on Kalman filtering, Dept. of Electronic and Electrical Eng., Univ. of Birmingham, U.K.

Silvey, S. D. (1980). "Optimal Design". Chapman & Hall, London.
Söderström, T. (1977). On model structure testing in system identification. *Int. J. Control* **26**, 1–18.
Solo, V. (1983). "Advanced Topics in Time Series Analysis". Springer-Verlag, Berlin and New York.
Stoica, P., and Söderström, T. (1982). On the parsimony principle. *Int. J. Control* **36**, 409–418.
van den Boom, A. J. W. and van den Enden (1974). The determination of the orders of process and noise dynamics. *Automatica* **10**, 245–256.
Wadsworth, G. P., and Bryan, J. G. (1974). "Applications of Probability and Random Variables", 2nd ed. McGraw-Hill, New York.
Wellstead, P. E. (1978). An instrumental product moment test for model order estimation. *Automatica* **14**, 89–91.
Wellstead, P. E., and Rojas, R. A. (1982). Instrumental product moment model-order testing: Extensions and applications. *Int. J. Control* **35**, 1013–1027.
Woodside, C. M. (1971). Estimation of the order of linear systems. *Automatica* **7**, 727–733.
Young, P. C. (1984). "Recursive Estimation and Time-Series Analysis". Springer-Verlag, Berlin and New York.
Young, P. C., Jakeman, A., and McMurtrie, R. (1980). An instrumental variable method for model order identification. *Automatica* **16**, 281–294.
Yuan, J. S-C., and Wonham, W. M. (1977). Probing signals for model reference identification. *IEEE Trans. Autom. Control* **AC-22**, 530–538.
Zarrop, M. B. (1979). "Optimal Experiment Design for Dynamic System Identification". Springer-Verlag, Berlin and New York.

PROBLEMS

9.1 Which of the coefficients y_v, n_v, n_r and n_ζ in Example 9.2.1 are likely to be hard to identify by an experiment in which the input signal $\zeta(t)$ is (i) of small bandwidth, (ii) narrowband, at a frequency high enough for the frequency-independent terms to have little effect on L/Z and R/Z? [Assume y_ζ is known in advance.]

9.2 Investigate the improvement in accuracy of $\hat{k}_{12}$ and $\hat{k}_{21}$ in Example 9.2.2, compared to the results of sampling at times 1, 2 and 3, due to (i) sampling at times 1, 2, 3 and 4; (ii) sampling at 1, 2, 3 and 10; (iii) adding a second sample at time 1, independent of the first; (iv) adding a second sample at time 3, independent of the first; (v) halving the sampling interval.

9.3 Three alternative model structures are

$$\mathcal{M}_1: \quad \dot{x} = ax + (fx/(x+g)) + bu$$

$$\mathcal{M}_2: \quad \dot{x} = ax + h + bu$$

$$\mathcal{M}_3: \quad \dot{x} = a_1x + a_2x^2 + \cdots + a_nx^n + bu$$

What combinations of two or more of these structures are hierarchical? [State any restrictions necessary.]

9.4 If two linear model structures differing by q terms are to be compared by an F test, and the records are 200 or more samples long, the F distribution relevant to the test is close to $F(q, \infty)$. The values of f exceeded by an $F(q, \infty)$ variate with probability 0.05 are

q	1	2	4	6	8	10
$f_{0.05}$	3.84	3.00	2.37	2.10	1.94	1.83

Tabulate the percentage excess of the smaller model's sum of squared output errors over that of the larger model when the hypothesis that the larger model is no better is just accepted by the F test, for these values of q and for records of length $N = 200$, 1000 and 5000. Ponder the likely practical significance of percentage differences of this size.

9.5 Tabulate the percentage excess of a smaller model's sum of squared output errors over that of a larger model when the Akaike information criterion says that two linear, time-invariant model structures differing by q terms are equally acceptable, for $q = 1, 2, 4, 6, 8$ and 10, and for records of length $N = 200$, 1000 and 5000. [Use the mean-square output error of the larger model as σ^2.] Compare this table with the one compiled in Problem 9.4.

Chapter 10

Model Validation

10.1 INTRODUCTION

10.1.1 Nature of Validation Tests

A model is validated by answering two questions: is it credible, and does it work? The first supposes we have some background knowledge and want the model to conform with it.

Example 10.1.1 The model

$$y_t = -a_1 y_{t-1} - a_2 y_{t-2} + b_1 u_{t-1} + b_2 u_{t-2} + c_1 v_{t-1} + c_2 v_{t-2} + c_3 v_{t-3} + v_t$$

is fitted to hourly recordings of rainfall $\{u\}$ in the catchment of the River Eden in north-west England and corresponding recordings of river flow $\{y\}$. The aim is to develop an on-line flow predictor.

After 100 steps of the e.l.s. algorithm, we have

$$[\hat{a}_1 \quad \hat{a}_2 \quad \hat{b}_1 \quad \hat{b}_2] = [-1.527 \quad 0.598 \quad -0.141 \quad 0.992]$$

Since b_1 is the first non-zero ordinate of the u.p.r., it cannot be negative, which would imply that the initial effect of rainfall is to reduce flow. Hence, $\hat{b}_1$ is implausible. The reason is that the model has too short a dead time; the latest input affecting y_t is u_{t-2}. In the absence of any influence of u_{t-1} on y_t, the value of $\hat{b}_1$ is determined by its effect on the subsequent u.p.r. shape, i.e. coefficients from h_2 on in

$$h_1 z^{-1} + h_2 z^{-2} + \cdots \infty = \frac{b_1 z^{-1} + b_2 z^{-2}}{1 + a_1 z^{-1} + a_2 z^{-2}}$$

When the u.p.r. of a model with negative $\hat{b}_1$ and too short a dead time is plotted, it normally approximates the u.p.r. of the model with the correct dead time quite well at lags beyond the correct dead time. △

A direct and revealing test of whether a model works is to try it on records different from those it was estimated from. However, there could well be a noticeable finite-sample difference between the performances of the model on the two sets of records, even if the model were optimal in some statistical sense and structurally well chosen. We might consider testing the significance of the difference in performance by comparing a statistic from the new records, for instance the m.s. output prediction error, with its theoretical value, computed from the estimated covariance P_N of the final parameter estimate $\hat{\mathbf{x}}_N$ of the original records. At sample instant i in the new records, the theoretical m.s. error in the output prediction $\hat{y}_i(\hat{\mathbf{x}}_N)$ obtained from an unbiased estimate $\hat{\mathbf{x}}_N$, for a system described by

$$y_i = \mathbf{h}_i^{\mathrm{T}}\mathbf{x} + v_i \tag{10.1.1}$$

is

$$\begin{aligned} E[(y_i - \hat{y}_i(\hat{\mathbf{x}}_N))^2] &= E[(\mathbf{h}_i^{\mathrm{T}}(\mathbf{x} - \hat{\mathbf{x}}_N) + v_i)(\mathbf{h}_i^{\mathrm{T}}(\mathbf{x} - \hat{\mathbf{x}}_N) + v_i)^{\mathrm{T}}] \\ &= \mathbf{h}_i^{\mathrm{T}} E[(\mathbf{x} - \hat{\mathbf{x}}_N)(\mathbf{x} - \hat{\mathbf{x}}_N)^{\mathrm{T}}]\mathbf{h}_i + E[v_i^2] \\ &= \mathbf{h}_i^{\mathrm{T}} P_N \mathbf{h}_i + \sigma_{vi}^2 \end{aligned} \tag{10.1.2}$$

Here $\mathbf{h}_i$ has been treated as deterministic as we are not averaging over a range of possible $\mathbf{h}_i$, and (10.1.1) has been assumed an adequate description of y_i. Also, v_i is assumed to be zero-mean and not correlated with $\hat{\mathbf{x}}_N$. The actual covariance P_N and variance σ_{vi}^2 would be replaced by their computed estimates to approximate the expected sample m.s. prediction error

$$E\left[\frac{1}{t}\sum_{i=1}^{t}(y_i - \hat{y}_i(\hat{\mathbf{x}}_N))^2\right] = \frac{1}{t}\sum_{i=1}^{t}(\mathbf{h}_i^{\mathrm{T}} P_N \mathbf{h}_i + \sigma_{vi}^2) \tag{10.1.3}$$

Such a statistic is easy to compute, for given $\{\mathbf{h}\}$ and $\{\sigma_v^2\}$, but has some unconvincing aspects. We should really take into account the uncertainty in P_N and the estimates of σ_{vi}^2. To assess the significance of a deviation of the statistic from its theoretical value given by (10.1.3) we need its sampling distribution. The distribution may be difficult to specify, as the prediction errors may well not form a stationary sequence. Formal tests of this sort may have a role in refining an already good model, but in earlier stages of model validation or when a good model is not realistically attainable, less formal checks less reliant on idealising assumptions are more to the point.

The remainder of this chapter illustrates a selection of validation checks applied to results from actual records. The tests are mostly quite informal, and bring out typical weaknesses in models. As usual, we do not pretend the tests are comprehensive or universally applicable.

10.1.2 What Do We Test?

The short answer is "everything we can", but more specifically we can test

(i) the records, before we do anything with them;
(ii) the parameter estimates, in the light of background knowledge;
(iii) the fit of the model to the records, through the residuals $\{y-\hat{y}\}$;
(iv) the estimated covariance of the parameter estimates; and
(v) the behaviour of the model as a whole, measured for instance by steady-state gain, u.p.r., poles and zeros.

We shall examine these tests entirely by examples, and try to resist too-sweeping conclusions.

10.2 CHECKS BEFORE AND DURING ESTIMATION

10.2.1 Checks on Records

Valuable information can be gained by looking at a plot of the records as soon as they are received.

Example 10.2.1 The rainfall and river flow records used in Example 10.1.1 are shown in Fig. 10.2.1. We see immediately that two distinct situations alternate: rapidly changing flow with frequent or continuous rainfall, and smooth monotonic flow decrease (recession) with little or no rainfall. The question arises whether one constant-parameter model can cater for both situations. We also see that the low-frequency gain from rainfall to flow varies from one flow peak to another; for instance, the peak near 213 h is higher than the one near 189 h, but preceded by less rainfall in the previous 10 h or so. We should not expect too much from a time-invariant linear model, and may well have to resort to a time-varying model for on-line prediction, the ultimate aim. A time-invariant non-linear model might be preferable if we knew enough to choose its form, but we shall not pursue that option.

The spread of time constants looks large, from two hours or so as indicated by the rises to perhaps ten hours or more for the slowest recession components. The flow record is quite smooth, so the sampling interval is probably short enough to avoid aliasing. The rainfall record is both uneven and heavily quantised, but each point is the integrated rainfall over an hour and the rainfall-to-flow dynamics are clearly of low bandwidth. Any increase in explanatory power of the rainfall record achieved by shortening the

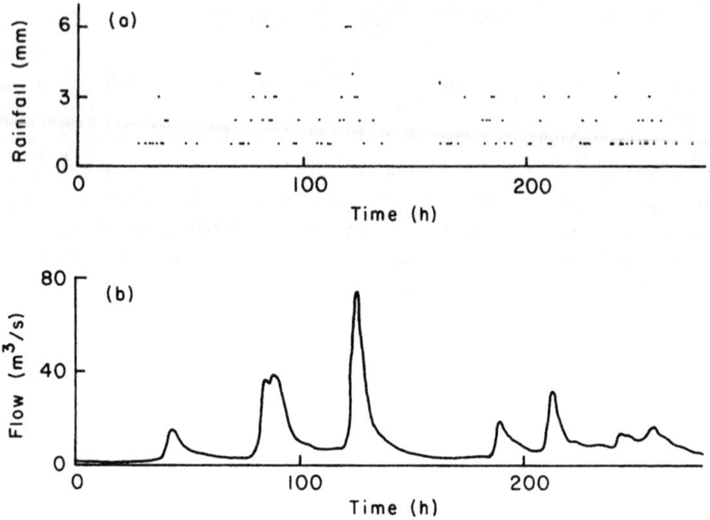

Fig. 10.2.1 (a) Rainfall and (b) river-flow records.

sampling interval would consequently be outweighed by the increase in quantisation error. The sampling interval may be too short to allow precise estimation of the longest time constant of the u.p.r.

The dead time cannot be assessed from the raw records in this instance, but is not more than a few hours.

We shall employ these records in many examples, since their behaviour is complex enough to be a good test of identification and validation methods.

△

Other things to look for, not evident in the example, are instrument breakdowns, transcription errors and patching-up of interrupted records. Errors are hard to avoid in records taken or transcribed manually, but they are easy to detect in smooth records and need not be detected if infrequent and comparable with noise from other sources. Breaks in records, particularly due to oversight, are sometimes repaired by crude interpolation without the recipient being informed, so any constant or straight-line section should be regarded with suspicion.

10.2.2 Checks on Parameter Estimates

In Example 10.1.1, unrealistic model behaviour was detected through the value of a single parameter. Quantities affected by all the parameters, such as u.p.r., steady-state (zero-frequency) gain and poles and zeros can be computed easily from a.r.m.a.x. parameter estimates, on line if necessary, and checked against background knowledge. The u.p.r. $\{\hat{h}\}$ is given by

$$\hat{h}_t = -\hat{a}_1\hat{h}_{t-1} - \cdots - \hat{a}_n\hat{h}_{t-n} + \hat{b}_1\delta_{t-k-1} + \cdots + \hat{b}_m\delta_{t-k-m} \quad (10.2.1)$$

where δ_{t-k-i} is 1 for $t = k + i$ and 0 otherwise. The steady-state gain is

$$g_s = \frac{\hat{b}_1 + \cdots + \hat{b}_m}{1 + \hat{a}_1 + \cdots + \hat{a}_n} = \sum_{t=k+1}^{\infty} \hat{h}_t \quad (10.2.2)$$

The poles are readily interpreted in terms of time constants. A positive real pole $z = \alpha$ corresponds to a sampled exponential component proportional to

$$\alpha^i = (\exp(-T/\tau))^i \equiv \exp(-iT/\tau), \qquad i = 0, 1, 2, \ldots \quad (10.2.3)$$

in the u.p.r. Hence the time constant τ is $-T/\ln \alpha$, where T is the sampling interval. Complex-conjugate poles can be interpreted as in the following example.

Example 10.2.2 We decided that the hydrological records in Example 10.2.1 exhibited a wide spread of time constants, and there was no evidence of oscillatory response in the flow, so we expect positive real poles between 0 and 1.

A model

$$y_t = -a_1 y_{t-1} - a_2 y_{t-2} + b_1 u_{t-2} + b_2 u_{t-3} + c_1 v_{t-1} + c_2 v_{t-2} + c_3 v_{t-3} + v_t$$

i.e. an a.r.m.a.x. model with $(n, m, q) = (2, 2, 3)$ and dead time 1, was fitted by e.l.s. to 280 input–output pairs from the hydrological records. The final a.r.m.a.x. coefficient estimates were

$$[\hat{a}_1 \quad \hat{a}_2 \quad \hat{b}_1 \quad \hat{b}_2] = [-1.401 \quad 0.5135 \quad 0.5399 \quad 0.8917]$$

The poles are complex conjugate, and since the z-transforms of

$$\exp(-t/\tau)\sin \beta t \quad \text{and} \quad \exp(-t/\tau)\cos \beta t$$

have denominator $1 - 2z^{-1}\exp(-T/\tau)\cos \beta T + z^{-2}\exp(-2T/\tau)$, the envelope time constant τ of the damped oscillatory u.p.r. is $-2T/\ln \hat{a}_2$, i.e. 3.0 h, and the oscillation period $2\pi/\beta$ is $2\pi T/\cos^{-1}(-\hat{a}_1/2\sqrt{\hat{a}_2})$, i.e. 29.6 h. The oscillation has no obvious physical explanation, and takes the u.p.r. negative for lags between 17 and 31 h, contrary to expectation.

We can permit a wider range of u.p.r. shapes by raising the m.a. order m to 6. The (2, 6, 3) model obtained with dead time zero (in case there is some small early response) has $[\hat{a}_1 \quad \hat{a}_2] = [-1.086 \quad 0.2629]$, implying time constants

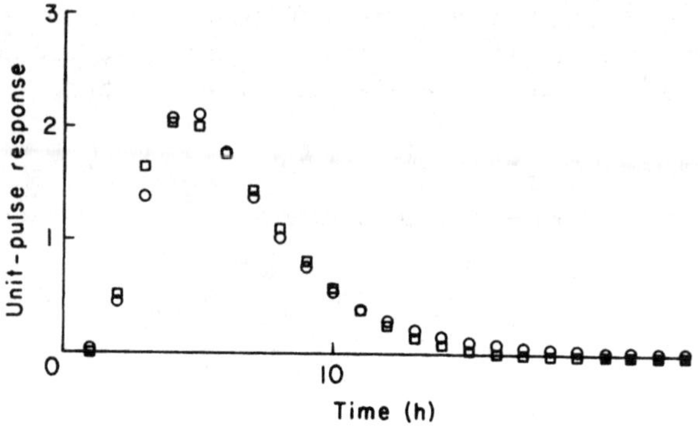

Fig. 10.2.2 Unit-pulse responses of (2, 2, 3) and (2, 6, 3) catchment models. □: $n = 2$, $m = 2$, $q = 3$; ○: $n = 2$, $m = 6$, $q = 3$.

3.07 and 0.99 h and a completely non-negative u.p.r. As Fig. 10.2.2 shows, the u.p.r.'s of the (2, 2, 3) and (2, 6, 3) models are quite close over the entire range of lags, so we should expect similar prediction performances. The smaller model in fact has r.m.s. prediction error 9% higher than the larger one over these records. △

The steady-state gain will be of most interest when we examine time-varying models, later.

10.2.3 Checks on Residuals

The recursive algorithms of Chapter 7 correct the parameter estimates at each update by an amount proportional to the most recent innovation $y_t - \mathbf{h}_t^T\hat{\mathbf{x}}_{t-1}$. Although e.l.s. and some other algorithms calculate the correction gain on a logical basis, the gain will still be poor if the calculation is based on inadequate information about the reliability of $\hat{\mathbf{x}}_{t-1}$ or the relation between y_t and $\mathbf{x}$. An awkward and easily overlooked point is that excessive correction may give $\hat{\mathbf{x}}_t$ such that $y_t - \mathbf{h}_t^T\hat{\mathbf{x}}_t$ is very small but $y_{t+1} - \mathbf{h}_{t+1}^T\hat{\mathbf{x}}_t$ is large; this is particularly likely when parameters are represented as time-varying and assigned too much short-term variability. We must therefore keep an eye on both the residuals sequence $\{y_t - \mathbf{h}_t^T\hat{\mathbf{x}}_t\}$ and the innovations sequence $\{y_t - \mathbf{h}_t^T\hat{\mathbf{x}}_{t-1}\}$. They should be similar in m.s. value if $\{\hat{\mathbf{x}}\}$ is about right (Norton, 1975).

Before we compare m.s. residuals and innovations, model deficiencies can be detected on line or off by noting large and time-structured residuals or innovations.

Example 10.2.3 Figure 10.2.3 shows the innovations produced by the (2, 2, 3)

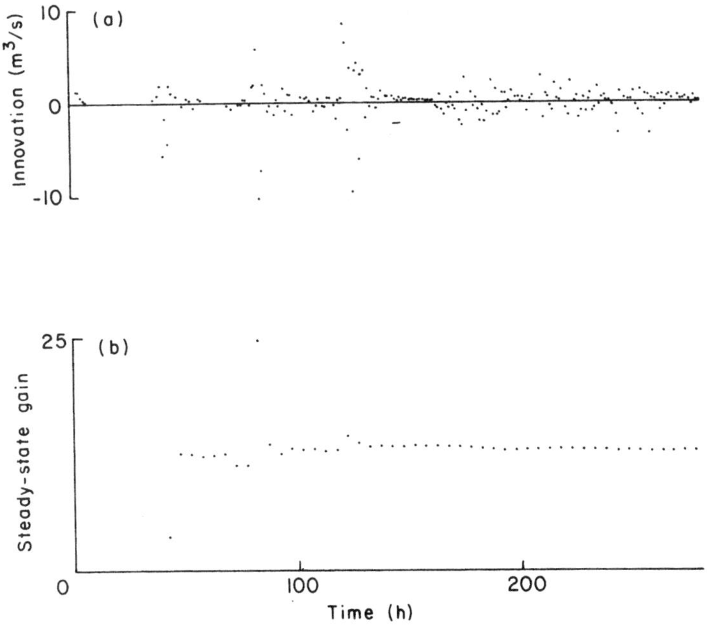

Fig. 10.2.3 (a) One-step prediction errors and (b) steady-state gain for (2, 2, 3) model of Example 10.2.2.

model of Example 10.2.2, and the accompanying parameter variation as reflected in the steady-state gain at intervals of 5 h. From about 130 h, parameter updating fails to respond to the fall in prediction accuracy during flow rises. The prediction performance over the less severely disturbed period from 160 h on is unimpressive, and there is sustained error even during flow recession between 140 and 160 h. It appears that the parameters should be treated as time-varying, by one of the methods of Section 8.1, since when they are represented as constant, the correction gain is too low. △

Although poor performance may be detected on line as in this example, the remedial action needed will vary. For that reason on-line adaptive identification will be difficult, whether it modifies the model structure and assumptions or adjusts the correction gain directly. Experience in state estimation, where adaptive recursive filtering is more an art than a science (Maybeck, 1982), bears this out. With enough prior experimentation, adaptive state-estimation algorithms can sometimes be made to work, but the best technique in a particular case is hard to predict. In short, recursive algorithms must be tuned off-line.

10.2.4 Checks on Covariance

An obvious way to assess the reliability of parameter estimates is to inspect their estimated covariance in the algorithms which, like e.l.s., compute it. The most convenient implementation of e.l.s. updates the normalised covariance

$$S_t \triangleq P_t/\sigma_t^2 \equiv (\text{cov}\,\hat{\mathbf{x}}_t)/E[v_t^2] \tag{10.2.4}$$

The algorithm is then

$$S_t = S_{t-1} - \frac{S_{t-1}\mathbf{h}_t\mathbf{h}_t^{\mathrm{T}}S_{t-1}}{1+\mathbf{h}_t^{\mathrm{T}}S_{t-1}\mathbf{h}_t}, \qquad \hat{\mathbf{x}}_t = \hat{\mathbf{x}}_{t-1} + S_t\mathbf{h}_t(y_t - \mathbf{h}_t^{\mathrm{T}}\hat{\mathbf{x}}_{t-1}) \tag{10.2.5}$$

The covariance of $\hat{\mathbf{x}}_t$ can be estimated as

$$P_t = \sigma_t^2 S_t \simeq \frac{S_t}{t-t_0}\sum_{i=t_0}^{t}(y_i - \mathbf{h}_i^{\mathrm{T}}\hat{\mathbf{x}}_i)^2 \tag{10.2.6}$$

if σ_t^2 is assumed independent of t. The summation should start late enough to miss the residuals greatly affected by the poor initial guess $\hat{\mathbf{x}}_0$.

Example 10.2.4 Figure 10.2.4 shows the estimated standard deviation $p_{33}^{1/2}$ of the estimate of the third parameter b_1 which became unrealistically negative in Example 10.1.1. It is found by (10.2.6) with $t_0 = 50$. Also shown are $s_{33}^{1/2}$ and $p_{33}^{1/2}/\hat{b}_1$. The normalised s.d. $s_{33}^{1/2}$ is uninformative about the reliability of $\hat{b}_1$

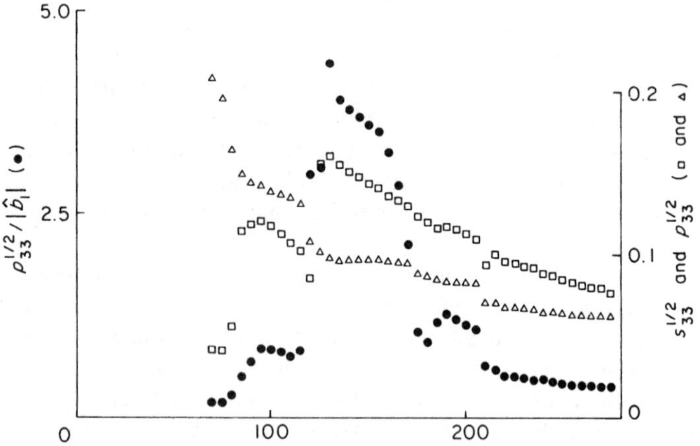

Fig. 10.2.4 Computed standard deviation of $\hat{b}_1$, standard deviation as proportion of $\hat{b}_1$, and square root of principal-diagonal element of S, in model of Example 10.1.1. ●: $p_{33}^{1/2}/|\hat{b}_1|$; △: $s_{33}^{1/2}$; □: $p_{33}^{1/2}$.

since σ_t^2 is far from constant. The estimated $p_{33}^{1/2}$ gives a better picture, showing for instance that the increase in σ_t^2 at about 120 h more than cancels the fall in $s_{33}^{1/2}$. Even $p_{33}^{1/2}$ is less than ideal, though, as it relies on a dubious estimate of σ_t^2, too small in disturbed periods and too large in smooth recessions. It can be seen from Fig. 10.2.4 that $p_{33}^{1/2}$ is initially erratic, partly because σ_t^2 is estimated from a small number of residuals. Later, $p_{33}^{1/2}$ is under-responsive, e.g. around 180 h, since σ_t^2 is estimated from many residuals, not all still relevant. Between 120 and 210 h, $p_{33}^{1/2}/|\hat{b}_1|$ warns clearly that $\hat{b}_1$ is unreliable. The warning is less clear from 85 to 120 h, when $\hat{b}_1$ is unrealistically negative. None of the quantities in Fig. 10.2.4 seems to be a trustworthy guide to reliability of $\hat{b}_1$.

Further evidence that the reliability of parameter estimates cannot always be gauged effectively by such quantities appears in the results of Example 10.2.2. The estimates $\hat{a}_1$ and $\hat{a}_2$ which implied unconvincing complex-conjugate poles in the (2, 2, 3) model have $p_{11}^{1/2}/|\hat{a}_1| = 0.025$ and $p_{22}^{1/2}/|\hat{a}_2| = 0.062$. We should therefore expect both estimates to be highly reliable, yet in the (2, 6, 3) model, which has credible poles, $p_{11}^{1/2}/|\hat{a}_1| = 0.083$ and $p_{22}^{1/2}/|\hat{a}_2| = 0.268$ (a result of spreading the information in the records over more parameter estimates). Evidently choices between model structures cannot be made on this basis.

A final comment is that at any one point in the recursion, all the process a.r. coefficients ($\hat{a}_1$, etc.) have similar estimated variances, seldom differing by more than a factor of 2. This empirical fact applies also to the process m.a. coefficients ($\hat{b}_1$, etc.) and the noise-model coefficients, and seems to be true in a wide range of examples, not only this one. Estimated variances appear not to be very sensitive to poor choices of model structure. △

We next examine results retrospectively (off-line) to find out whether and how the model should be modified. Off-line working allows us to reprocess the records and estimates at leisure, which will prove especially helpful in models with strongly time-varying parameters.

10.3 POST-ESTIMATION CHECKS

10.3.1 Employment of Time-Varying Models

The markedly non-stationary innovations sequence in Example 10.2.3 suggested a need for a higher-order or time-varying or non-linear model. As it is even less feasible to draw general conclusions about identification from non-linear examples than from linear ones, we shall focus on time-varying and higher-order models. By time-varying we mean with the dynamics represented as time-varying in the model structure; we are not thinking of the incidental time variation of recursive estimates of time-invariant dynamics as they settle from inaccurate initial guesses.

As we saw in Section 8.1.5, random walks provide a flexible representation of time-varying model coefficients $\mathbf{x}$. We model $\mathbf{x}$ by

$$\mathbf{x}_t = \mathbf{x}_{t-1} + \mathbf{w}_{t-1}, \qquad E\mathbf{w}_{t-1} = \mathbf{0}, \qquad \operatorname{cov} \mathbf{w}_{t-1} = Q \tag{10.3.1}$$

where Q is diagonal and each principal-diagonal element controls the variation of one coefficient. In e.l.s., Q merely increases P before each new observation is processed.

The estimation of a time-varying model is not only of interest when the final model will be time-varying, but is also valuable as a bridge between a very simple first-attempt model and a refined and extended time-invariant final model. The nature of the time variation in a simple model is a good pointer to the extra or modified features the final model should have.

To get the most benefit from the random-walk representation, we must distinguish genuine time-variation in $\mathbf{x}$ from the initial variation of $\hat{\mathbf{x}}$ as it converges from a poor $\hat{\mathbf{x}}_0$. The only way we can do so is by improving early estimates retrospectively, by optimal smoothing which brings into $\hat{\mathbf{x}}_t$ all the information about $\mathbf{x}_t$ contained in later observations, as outlined in Section 8.1.6.

We employ time-varying models and retrospective updating extensively from now on. However, some of the validation checks can equally be applied to models with coefficients represented as constant.

10.3.2 Checks on Parameter Estimates

The detailed variation of parameter estimates or derived quantities such as u.p.r. and steady-state gain can be checked against qualitative knowledge of the physics underlying the dynamics.

Example 10.3.1 Figure 10.3.1 shows the time-varying steady-state gain $\hat{g}_s$ and coefficient $\hat{b}_1$ estimated by e.l.s. and optimal smoothing from the records of Example 10.2.1. The (2, 6, 3) model has dead time 1. The principal-diagonal elements of Q/σ_v^2 are 10^{-4} and 10^{-2} for the a.r. and m.a. coefficients in $A(z^{-1})$ and $B(z^{-1})$, respectively, and zero for the noise polynomial $C(z^{-1})$. These values result from trial-and-error adjustment by factors of 10 to achieve the smallest m.s. residual obtainable without excessive time variation of the parameters and consequent inflation of the m.s. innovation (Norton, 1975). There is a substantial rise in $\hat{g}_s$ during each flow rise (Fig. 10.2.1), and a fall during dry spells. The interpretation is that more of the rainfall is absorbed into the ground after dry spells and less after heavy rain. The increases in $\hat{b}_1$

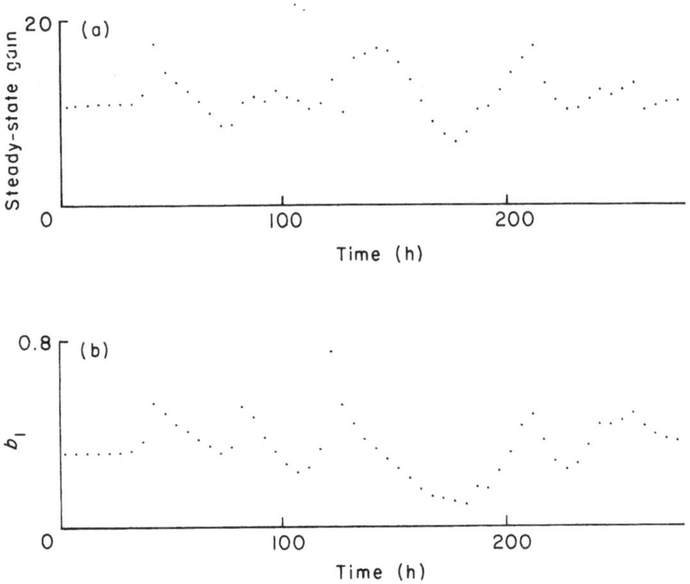

Fig. 10.3.1 Estimates of time-varying (a) steady-state (zero-frequency) gain and (b) b_1, Example 10.3.1.

during flow rises are even sharper than those in $\hat{g}_s$, suggesting that the dead time shortens.

Changes in the initial part of the u.p.r., as the dead time and shortest time constant vary, are made clear in Fig. 10.3.2 by plotting the u.p.r. normalised by its peak value, for various significant instants. Shortening of the dead time and shortest time constant occurs during heavy rainfall between 110 and

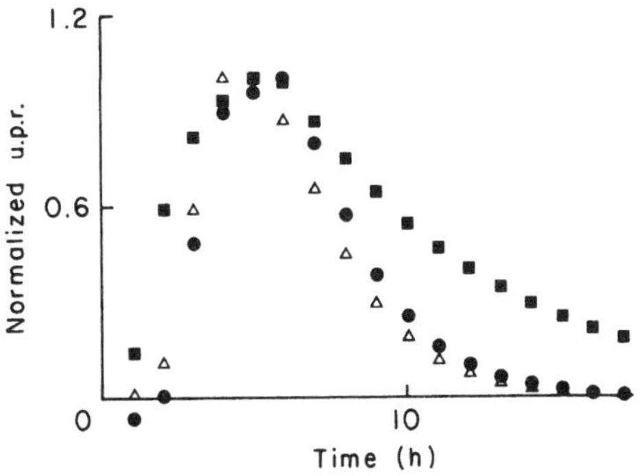

Fig. 10.3.2 Unit-pulse responses at various instants, normalised to have unity peak values, Example 10.3.1. Unit pulse response is for model at ●: 110 h; △: 130 h; ■: 250 h.

130 h. Comparison of the u.p.r.'s at 130 and 250 h finds a lengthening of the longest time constant and a further reduction in dead time, both due to cumulative wetting of the catchment. We conclude that ideally a non-linear model should give rising gain, shortening dead time and lengthening overall response as the catchment gets wetter. △

The following example illustrates (in a constant-parameter model) how a deficiency in model structure may show up clearly in unlikely parameter values although it is not obvious from the residuals.

Example 10.3.2 In a methionine tolerance test (Brown *et al.*, 1979), the response of methionine concentration in the blood of a subject following a rapid dose was

time t (h)	0	0.5	0.75	1.25	1.75	2.25
response y	0	90	115	85	55	40

A plot of this response and several others suggested that they might be fitted by $y(t) = a(\exp(-t/\tau_1) - \exp(-t/\tau_2))$, with the aim of investigating whether the response parameters are related to clinical condition.

Because the samples are unevenly spaced, parameters a, τ_1 and τ_2 were estimated directly by non-linear least squares rather than trying to fit a transfer-function model. The Levenberg–Marquardt method of Section 4.3.3 produced

$$\hat{y}(t) = 5.41 \times 10^4(\exp(-t/0.714) - \exp(-t/0{\cdot}710)).$$

The fit is moderate, with an r.m.s. error 6.03, about 16% of the r.m.s. deviation of the samples from their mean. However, $\hat{a}$ is implausibly large and the time constants suspiciously similar. A possible reason is omission of dead time. Figure 10.3.3 shows how r.m.s. error, $\hat{a}$, $\hat{\tau}_1$ and $\hat{\tau}_2$ were found to vary with dead time. The optimal dead time 0.35 h reduces the r.m.s. error to 1.13 and gives credible values for $\hat{a}$, $\hat{\tau}_1$ and $\hat{\tau}_2$, reassuring us that the improvement in fit is not merely due to increasing by one the number of parameters estimated from a very short record. △

10.3.3 Checks on Residuals

A straightforward plot of the residuals or innovations can say a great deal about the adequacy of the model, as in Example 10.2.3. Another post-estimation check often proposed is to test whether the sample autocorrelation

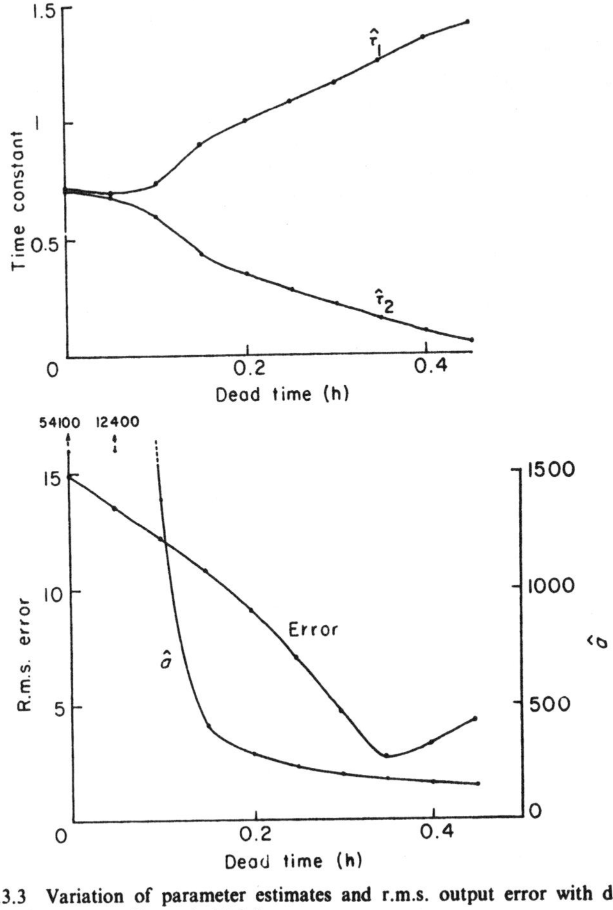

Fig. 10.3.3 Variation of parameter estimates and r.m.s. output error with dead time, methionine tolerance test, Example 10.3.2.

function of the innovations differs significantly from that of a white sequence, i.e. zero at all non-zero lags. The idea is that a good model predicts all the systematic part of the output, leaving an unstructured innovations sequence. Unfortunately, sample autocorrelation is not always a good measure of whether a sequence is structured, as we shall now see.

Example 10.3.3 Time-varying (2, 2, 3) models with various dead times were estimated from the records shown in Fig. 10.2.1, with Q as in Example 10.3.1. The sample a.c.f. of the innovations is given in Fig. 10.3.4 for dead times 1 and 9. The a.c.f. for dead time 9 is quite compatible with the assumption that the innovations sequence is white, with only one value outside the ± 2 standard-deviation lines for an uncorrelated sequence, and that only marginally. For dead time 1 the a.c.f. looks more structured and is beyond the ± 2 standard-deviation lines at two lags and close at a third. Nevertheless, dead time 1 is far closer to the truth. It gives smaller residuals, better predictions and smaller estimated standard deviations for the parameter estimates. It is also consistent with the look of the records; dead time 9 is not.

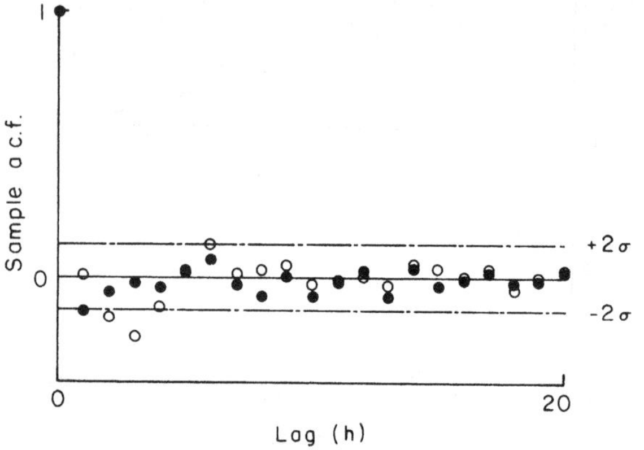

Fig. 10.3.4 Sample autocorrelation function of innovations for models with different dead times, Example 10.3.3. Model dead time ○: 1 h; ●: 9 h.

The explanation for the failure of the test is that a few large innovations dominate the sample a.c.f., the reliability of which is therefore low. For instance, the five largest innovations for dead time 9 account for 53.6% of the sum of squares. Incidentally, none of the large innovations is near the start of the records and so avoidable by a later start for the a.c.f. calculation. We conclude that the sample a.c.f. is helpful only if the innovations sequence is reasonably stationary, a rare event in practice. △

We next ask whether the r.m.s. innovation is an adequate guide to the best model. At the same time we see the effects on the innovations of retrospective re-estimation of $\{\mathbf{x}\}$ by optimal smoothing.

Example 10.3.4 Root mean square innovations are plotted in Fig. 10.3.5a for (2, 2, 3) and (2, 6, 3) models, with and without optimal smoothing of $\{\hat{\mathbf{x}}\}$, for a range of model dead times. In all cases $\mathbf{x}$ is represented as a random walk, with Q the same as before. The optimally smoothed estimates are retrospective, being based on the entire record of N input–output pairs rather than the observations received up to each point in the recursion. The one-step "predictions" they yield are not therefore available on line. However, the smoothed "predictions" $\{\mathbf{h}_t^T\hat{\mathbf{x}}_{t-1|N}\}$ should give a better indication of long-term one-step prediction performance than $\{\mathbf{h}_t^T\hat{\mathbf{x}}_{t-1|t-1}\}$, because early in the records the latter is still much affected by the error in the initial estimate.

In this example, optimal smoothing reduces the r.m.s. value of the innovations by about a factor of 2. The variation with dead time and process m.a. order is, however, much the same with and without smoothing.

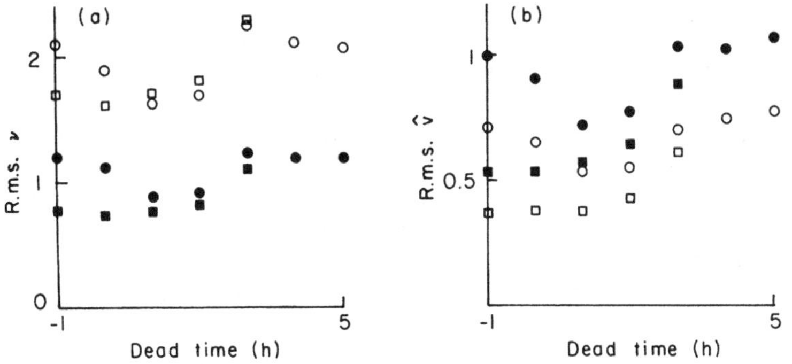

Fig. 10.3.5 (a) Variation of r.m.s. innovation with dead time, for time-varying models obtained with and without optimal smoothing, Example 10.3.4, and (b) variation of r.m.s. residual with dead time. Without smoothing: ○: (2, 2, 3) model; □: (2, 6, 3) model. With optimal smoothing: ●: (2, 2, 3) model; ■: (2, 6, 3) model.

From Fig. 10.3.5a, u_{t-3} is the latest input essential for predicting y_t; the (2, 2, 3) and (2, 6, 3) models perform well only when u_{t-3} is included in the explanatory variables. Only a small deterioration in prediction performance results when too short a dead time is specified in the (2, 6, 3) model, even though the redundant leading m.a. terms $b_1 u_{t-k-1}$, etc., contribute nothing. Evidently fewer than six m.a. terms are necessary if the dead time is well chosen. The closeness of the r.m.s. innovation values of the (2, 2, 3) and (2, 6, 3) models at dead time 1 or 2 confirms this.

Figure 10.3.5 gives the r.m.s. values of residuals $\{y_t - \mathbf{h}_t^T\hat{\mathbf{x}}_{t|t}\}$ and $\{y_t - \mathbf{h}_t^T\hat{\mathbf{x}}_{t|N}\}$. They demonstrate how effective optimal smoothing is in reducing noise-induced spurious short-term variation of the parameter estimates. Such variation makes the r.m.s. innovation three to four times the size of the r.m.s. residual in the forward recursion. In the backward recursion which produces $\{\hat{\mathbf{x}}_{t|N}\}$, $\{y_t - \mathbf{h}_t^T\hat{\mathbf{x}}_{t-1|N}\}$ and $\{y_t - \mathbf{h}_t^T\mathbf{x}_{t|N}\}$, the r.m.s. innovation is only 20 to 30% larger. We conclude that the optimally smoothed estimates $\{\hat{\mathbf{x}}_{t|N}\}$ contain considerably less spurious variation than the on-line estimates $\{\hat{\mathbf{x}}_{t|t}\}$. △

10.3.4 Simulation-Mode Runs

A severe and informative test of a model is to run it over records in simulation mode, that is, with earlier *model*-output samples in place of the observed output in the explanatory variables, generating

$$\hat{y}_t = -a_1\hat{y}_{t-1} - \cdots - \hat{a}_n y_{t-n} + \hat{b}_1 u_{t-k-1} + \cdots$$
$$+ \hat{b}_m u_{t-k-m} + \hat{c}_1\hat{v}_{t-1} + \cdots + \hat{c}_q\hat{v}_{t-q} \tag{10.3.2}$$

The recursion is started with y_0 to y_{n-1} for $\hat{y}_0$ to $\hat{y}_{n-1}$. Deficiencies in model structure and poor parameter estimates or dead time give rise to obvious systematic error in the simulation-mode output sequence.

Example 10.3.5 A (2, 6, 3) model with dead time 1 was estimated from the records shown in Fig. 10.2.1, with the parameters represented as constant. Figure 10.3.6 compares the observed and simulation-mode flows over part of the record. The shortcomings of the model are plain to see: It overestimates the lower peak flows and underestimates the largest peak, misses the slowest dynamics in the recessions and gives too large and rapid a response to the start of rain after a long dry spell. These features appear much more clearly than in the on-line residuals or innovations, for several reasons. The simulation-mode computation deliberately omits the noise model, which on-line takes up some of the output behaviour not captured by the input–output part of the model. Missing or inaccurate dynamics in the model causes cumulative output error in the simulation mode, but the error is constantly removed on-line by use of observed flows in predicting the present flow. Finally, the parameter estimates evolve on-line even if represented as constant with Q zero, and can follow time-varying dynamics to some extent. This is so until the covariance P and

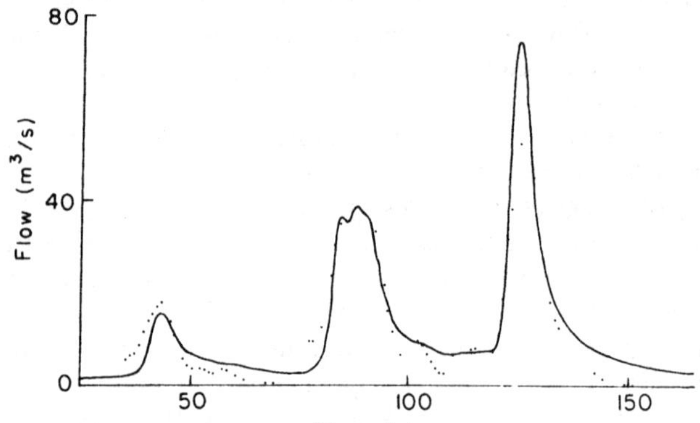

Fig. 10.3.6 Observed and simulation-mode flows, Example 10.3.5. Observed flow is a continuous line.

correction gain have become small. Simulation-mode results are pessimistic if the correction gain does indeed stay large enough to allow significant variation of $\hat{\mathbf{x}}$.

Raising the model orders to (6, 6, 3) in this example improves the simulation-mode performance very little, reinforcing the conclusion that only time-varying representation of the a.r.m.a. coefficients will match the dynamics better. △

Simulation-mode results for short sections of record where the parameter estimates of a time-varying model change rapidly help us to understand the changes and assess whether they are sufficient. The "before change" and "after change" models are compared according to how the simulation-mode $\{\hat{y}\}$ fits $\{y\}$ in the vicinity of the change.

Example 10.3.6 A (2, 2, 3) model with dead time 1 and coefficients represented as time-varying was estimated from the records of Fig. 10.2.1. The estimates change rapidly during the flow rise at about 120 h. Figure 10.3.7 plots the simulation-mode $\{\hat{y}\}$ generated by the model as it stood at 115 and 125 h, for a period covering the rise, peak and recession. The model at 115 h has far too low a gain and is too slow; the flow peak is 2 h late. The 125 h model, by contrast, has about the right gain and places the peak at the correct time. Its recession time constant is too short, but that is of little consequence for on-line flow prediction; any fairly long time constant ensures good one-step prediction while the flow changes slowly, and predictions during a recession are in any case of little importance. We can be satisfied with the change in $\hat{\mathbf{x}}$ during the rise, it seems. △

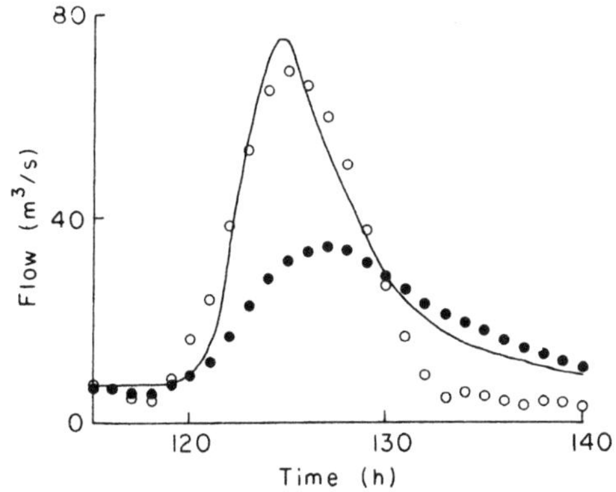

Fig. 10.3.7 Observed flow and simulation-mode flows given by models before and after changes during flow rise, Example 10.3.6. Observed flow is a continuous line. Flows modelled at ●: 115 h; ○: 125 h.

10.3.5 Informal Checks

Worthwhile checks after an estimation run are for:

(i) isolated large residuals (outliers), which may reveal transcription or instrument errors;

(ii) short periods of large and highly structured, or anomalously small, residuals, which may show up doctored records;

(iii) abrupt and unexpected changes in the parameter estimates or residuals, which may point to unrecorded incidents such as a feed-stock change, unrecorded control action, change or inconsistency in the way measurements are made or shift-to-shift variations in process-operating practice;

(iv) input features with no apparent output consequences or output features with no apparent cause, as shown by large residuals over short periods, suggesting that more extensive or better measurements or a higher sampling rate may be required;

(v) periodicity: specialised models for periodic phenomena may be necessary, an important topic we have too little space to pursue (Box and Jenkins, 1970).

Simulation-mode runs are effective in bringing out such features.

Example 10.3.7 Hourly records were taken of flow in the Mackintosh River in Western Tasmania, together with rainfall at two gauges, one not far above the stream-gauging point and the other diametrically across the catchment

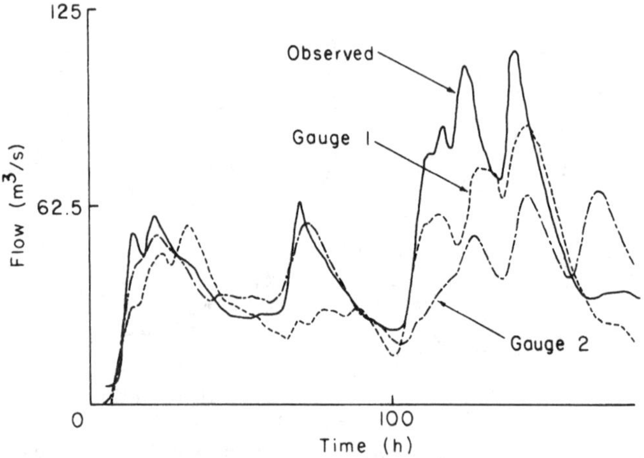

Fig. 10.3.8 Observed flow and simulation-mode flows given by two separate rain-gauge records, Example 10.3.7.

and just outside it. Figure 10.3.8 gives 180 h of observed flow and the simulation-mode flows calculated from rainfall at each gauge on its own, using (2, 6, 3) constant-parameter models fitted to 217 h of record. Gauge 1 predicts a flow peak at 32 h absent from the observed flow, and misses the peak at 70 h almost completely. A peak at 168 h predicted by gauge 2 fails to eventuate, and the observed peak at 114 h is missed by gauge 2. Apparently the catchment is too large and the rainfall too local for two rain gauges to be enough. Furthermore, the absolute as well as relative success of the two gauges varies greatly, so a weighted sum of their readings would not do, either. △

10.4 EPILOGUE

Let us finish with a story.

Once upon a time, a researcher interested in identification was invited by an industrial research association to identify the dynamics of a certain process. He was told records could be provided but were likely to be very noisy. He agreed to take the job on. There was some delay in producing the records, and the researcher moved overseas. He was keen to do the work, even though contact with the people producing the records was now less easy, so the agreement stood. In due course the records arrived. Day after day he laboured to fit a convincing model, but with no success. Finally he gave up and, rather shamefaced, wrote to the research association to admit defeat. Many weeks passed, until one day a letter arrived from the research association. It read:

"Dear ________,

Thank you for your recent letter. We appreciate your efforts on our behalf, and were sorry that they met with no success. You will be pleased to hear that we can now account for the difficulty. The input record we sent you was for Tuesday 23rd March. The output record was for Tuesday 16th March.

Yours sincerely,

________"

FURTHER READING

Tests for absence of structure in residual sequences are discussed by Box and Jenkins (1970), Kendall (1976) and the regression texts mentioned at the end of Chapter 4.

Identification case studies are given by Bohlin (1976), Ljung and Söderström (1983), Olson (1976), Söderström and Stoica (1983) and Young (1984). Gustavsson (1975) surveys applications in the process industries and gives 143 references.

REFERENCES

Bohlin, T. (1976). Four cases of identification of changing systems. *In* "System Identification: Advances and Case Studies" (R. K. Mehra and D. G. Lainiotis, eds.). Academic Press, New York and London.

Box, G. E. P., and Jenkins, G. M. (1970). "Time Series Analysis, Forecasting and Control". Holden-Day, San Francisco, California.

Brown, R. F., Godfrey, K. R., and Knell, A. (1979). Compartmental modelling based on methionine tolerance test data: A case study. *Med. Biol. Eng. Comput.* **17**, 223–229.

Gustavsson, I. (1975). Survey of applications of identification in chemical and physical processes. *Automatica* **11**, 3–24.

Kendall, M. G. (1976). "Time-Series", 2nd ed. Griffin, London.

Ljung, L., and Söderström, T. (1983). "Theory and Practice of Recursive Identification". MIT Press, Cambridge, Massachusetts.

Maybeck, P. S. (1982). "Stochastic Models, Estimation and Control", Vol. 2. Academic Press, New York and London.

Norton, J. P. (1975). Optimal smoothing in the identification of linear time-varying systems. *Proc. IEE* **122**, 663–668.

Olsson, G. (1976). Modeling and identification of a nuclear reactor. *In* "System Identification: Advances and Case Studies" (R. K. Mehra and D. G. Lainiotis, eds.). Academic Press, New York and London.

Söderström, T., and Stoica, P. G. (1983). "Instrumental Variable Methods for System Identification." Springer-Verlag, Berlin and New York.

Young, P. C. (1984). "Recursive Estimation and Time-Series Analysis." Springer-Verlag, Berlin and New York.

Index

K

L

M

Q

R

S

T

U